10 0221950 X

KU-538-989

NOT TO
BE TAKEN
OUT OF
THE
LIBRARY

GELS HANDBOOK

Volume 3

GELS HANDBOOK

Volume 3
Applications

Editors-in-Chief

Yoshihito Osada and Kanji Kajiwara

Associate Editors

Takao Fushimi, Okihiko Hirasa,
Yoshitsugu Hirokawa, Tsutomu Matsunaga,
Tadao Shimomura, and Lin Wang

Translated by

Hatsuo Ishida

ACADEMIC PRESS

A Harcourt Science and Technology Company

San Diego   San Francisco   New York   Boston
London   Sydney   Tokyo

This book is printed on acid-free paper. ∞

Copyright © 2001 by Academic Press

All rights reserved.
No part of this publication may be reproduced or transmitted in any form or by any means, electronic or mechanical, including photocopy, recording, or any information storage and retrieval system, without permission in writing from the publisher.

Requests for permission to make copies of any part of the work should be mailed to: Permissions Department, Harcourt, Inc., 6277 Sea Harbor Drive, Orlando, Florida, 32887-6777.

ACADEMIC PRESS
A Harcourt Science and Technology Company
525 B Street, Suite 1900, San Diego, CA 92101-4495, USA
http://www.academicpress.com

Academic Press
Harcourt Place, 32 Jamestown Road, London, NW1 7BY, UK

Library of Congress Catalog Number: 00-107106
International Standard Book Number: 0-12-394690-5 (Set)

International Standard Book Number, Volume 3: 0-12-394963-7

Printed in the United States of America
00 01 02 03 04 IP 9 8 7 6 5 4 3 2 1

UNIVERSITY OF NOTTINGHAM
1002219508
WITHDRAWN
FROM THE LIBRARY
UNIVERSITY LIBRARY NOTTINGHAM

Contents

Chapter 2   Daily Commodities   35

Section 1   Cosmetics

Section 2   Air Fresheners and Deodorizers

Section 3   Disposable Portable Heaters

Section 4   Sanitary products for Pets

Section 5   Photographic Films

Chapter 3   Foods and Packaging   109

Chapter 4   Medicine and Medical Care   141

Chapter 5   Farming and Agriculture   259

Chapter 8   Electric and Electronic Industries   409

Section 1   Communication Cables

Section 2   Batteries

Section 3   Fuel Cells

Chapter 9   Sport and Leisure Activity Industries   479

Preface

The development, production, and application of superabsorbent gels is increasing at a remarkable pace. Research involving functional materials in such areas as medical care, medicine, foods, civil engineering, bioengineering, and sports is already widely documented. In the twenty-first century innovative research and development is growing ever more active. Gels are widely expected to be one of the essential solutions to various problems such as limited food resources, environmental preservation, and safeguarding human welfare.

In spite of the clear need for continued gel research and development, there have been no comprehensive references involving gels until now. In 1996, an editorial board led by the main members of the Association of Polymer Gel Research was organized with the primary goal of collecting a broad range of available information and organizing this information in such a way that would be helpful for not only gels scientists, but also for researchers and engineers in other fields. The content covers all topics ranging from preparation methods, structure, and characteristics to applications, functions, and evaluation methods of gels. It consists of Volume 1, The Fundamentals; Volume 2, Functions; Volume 3, Applications; and Volume 4, Environment: Earth Environment and Gels, which consists of several appendices and an index on gel compounds.

Because we were fortunate enough to receive contributions from the leading researchers on gels in Japan and abroad, we offer this book with great confidence. We would like to thank the editors as well as the authors who willingly contributed despite their very busy schedules.

This handbook was initially proposed by Mr. Shi Matsunaga. It is, of course, due to the neverending effort by him and the editorial staff that this handbook was successfully completed. We would also like to express great appreciation to the enthusiasm and help of Mr. Takashi Yoshida and Ms. Masami Matsukaze of NTS Inc.

Yoshihito Osada
Kanji Kajiwara
November, 1997

Contributors

Editors-in-Chief

Yoshihito Osada, *Professor, Department of Scientific Research, Division of Biology at Hokkaido University Graduate School*

Kanji Kajiwara, *Professor, Department of Technical Art in Material Engineering at Kyoto University of Industrial Art and Textile*

Principal Editorial Members

Tadao Shimomura, *President, Japan Catalytic Polymer Molecule Research Center*

Okihiko Hirasa, *Professor, Department of Education and Domestic Science at Iwate University*

Yoshitsugu Hirokawa, *Technical Councilor, Science and Technology Promotional Office, Hashimoto Phase Separation Structure Project*

Takao Fushimi, *Examiner, Patent Office Third Examination Office at Ministry of International Trade and Industry*

Tsutomu Matsunaga, *Director, Chemistry Bio-Tsukuba*

Lin Wang, *Senior Scientist, P&G Product Development Headquarters*

Ito Takeshi, *Assistant Manager, Tokyo Office Sales and Development Division of Mitsubishi Chemical Co.*

Seigo Ouchi, *Head Researcher, Kanishi Test Farm at Agricultural Chemical Research Center of Sumitomo Chemical Co.*

Mitsuo Okano, *Professor, Tokyo Women's Medical College*

Masayoshi Watanabe, *Assistant Professor, Yokohama National University Department of Engineering, Division of Material Engineering*

Contributors

Aizo Yamauchi, *President, International Research Exchange Center of Japan Society of Promotion for Industrial Technology*

Yoshihito Osada, *Professor, Department of Scientific Research in Biology at Hokkaido University Graduate School*

Hidetaka Tobita, *Assistant Professor, Department of Engineering, Material Chemistry Division at Fukui University*

Yutaka Tanaka, *Research Associate, Department of Engineering, Material Chemistry Division at Fukui University*

Shunsuke Hirotsu, *Professor, Department of Life Sciences and Engineering, Division of Organism Structures at Tokyo Institute of Technology*

Mitsuhiro Shibayama, *Professor, Department of Textiles, Polymer Molecule Division at Kyoto University of Industrial Art and Textile*

Hidenori Okuzaki, *Assistant, Department of Chemistry and Biology, Division of Biological Engineering at Yamanashi University*

Kanji Kajiwara, *Professor, Department of Technical Art in Material Engineering at Kyoto University of Industrial Art and Textile*

Yukio Naito, *Head of Research, Biological Research Center for Kao*

(the late) Kobayashi Masamichi, *Honorary Professor, Department of Science, Division of Polymer Molecular Research at Osaka University Graduate School*

Hidetoshi Oikawa, *Assistant Professor, Emphasis of Research on Higher Order Structural Controls in Department of Reactive Controls at Reactive Chemistry Research Center at Tohoku University*

Yositsugu Hirokawa, *Technical Councilor, Science and Technology Promotional Office, Hashimoto Phase Separation Structure Project*

Makoto Suzuki, *Professor, Department of Engineering, Division of Metal Engineering at Tohoku University Graduate School*

Ken Nakajima, *Special Research, Division of Basic Science in International Frontier Research System Nano-organic Photonics Material Research Team at Physics and Chemistry Research Center*

Toshio Nishi, *Professor, Department of Engineering Research, Division of Physical Engineering at Tokyo University Graduate School*

Hidemitsu Kuroko, *Assistant Professor, Department of Life Environment, Division of Life Environment at Nara Women's University*

Shukei Yasunaga, *Assistant, Department of Technical Art in Material Engineering at Kyoto University of Industrial Art and Textile*

Mitsue Kobayashi, *Special Researcher, Tokyo Institute of Technology*

Hajime Saito, *Professor, Department of Science, Division of Life Sciences at Himeji Institute of Technology*

Hazime Ichijyo, *Manager of Planning Office, Industrial Engineering Research Center in Department of Industrial Engineering, Agency of Industrial Science and Technology at Ministry of International Trade and Industry*

Masayoshi Watanabe, *Assistant Professor, Yokohama National University Department of Engineering, Division of Material Engineering*

Kunio Nakamura, *Professor, Department of Agriculture, Division of Food Sciences at College of Dairy Agriculture*

Hideo Yamazaki, *Shial, Inc. (Temporarily transferred from Tonen Chemical Co.)*

Koshibe Shigeru, *Shial, Inc. (Temporarily transferred from Tonen Chemical Co.)*

Hirohisa Yoshida, *Assistant, Department of Engineering, Division of Industrial Chemistry at Tokyo Metropolitan University*

Yoshiro Tajitsu, *Professor, Department of Engineering at Yamagata University*

Hotaka Ito, *Instructor, Division of Material Engineering at National Hakodate Technical High School*

Toyoaki Matsuura, *Assistant, Department of Opthamology at Nara Prefectural Medical College*

Yoshihiko Masuda, *Lead Researcher, Third Research Division of Japan Catalytic Polymer Molecule Research Center*

Toshio Yanaki, *Researcher, Shiseido Printed Circuit Board Technology Research Center*

Yuzo Kaneko, *Department of Science, Division of Applied Chemistry at Waseda University*

Kiyotaka Sakai, *Professor, Department of Science, Division of Applied Chemistry at Waseda University*

Teruo Okano, *Professor, Medical Engineering Research Institute at Tokyo Women's Medical College*

Shuji Sakohara, *Professor, Department of Engineering, Chemical Engineering Seminar at Hiroshima University*

Jian-Ping Gong, *Assistant Professor, Department of Scientific Research, Division of Biology at Hokkaido University Graduate School*

Akihiko Kikuchi, *Assistant, Medical Engineering Research Institute at Tokyo Women's Medical College*

Shingo Matukawa, *Assistant, Department of Fisheries, Division of Food Production at Tokyo University of Fisheries*

Kenji Hanabusa, *Assistant Professor, Department of Textiles, Division of Functional Polymer Molecules at Shinshu University*

Ohhoh Shirai, *Professor, Department of Textiles, Division of Functional Polymer Molecules at Shinshu University*

Atushi Suzuki, *Assistant Professor, Department of Engineering Research, Division of Artificial Environment Systems at Yokohama National University Graduate School*

Junji Tanaka, *Department of Camera Products Technology, Division Production Engineering, Process Engineering Group at Optical Equipment Headquarters at Minolta, Inc.*

Eiji Nakanishi, *Assistant Professor, Department of Engineering, Division of Material Engineering at Nagoya Institute of Technology*

Ryoichi Kishi, *Department of Polymer Molecules, Functional Soft Material Group in Material Engineering Technology Research Center in Agency of Industrial Science and Technology at Ministry of International Trade and Industry*

Toshio Kurauchi, *Director, Toyota Central Research Center*

Tohru Shiga, *Head Researcher, LB Department of Toyota Central Research Center*

Keiichi Kaneto, *Professor, Department of Information Technology, Division of Electronic Information Technology at Kyushu Institute of Technology*

Kiyohito Koyama, *Professor, Department of Engineering, Material Engineering Division at Yamagata University*

Yoshinobu Asako, *Lead Researcher, Nippon Shokubai Co. Ltd., Tsukuba Research Center*

Tasuku Saito, *General Manager, Research and Development Headquarters, Development Division No. 2 of Bridgestone, Inc.*

Toshihiro Hirai, *Professor, Department of Textiles, Division of Raw Material Development at Shinshu University*

Keizo Ishii, *Manager, Synthetic Technology Research Center at Japan Paints, Inc.*

Yoshito Ikada, *Professor, Organism Medical Engineering Research Center at Kyoto University*

Lin Wang, *Senior Scientist, P&G Product Development Headquarters*

Rezai E., *P&G Product Development Headquarters*

Fumiaki Matsuzaki, *Group Leader, Department of Polymer Molecule Science Research, Shiseido Printed Circuit Board Technology Research Center*

Jian-Zhang (Kenchu) Yang, *Researcher, Beauty Care Product Division of P&G Product Development Headquarters*

Chun Lou Xiao, *Section Leader, Beauty Care Product Division of P&G Product Development Headquarters*

Yasunari Nakama, *Councilor, Shiseido Printed Circuit Board Technology Research Center*

Keisuke Sakuda, *Assistant Director, Fragrance Development Research Center at Ogawa Perfumes, Co.*

Akio Usui, *Thermofilm, Co.*

Mitsuharu Tominaga, *Executive Director, Fuji Light Technology, Inc.*

Takashi Naoi, *Head Researcher, Ashikaga Research Center of Fuji Film, Inc.*

Makoto Ichikawa, *Lion, Corp. Better Living Research Center*

Takamitsu Tamura, *Lion, Corp. Material Engineering Center*

Takao Fushimi, *Examiner, Patent Office Third Examination Office at Ministry of International Trade and Industry*

Kohichi Nakazato, *Integrated Culture Research Institute, Division of Life Environment (Chemistry) at Tokyo University Graduate School*

Masayuki Yamato, *Researcher, Doctor at Japan Society for the Promotion of Science, and Japan Medical Engineering Research Institute of Tokyo Women's Medical College*

Toshihiko Hayasi, *Professor, Integrated Culture Research Institute, Division of Life Environment (Chemistry) at Tokyo University Graduate School*

Naoki Negishi, *Assistant Professor, Department of Cosmetic Surgery at Tokyo Women's Medical College*

Mikihiro Nozaki, *Professor, Department of Cosmetic Surgery at Tokyo Women's Medical College*

Yoshiharu Machida, *Professor, Department of Medical Pharmacology Research at Hoshi College of Pharmacy*

Naoki Nagai, *Professor, Department of Pharmacology at Hoshi College of Pharmacy*

Kenji Sugibayashi, *Assistant Professor, Department of Pharmacology at Josai University*

Yohken Morimoto, *Department Chair Professor, Department of Pharmacology at Josai University*

Toshio Inaki, *Manager, Division of Formulation Research in Fuji Research Center of Kyowa, Inc.*

Seiichi Aiba, *Manager, Department of Organic Functional Materials, Division of Functional Polymer Molecule Research, Osaka Industrial Engineering Research Center of Agency of Industrial Science and Technology at Ministry of International Trade and Industry*

Masakatsu Yonese, *Professor, Department of Pharmacology, Division of Pharmacology Materials at Nagoya City University*

Etsuo Kokufuta, *Professor, Department of Applied Biology at Tsukuba University*

Hiroo Iwata, *Assistant Professor, Organism Medical Engineering Research Center at Kyoto University*

Seigo Ouchi, *Head Researcher, Agricultural Chemical Research Center at Sumitomo Chemical Engineering, Co.*

Ryoichi Oshiumi, *Former Engineering Manager, Nippon Shokubai Co. Ltd. Water-absorbent Resin Engineering Research Association*

Tatsuro Toyoda, *Nishikawa Rubber Engineering, Inc. Industrial Material Division*

Nobuyuki Harada, *Researcher, Third Research Division of Japan Catalytic Polymer Molecule Research Center*

Osamu Tanaka, *Engineering Manager, Ask Techno Construction, Inc.*

Mitsuharu Ohsawa, *Group Leader, Fire Resistance Systems Group of Kenzai Techno Research Center*

Takeshi Kawachi, *Office Manager, Chemical Research Division of Ohbayashi Engineering Research Center, Inc.*

Hiroaki Takayanagi, *Head Researcher, Functional Chemistry Research Center in Yokohama Research Center of Mitsubishi Chemical, Inc.*

Yuichi Mori, *Guest Professor, Department of Science and Engineering Research Center at Waseda University*

Tomoki Gomi, *Assistant Lead Researcher, Third Research Division of Japan Catalytic Polymer Molecule Research Center*

Katsumi Kuboshima, *President, Kuboshima Engineering Company*

Hiroyuki Kakiuchi, *Mitsubishi Chemical, Inc., Tsukuba Research Center*

Baba Yoshinobu, *Professor, Department of Pharmacology, Division of Pharmacological Sciences and Chemistry at Tokushima University*

Toshiyuki Osawa, *Acting Manager, Engineer, Thermal Division NA-PT at Shotsu Office of Ricoh, Inc.*

Kazuo Okuyama, *Assistant Councilor, Membrane Research Laboratory, Asahi Chemical Industry Co., Ltd.*

Takahiro Saito, *Yokohama National University Graduate School, Department of Engineering, Division of Engineering Research*

Yoshiro Sakai, *Professor, Department of Engineering, Division of Applied Chemistry at Ehime University*

Seisuke Tomita, *Managing Director, Development and Production Headquarters at Bridgestone Sports, Inc.*

Hiroshi Kasahara, *Taikisha, Inc. Environment System Office*

Shigeru Sato, *Head Researcher, Engineering Development Center at Kurita Engineering, Inc.*

Okihiko Hirasa, *Professor, Iwate University*

Seiro Nishio, *Former Member of Disposable Diaper Technology and Environment Group of Japan Sanitary Material Engineering Association*

VOLUME III
Applications

CHAPTER 1

Sanitary Products

Chapter contents

Section 1
Disposable Diapers

LING WANG AND E. REZAI

1 INTRODUCTION

The most important property of superabsorbent polymers is their ability to absorb water. Traditional water-absorbing materials such as cotton, pulp and sponges absorb water into interstices by capillary phenomenon. By contrast, superabsorbent polymer absorbs water into three-dimensional (3D) networks of a crosslinked polymer by the compatibility of polymer chains and water and by osmotic pressure. Hence, superabsorbent polymers lack absorption speed but have a much better water-absorption capacity compared to cotton or pulp. Products that have a high water-absorption property include sanitary products, such as disposable diapers, construction materials, mulch, water sealant for electric cables, freshness maintenance materials in the food industry. Of the sanitary products, of the disposable diaper is the best example of a product that has a high water-absorbing capacity. The superabsorbent polymer is indispensable for disposable diapers, and it drastically improved the performance of disposable diapers. Here, the development and current status of disposable diapers, the structure of disposable diapers in which a superabsorbent polymer is incorporated, and advances in the development of super-absorbent polymers will be described.

4

2   EVALUATION OF DISPOSABLE DIAPERS

Disposable diapers were developed in Sweden in the 1940s. During the 1960s, Procter & Gamble introduced a disposable diaper under the brand name Pampers, which contributed to the widespread use of disposable diapers and subsequently was exported to Europe, where it became a synonym for disposable diapers. Pampers was imported from the United States and introduced to the Japanese market around 1977. Around 1980, superabsorbent polymers began to be used for disposable diapers, which improved their performance drastically. In the 1980s, disposable diapers were distinguished based on performance as new concepts and technologies began to have a great impact on the disposable diaper market. As a result, ultrathin diapers using a high concentration fo superabsorbent polymer were introduced along with such innovations as elastic gathering at the waist and leg, the use of an air-permeable back panel, and tapeless panty-type diapers. In particular, since its introduction in 1990, the panty-type diaper has become very popular for its ease of use, and its market share has increased rapidly.

3   MARKET FOR DISPOSABLE DIAPERS FOR CHILDREN

The market for disposable diapers worldwide is approximately 56 billion diapers [1, 2], which amounts to approximately 17 billion dollars. As shown in Fig. 1, consumption in North America is about 19 billion diapers

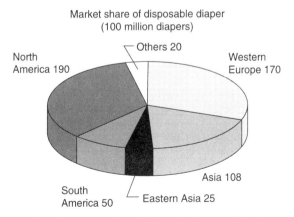

Fig. 1   Regional market share for children's disposable diapers worldwide.

(33.7 % of the total consumed), in South America it is about 5 billion diapers (8.9 %), 17 billion in Western Europe (30.2 %), in Eastern Europe 2.5 billion diapers (4.4 %), and in Asia 10.8 billion diapers (19.1 %).

In recent years, Asia has experienced rapid economic growth and a remarkable improvement in the level of living standards. Consequently, there has been a rapid expansion in the disposable diaper market.

Table 1 lists the disposable diaper market share of thirteen countries and regions in Asia. As seen in the table, other than Japan, the market share in Asia is low. Nevertheless, the growth has been remarkable. For example, the market share in South Korea was approximately 25 % distribution and reached about 55 % in 1994, an annual average growth rate of over 25 % [3]. In the countries or regions with high economic growth, such as Singapore, Taiwan, and Hong Kong, the annual growth rate in the past several years has been remarkable, and this trend is expected to continue until the present.

The market share for disposable diapers in China is less than 1 %. However, with the rapid increase in its urban population, has been the establishment of a market for disposable diapers. Assuming that 5 % of China's 1.1 billion population is represented by children under three years of age, the consumption of disposable diapers will be 10 billion even if the market share is 10 %. This is equivalent to the consumption of all the Asian countries combined. The major disposable diaper producers in the world are paying serious attention to this huge market and are planning active entry into it.

Table 1   Market share for children's disposable diapers for Asian countries.

Country or region	Market size (1 milltion diapers)	Population (1 million people)
Japan	5,000	125
South Korea	1,600	43
China	1,500	1,150
Taiwan	1,200	21
Australia	600	17
Hong Kong	250	6
Malaysia	150	18
Singapore	100	3
New Zealand	100	4
Philippines	100	63
Thailand	70	57
Indonesia	30	180
India	20	865

At the moment, Japan is the largest market in Asia and occupies 45 % of all Asian markets. The consumption was 1.281 billion yen in 1991, 1.320 billion yen in 1992, 1.395 billion yen in 1993, 1.402 billion yen in 1994, 1.385 billion yen in 1995, and 1.420 billion yen in 1996. Although the growth rate was slow, it was steady. In the early 1990s, introduction of the panty-type diaper in toilet training of children prolonged the time of diaper use and perhaps contributed to the increase of activity in the market. However, the consumption in 1995 was 1.385 billion yen, which was a reduction from 1994 by 1.2 %. In recent years, many panty-type diapers for older children have been introduced, and it was expected that the growth rate based on the dollar amount would be high. However, the consumption of inexpensive diapers continues to be strong, which suggests the existence of a dipolar trend [4].

The makers of disposable diapers in Japan are Unicharm, Procter & Gamble, Kao, Nepia, Shiseido, Daio Seishi, and others totalling more than ten companies. However, the top three (Unicharm, Procter & Gamble, and Kao) occupy more than 80 % of market share.

4   DEVELOPMENT OF TREND OF CHILDREN'S DISPOSABLE DIAPERS

The usual disposable diaper consisted of: (i) a top sheet through which water passes; (ii) a water-absorbing core; and (iii) a back panel that is impermeable. Until the beginning of the 1980s, approximately 70 g of pulp was used as the absorption core. Later, the core weight was reduced to 40 g by the introduction of a superabsorbent polymer. The use of the superabsorbent polymer provided such advantages as: (i) no leakage; (ii) no build up of heat; (iii) the prevention of diaper rash; and (iv) thin size.

However, if the concentration of a superabsorbent polymer is further increased, the use efficiency reduces due to gel blocking. In order to solve this problem, investigations are made from the viewpoints of: (i) improvement of superabsorbent polymer performance; (ii) development of new materials that transport water effectively and separately; and (iii) application of revolutionary product design.

With regard to item (i), the gel strength and water absorption under stress have been improved by crosslinking the surface of the superabsorbent polymer [5]. This has made it possible to develop ultrathin products. The development of superabsorbent polymers for disposable diapers will be described later.

Many new materials have been developed for transporting water effectively and separately [6–9], including crosslinked cellulose called curly fibers, thermally fused PE/PP fibers, and a special nonwoven rayon cloth. These new materials absorb a liquid quickly and distribute it to the superabsorbent polymer. In addition, these new materials show good integrity to moisture. In particular, there is no corruption of the structure even after wetting, allowing the maintenance of channels through which the liquid can be transferred.

Finally, with regard to revolutionary design, many systems adopt an acquisition layer or a combination of an acquisition layer and transport layers between the absorption core (Fig. 2) and the top sheet [9]. For the acquisition layer or the transport layer, either a crosslinked cellulose with a density of 0.04–0.1 g/cm^3, thermally fused PE/PP fibers, or a nonwoven cloth is used. The absorption core consists of surface-crosslinked superabsorbent polymer and pulp. In general, the concentration of the superabsorbent polymer is 40 to 60 wt %, and the weight of the polymer used per diaper is approximately 10 to 12 g [9, 10].

As an example of ultrathin disposable diapers, Fig. 3 shows a birdseye view [12]. This is a so-called multilayer design, which consists of a front panel first diffusion layer/absorption core/second diffusion layer/back panel. The first diffusion layer utilizes a web made of a synthetic polymer with its unit area weight is 20 to 200 g/cm^2 and the

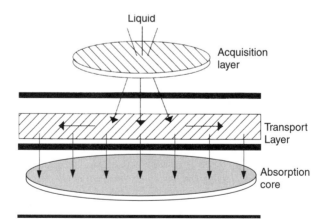

Fig. 2 Design of the absorption core of an ultrathin disposable diaper [9].

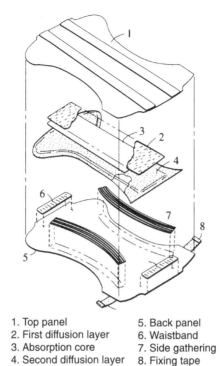

1. Top panel   5. Back panel
2. First diffusion layer   6. Waistband
3. Absorption core   7. Side gathering
4. Second diffusion layer 8. Fixing tape

Fig. 3   Structure of disposable diaper [12].

density is 0.01 to $0.12 \, \text{g/m}^3$. The absorption core is made of a super-absorbent polymer and pulp in which the polymer occupies more than 50 wt %. The second diffusion layer uses hydrophilic fibers such as cellulose or rayon fibers with the unit area weight is the same as the first layer at 20 to $200 \, \text{g/m},^2$ and slightly higher density of 0.04 to $0.2 \, \text{g/cm}^3$.

5   MARKET SHARE AND DEVELOPMENT OF ADULT DISPOSABLE DIAPERS

The size of hew adult disposable diaper market is still small compared with the children's market. However, the potential is great because of the increasing seniors population in developed countries, particularly Japan, where social security is an important issue. The current adult disposable diaper market share worldwide is approximately 360 billion yen. With sanitary products occupying 10 %. The Japanese market share was 24.9

billion yen in 1992, 25.5 billion yen in 1993, 26.2 billion yen in 1994, 27.1 billion yen in 1995, and 28.5 billion yen in 1996, a 5 % increase over the previous year. As can be seen, the market share grew every year.

Adult disposable diapers can be divided into four types—the panty-type, the flat-type, the supplemental pad-type, and the pad-type. The panty-type is the adult disposable diaper in widest use and occupies a 40 % share of the market. The flat-type is used along with diaper cover or a net panty. The supplemental pad is placed onto a diaper in order to improve the absorption capability of the diaper. Thus as these three types exhibit strong absorption and are used for those who require serious protection, the pad-type disposable diaper has a weak absorption capability and is used for those who do not require serious protection.

The adult panty-type diaper is similar in structure to the disposable diaper for children and consists of a top panel, acquisition/transport layer, absorption core, and back panel. The amount of superabsorbent polymer used is 13 to 20 g and the pulp is 50 to 70 g. The absorption core of the flat-type diaper is generally made of 3 to 10 g of superabsorbent polymer and 20 to 50 g of pulp. The concentration of superabsorbent polymer for adult diapers is lower than the ultrathin disposable diapers for children.

The adult disposable diaper may become thinner in a similar manner as those disposable diapers for children, however the reduction of thickness could cause problems due to the greater demand placed on the adult diapers. If the concentration of superabsorbent polymer is increased, the absorption core tends to exhibit gel block. In addition, an absorption core, which has high absorption and diffusion rates, is required as the urination speed and amount of urine is high for adults. If the amount of pulp is excessively reduced to reduce the thickness of the diaper, the absorption and diffusion rates decrease and leakage will result.

The two areas of development for adult disposable diapers are better leakage prevention and further reduction of thickness through improvement in the absorption core, and improvement in product design to make the use of diapers more comfortable. A high gel strength and fast absorption speed are two indispensable properties of superabsorbent polymers for high-performance cores in adult diapers, and development of such polymer gels is strongly desired. Improvement in comfort level reduces the load on those involved in elder care and further protects the privacy of the users. Hence, further effort in this direction is also needed.

6 SUPERABSORBENT POLYMERS FOR DISPOSABLE DIAPERS

Superabsorbent polymers were developed in 1974. Initially, they were mostly graft copolymers of polysaccharides such as starch and cellulose. However, the majority of currently used superabsorbent polymers are lightly crosslinked acrylic acid polymers.

The most important property for the superabsorbent polymers used for disposable diapers is absorbency, which here includes free-swelling absorbency, maintenance capacity, and absorbency under load (AUL). Free-swelling absorbency indicates the absorbency per gram of superabsorbent polymer (g/g) in the presence of an excess amount of a solution such as saline solution. It is also called gel volume, and it can be determined either by the centrifuge capacity method or the blue dextrin (BD) method. The maintenance capacity is the sustained liquid per gram of the polymer (g/g) after applying a constant pressure to the free-swelled gel. On the other hand, absorbency under load is the absorbency (g/g) when the superabsorbent polymer under a constant pressure is made to contact with a liquid. Gel volume, maintenance capacity, and absorbency under load are strongly influenced by crosslink density and surface crosslinking.

6.1 Crosslink Density and Absorbency

A gel volume of a superabsorbent polymer is determined by the balance of forces acting upon the gel. The greater the hydrophilicity and charge density, the larger the gel volume. However, in gels with a constant chemical composition, for example, a crosslinked poly(acrylic acid) salt, the higher the crosslink density, the smaller the gel volume. The most widely used poly(acrylic acid)-type superabsorbent polymers are synthesized by adding a crosslinking agent at the time of polymerization. The crosslinking by this method is chemical crosslinking and its crosslink density greatly influences gel volume, gel strength, and extractable level. The gel strength of a superabsorbent polymer is the strength of a swollen gel to resist deformation and flow under a pressure. Because the shear modulus of a swollen superabsorbent polymer is proportional to the gel strength, it is used to evaluate the strength of a gel. The extractable level is the ratio of the uncrosslinked water-soluble polymer against the total weight.

By reducing the amount of crosslinking agent, the gel volume of superabsorbent polymers can be increased. However, as for the reduction of crosslink density, the gel strength also decreases and the extractable level is increased. In reality, the swelling of superabsorbent polymers is not free but occurs under pressure. The superabsorbent polymers with tow gel strength tend to deform and flow under pressure, leading to undesirable gel blocking. When a superabsorbent polymer swells, the swollen particles deform and eliminate the gap between particles As a result, the passage of the liquid through those gaps will be determined as shown in Fig. 4.

The uncrosslinked water-soluble components from a high extractable level superabsorbent polymer can change the properties of the liquid and increase the adhesion between gel particles. This leads to reduced absorption rate and efficiency. For superabsorbent polymers used for sanitary products, not only the gel volume but also high absorbency under load, high gel strength, and a low extractable level are required. Therefore, the influence of crosslink density on gel volume, absorbency under load, and extractable level are actively evaluated. For example, polymerization of acrylic acid salt using trimethylolpropanetriacrylate (TMPTA) as a crosslinking agent has recently been reported [13]. In this case, the absorbency under load can be increased but the gel volume greatly decreased by the increased concentration of the crosslinking agent The influence of an oxidizing agent such as perchlorate along with a

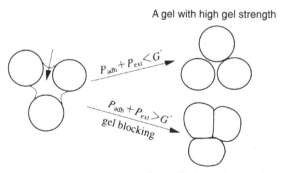

A gel with high gel strength

$$P_{adh} + P_{ext} < G'$$

$$P_{adh} + P_{ext} > G'$$
gel blocking

A gel with low gel strength

Forces that cause gel blocking:
– Interparticle adhesive force P_{adh}
– External pressure P_{ext}

Fig. 4   Gel blocking of a superabsorbent polymer.

crosslinking agent on the gel volume has also been reported [13]. Reduction of the gel volume by the increased crosslinking agent is known to be inhibited by the presence of an oxidizing agent. If the obtained polymer is heated at 170 to 250 °C for 1 to 60 min, absorbency under load can further be improved.

In theory, it is possible to synthesize ultrahigh molecular weight polymers to reduce the amount of crosslinking agents and the extractable level. There is a report on the excellent absorption of water of ultrahigh molecular weight poly(2-acrylamide-2-methyl-l-propane sulfonic acid) (poly-AMPS) that supports this statement [14]. However, there has been no other report on this subject.

6.2   Surface Crosslinking and Absorbency

Surface crosslinking of superabsorbent polymers is widely used as an effective method to improve absorbency [15]. As shown in the schematic diagram of Fig. 5, surface crosslinking creates a shell with high crosslink density (high gel strength) on a core with low crosslink density (low gel strength). As a result, properties such as absorbency under load, extractable level, and gel strength can be improved. Compounds often used as surface crosslinking agents are polyhydroxy, epoxy and polyamine compounds. More specifically, for example, diethylene glycol, glycerol, and ethylene glycol diglycidyl ether are used. In particular, because glycerol is safe and inexpensive, it is widely used. In order to avoid deep penetration of a crosslinking agent into gel particles, the type, amount and crosslinking process must be carefully chosen. The amount of surface crosslinking agent is usually 0.5 to 3 % of the superabsorbent polymer. Surface crosslinking can be achieved by spraying a surface crosslinking agent onto gel particles made of crosslinked poly(acrylic

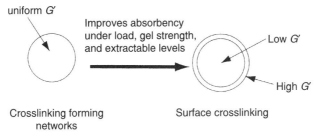

uniform G'

Improves absorbency under load, gel strength, and extractable levels

Low G'

High G'

Crosslinking forming networks

Surface crosslinking

Fig. 5 Surface crosslinking of a superabsorbent polymer.

acid) salt followed by heating at 150 to 200 °C for 10 to 30 min. The surface-crosslinked superabsorbent polymer exhibits a high gel strength and low extractable level. Hence, aggregation of particles can be minimized. As a result, absorption rate under load is high and high absorbency can be achieved.

Figure 6 shows absorbency under load before and after surface crosslinking. In general, the surface-crosslinked superabsorbent polymer shows a reduction of gel volume by 2 to 5 g. However, the absorbency under load is drastically improved. The increased absorbency is considered to be due to the improvement in gel strength and extractable level.

Recently, the surface-crosslinking agent of a porous superabsorbent polymer was studied [16]. The porous superabsorbent polymer is prepared by polymerizing acrylic acid/acrylic acid salt (30/70) using triarylamine as the crosslinking agent and sodium carbonate as a blowing agent. As a surface-crosslinking agent, glycerol is used. Gel strength and absorbency under load are significantly improved.

Figure 7 shows the gel volume on the abscissa and absorbency under load at 4100 Pa on the ordinate for the polymers reported in Reference 13. The majority of these polymers show absorbency under load of 10 to 30 g/g, which is smaller than the gel volume by 5 to 20 g/g. Among commercially available superabsorbent polymers, there are many that show this trend, and a few that show more than 60 g/g gel volume. However, these few exhibit low absorbency under load at 10 g/g. High-

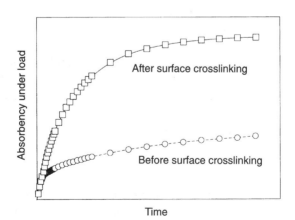

Fig. 6 Absorbency under load before and after surface crosslinking.

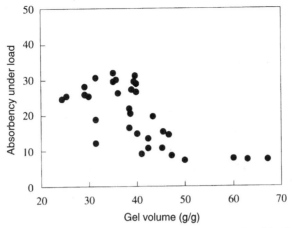

Fig. 7 Gel volume and absorbency under load [13].

superabsorbent polymers used for disposable diapers must have high gel volume and absorbency under load.

6.3 Particle Sizes and Absorbency

In a superabsorbent polymer without surface crosslinking the gel volume is not influenced by the particle size. On the contrary, the surface-crosslinked gel shows reduction of gel volume as the particle size reduces. The higher the degree of surface crosslinking, the more apparent is this trend. For the surface-crosslinked polymers, the surface area increased as the particle size decreases, leading to the increased degree of crosslinking per unit weight. This is probably the reason for the regulation of gel volume.

Distribution of particle sizes does not significantly influence the gel volume. However, those gels with wide particle-size distributions have poorer passage of liquid through particles than those with narrow distributions and thus are not desirable. The particle sizes used for disposable diapers are usually 100 to 800 µm.

6.4 Permeability

One of the important properties for disposable diapers is permeability. Permeability of a porous material includes the flow of a liquid by capillary flow and pressure gradient, and diffusion of the liquid by concentration gradient. An ordinary superabsorbent polymer can be viewed as a porous body with particle diameters of 100 to 800 μm. The transfer speed of water in superabsorbent polymers under gravity and external pressure is much faster than the diffusion speed by concentration gradient. Hence the permeation described here only indicates the former process.

Permeation through porous media can be expressed by the Darcy's empirical equation [17]:

$$Q = -K \frac{\Delta P}{L_0} = -\frac{k \Delta P}{\eta L_0} \tag{1}$$

where Q is the flux per unit cross-sectional area, ΔP is the pressure difference, and L_0 is the thickness of the porous media; K is a proportionality constant and is sometimes called the flow conductivity, and in addition it can be expressed as $K = k/\eta$, where k is the permeability and η is the viscosity of the fluid.

The permeability of swollen gel layers of superabsorbent polymers can be expressed by the Carmen-Kozeny equation, which is based on the Darcy equation when the porosity and surface area are to be considered:

$$Q = -\frac{\phi^3}{k' S_0^2 (1 - \phi)^2} \frac{\Delta P}{\eta L_0} \tag{2}$$

where ϕ is the porosity of swollen gel layers, S_0 is the surface area per unit volume of the swollen gel, and k' is the constant that relates to the tortuosity (geometrical shape and length change). Porosity is the ratio of the air volume to the total volume. The larger the porosity, the greater the permeability of water. From Eqs. (1) and (2), the permeability k of a swollen gel is expressed as follows:

$$k = \frac{\phi^3}{k' S_0^2 (1-)^2} \tag{3}$$

A monodisperse hard sphere is packed in a face-centered cube with the porosity of 0.295 and orthorhombic system with the porosity of 0.395. The porosities of dry superabsorbent polymers are generally 0.40 to 0.75, and are higher than a hard sphere, for example, a glass bead. However, the permeability of superabsorbent polymers is much lower than glass bends of a similar thickness.

Particles of superabsorbent polymers start swelling upon contact with water by the diffusion of water at the wet surface and tend to aggregate by increased adhesion. Especially under pressure, the aggregation of particles is accelerated by the deformation of the particles and the presence of soluble extracts. This is the so-called blocking phenomenon. As a result, the swollen surface layer becomes impermeable to water, which cannot enter the inner part of the gel powder. In order to increase the permeability of superabsorbent polymers, properties to minimize gel blocking such as high gel strength, high porosity and low extractable level are required.

The permeability of superabsorbent polymers can be obtained from Eq. (1) by measuring the flow flux during the steady-state flow through swollen gel layers under hydrostatic pressure. The porosity of a swollen gel ϕ can be calculated as follows by measuring the density:

$$\phi = 1\frac{\rho_s}{\rho_v} \qquad (4)$$

where ρ_s is the bulk density and ρ_r is the true density of the swollen gel, respectively.

7   ENVIRONMENTAL PROBLEMS

The use of forest resources to obtain pulp is often pointed out as an environmental problem with regard to disposable diapers [18–20]. As a result, most of those involved in the production of disposable diapers have already introduced plans to address environmental concerns. The pulp used for disposable diapers is obtained from needle-leaf trees which grow rapidly. The leading companies have instituted a policy of planting twice the number of trees than are cut. Thus, systematically planted needle-leaf trees are used for disposable diapers, thereby eliminating wasteful use of forest resources [18, 19]. In addition, those companies use a high concentration of superabsorbent polymers, hence, the consumption of pulp is now less than half that of the original disposable diapers.

Another problem related to disposable diapers is disposal after use [18, 20]. The weight of a disposable diaper for children is approximately 40 to 50 g, which increases to about 200 g after the use. The annual consumption of disposable diapers in the world is approximately 56 billion, which results in 11.2 million tons of waste. In Japan, the annual consumption is 4.6 billion diapers, which becomes 0.92 million tons of waste. In the Western countries, waste sanitary products are used for landfill whereas in Japan they are incinerated. In the case of landfill, superabsorbent polymers are considered to be safe based on animal tests [21]. Incineration is also safe as the concentrations of toxic materials such as nitroxides, sulfur dioxide, and heavy metals are much lower than the regulation requirements [18, 21].

In the early 1990s, composting and recycling of disposable diapers were actively considered in the United States, Canada and Europe. For composting, a biodegradable component such as pulp and superabsorbent polymer are used after removing the plastic back panel and tape. The compost has excellent water retention and can be used for soil improvement. Techniques to remove those plastics are now under development.

With regard to recycling, Knowaste Technologies, Inc. of Canada is investigating a process [20] in which three materials, the pulp, plastics, and superabsorbent polymer, are recovered after washing, sterilization, separation, and drying. The recycled plastics can be used for industrial oil-absorbing resins, the pulp and paper can be used for toilet paper or sanitary products, and the superabsorbent polymer can be used for water-retention material in horticulture.

8   FUTURE DIRECTIONS

Despite the worldwide recession in the past several years, the worldwide market for disposable diapers has steadily increased. This growth may be sustained on an annual increase of approximately 4 to 5 % until the present due to the rapid economic growth developing countries.

The market share of the panty-type diaper, especially the ultrathin disposable diapers, which are not only comfortable and convenient, but also reduce transpiration because of their compact size, which also requires less space on a store shelf, is expected to grow. There are three requirements for the successful progress of ultrathin diapers.

First, high-performance superabsorbent polymers must be developed. Such performance includes high gel volume, high absorbency under load, low extractable level, and high permeability. Furthermore, in addition to the traditional role of water absorption and holding, it is desirable that superabsorbent polymers perform the same as pulp in transport and distribution.

Second, the development of high-performance acquisition and transport materials is needed. A new material that can transport bodily fluids quickly from the diaper surface to the absorption core and yet minimize reversal of the flow is indispensable for ultrathin products. A capillary fiber is an example of such a material.

Finally, high performance per cost fibrous or sheet superabsorbent polymers instead of particulate polymers can be alternatives. The merits of these fibrous or sheet polymers are applicability at higher concentration, good core integrity, high absorption range, and flexibility of manufacturing processes. Although various superabsorbent fibers and sheets have already been developed, their performance or cost cannot yet compete with the currently available particulate products.

Further advances in the preservation of natural resources, composting and recycling will be made. The ultrathin diapers that were introduced during the 1990s successfully reduced the use of pulp. However, pulp, superabsorbent polymers, tissue, and nonwoven cloth must also be used only in required amounts at the appropriate location in the product. In order to reduce the use of superabsorbent polymers that are troublesome materials during waste treatment, development of polymers with high absorbency under load and biodegradability is needed. Furthermore, advances in the establishment of recovery systems and separation technologies for the back panel, pulp and superabsorbent polymer for composting and recycling are expected. In addition, a biodegradable back panel and a top panel is desired.

REFERENCES

1 Uggla, E. (1995). *Nonwovens World*, Spring 71.
2 Nanna, P. (1993). *The New Nonwovens World*, Winter, 110.
3 Lee, H.J. (1994). *Nonwovens World*, April, 53.
4 (1996). *Kokusai Shogyo* **5**:117.
5 EP 0 509. 708 A1; US Pat. 5,314,420 (1994); Tokkyo Kokai 39487 (1994); Tokkyo Kokai 74331 (1994).
6 (1992). *Nonwovens Markets* **7**(4):7, 2.
7 (1994). *Nonwovens Industry* **25**:51.
8 (1994). US Pat. 5,324,575.
9 Abitz, P.R. (1994). *Nonwovens World*, Fall 94.
10 Suzuki, T. (1994). *Nonwovens Cloth Report*, p. 4.
11 (1992). US Pat. 5,294,478.
12 WO 94/20547.
13 WO 94/20547.
14 Osada, Y. and Takase, M. (1983). *J. Chem. Soc., Jpn.* **3**:439.
15 (1989). Tokkyo Kokai 17411, Tokkyo Kokai 39487 (1994); Tokkyo Kokai 74331 (1994); Tokkyo Kokai 156034 (1993); EP 0605150 A1 (1993); Tokkyo Kokai 70694 (1979); Tokkyo Kokai 30336 (1990); and Tokkyo Kokai 62745 (1991).
16 (1994). US Pat. 5,314,420.

17   Chartterjee, P.K. (1985). *Absorbency,* Amsterdam: Elsevier Science, p. 43.
18   (1994). *Nonwoven Cloth Report,* 17.
19   (1994). *Steady Eyes* **95**:60.
20   (1991). *Managing Nonwoven Products Waste,* Miller Freeman Inc., pp. 44, 114, 142.
21   Goldstein, J.E. (1994). *The New Nonwovens World,* Winter, p. 104.

Section 2
Sanitary Napkins

LING WANG

1 INTRODUCTION

Major application areas for superabsorbent polymers (SAP) are disposable sanitary products, which include sanitary napkins, and disposable diapers for children and adults. Of these sanitary products, superabsorbent polymers were first used in sanitary napkins. Later, superabsorbent polymers were adopted for disposable diapers for children and adults. Accordingly, the performance of sanitary products has been dramatically improved. At the present time, more than 90 % of superabsorbent polymers produced are considered to be used for sanitary products. Here, the history, current status and recent trends and advances in the sanitary napkin will be given, and development of new superabsorbent polymers for sanitary napkin applications will be described.

2 HISTORY OF SANITARY NAPKIN

Prior to the 1950s, cotton or cloth was used to cope with menstrual blood. Sanitary napkins have undergone a revolutionary change since they were introduced in Japan in 1961. It has successfully changed the previously held negative attitudes of Japanese women regarding menstruation.

As listed in Table 1, there were a number of technological developments during the 35-year history of the sanitary napkin [1]. Many of the technological developments during the 35 years occurred as women became more active, which created the demand for a sanitary napkin that afforded comfort and convenience. The result has been a thinner, comfortable product that could be used for a longer time. One such technological development, which deserves special attention, occurred around 1978 when superabsorbent polymers were used for sanitary napkins. As a result, the thickness of sanitary napkins was reduced to less than half that of the original products and performance was also drastically improved. Triggered by this event, superabsorbent polymers were applied to disposable diapers for children two years later and somewhat later for adults. During 1979 and 1985, the position and size of the tape to prevent shifting were changed.

Table 1   History of sanitary napkins.

Years	Change of sanitary products
1960	Cotton and cloth were used as absorbents
1961	Introduction of sanitary napkins in Japan
1965	Introduction of both side cut, nonwoven roll-up type
1968	Fluff pulp was used as an absorbent and wet-produced nonwoven cloth was used as the top
1974	Introduction of adhesive tape for slippage prevention
1976	Dry-produced nonwoven cloth, span bond, was used as the top sheet. Products became thinner and more compact.
1978	A superabsorbent polymer was used as an absorbent for the first time. It is two years prior to the introduction of disposable diapers for children.
1982	Introduction of 3D cut, and increased variety of adhesive tape positions
1985	Introduction of longer-size products for long-term and nighttime use
1986	Market was activated by the introduction of a new surface-treated polyethylene dry mesh
1987	Introduction of a flapper-type that does not slip or leak
1989	Introduction of an ultrathin product 2 mm thick
1992	Adoption of one-step wrapping technology that allows both inside material and wrapping material to be peeled off at once
1993	Adoption of 3D side gathering
1994	A special sheet was added between the top sheet and the absorbent. They are made into a single part by quilting process.
1994	Introduction of a napkin with presslines to prevent shifting of the napkin by adding gap-like lines
1996	Introduction of a curved napkin emphasizing good fit

However, such technological developments had little impact on the market, which was inactive. In 1986, Procter & Gamble Co. Introduced an innovation, the dry mesh sheet, which was radical at that time but did succeed in activating the market. Until then, nonwoven cloths were used for the top sheet, but they absorb water and the portion in contact with the skin is always wet, thus creating discomfort. The dry mesh sheet is a poly-ethylene sheet with holes, and it does not absorb or reverse moisture but provides a dry and clean environment. As a result, the Procter & Gamble product became the leader in the market. Introduction in 1987 of a napkin with flaps that does not slip or leak and an ultrathin napkin 2 mm thick further accelerated the development of high value-added sanitary napkins. In the 1990s, various technologies were introduced, including one-step wrapping that can simultaneously peel off the interior and wrapping, three-dimensional (3D) side gatherings, a quilting process to prevent shifting by combining the top sheet and absorbent, and presslines that also prevent shifting by adding gap-like lines on both sides. In the near future, the competition for high value-added products will continue to intensity.

3   MARKET SHARE SANITARY NAPKINS

3.1   Worldwide Market

The worldwide market for sanitary napkins and disposable diapers for children and adults is approximately 30 billion dollars [2]. Of this amount, the market for sanitary napkins is about 10 billion dollars, around 33 % of all sanitary products. Based on the number of products worldwide, the total number of sanitary products comprises approximately 142 billion, which includes about 80 billion sanitary napkins (about 56 % of all sanitary products). Figure 1 illustrates the market shares based on the world regions.

Assuming that approximately half the world population of 5.6 billion is women and about 50 % of the women are in the age range of 15 to 50 and are users of sanitary napkins, the world market will be 252 billion napkins if 100 % of those who need them, use them. At the present time, only less than 32 % of women in this age bracket use sanitary napkins. Hence, emphasis on market development is believed to shift to developing countries.

The market share of sanitary napkins in the United States, Western Europe and Japan is nearly 100 % and no significant growth can be expected. Hence, even by introducing high value-added products, the

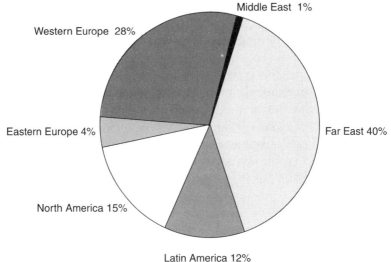

Fig. 1   Market share of sanitary napkins in various regions of the world.

future growth rate is expected to be around 1 to 3 %. On the other hand, in the developing countries of the Asia/Pacific region other than Japan and Latin American countries, the market share is low. Thus, it is expected that an annual growth of 10 to 15 % can be maintained until the present by cultivating new users.

The major producers of sanitary napkins in the world are Procter & Gamble, Kimberly-Clark, Molnlycke/Peaudouce, Johnson & Johnson, Kao and Unicharm. In addition to these, various regional producers are joining the market. However, the world market is dominated by the major producers and such trend is expected to intensify in 2001. It is noteworthy to recognize the healthy market share of low-cost products, that is, the so-called generic brand products, despite the worldwide recession in recent years.

3.2   Asian Market

In recent years, Asia has become the center of worldwide economic growth and has begun experiencing a rapidly improving living standard. Accompanying this has been a rapidly growing market for sanitary napkins.

The sanitary napkin market of all Asian countries combined is expected to be about 38 billion napkins. At first glance, this might seem large, but based on the previously mentioned assumptions, the market share is a mere 32 %. Due to the high economic growth and low market share, nevertheless, the growth in Asian countries other than Japan is remarkable.Using South Korea and China as examples, in South Korea, the sanitary napkin was first introduced on the market in South Korea in 1971. By 1980, the market share had increased to 40 % [3]. In 1993, the market was 1.9 billion napkins and the market share was 87 %. In the case of China, since its introduction in the 1980s, the market has rapidly expanded. By 1993, already a market of 21.6 billion napkins was established and the market share was 40 % [4]. In the next several years, an annual growth of 15 % is projected. In Thailand and the Philippines, the market is still growing. By 2001, the growth rate of Asian countries other than Japan is expected to be 10 to 15 %. At present, the leading makers of sanitary napkins in the Asian market are Procter & Gamble, Kimberly-Clark, Johnson & Johnson, Kao and Unicharm.

3.3   Japanese Market

The total sanitary product market, which includes sanitary napkins and disposable diapers for children and adults, is approximately 260 billion yen [5], of which sanitary napkins account for 90 billion yen. The market share among sanitary products is about 34 %. Japan is the largest market in Asia and occupies 45 % of the Asian market in sales. As many women already use sanitary napkins, and the number of products sold may not increase. Thus, the only way to increase the market is to rely on the development of high value-added products. As shown in Fig. 2, the consumption was 80.5 billion yen in 1991, 82.1 billion yen in 1992, 83.3 billion yen in 1993, and 87.1 billion yen in 1994 with low growth rate. These figures indicate market saturation [5]. Each producer attempts to increase their market share by introducing even more functional innovations such as three-dimensional side gathering, quilting processed top sheet, and one-step wrapping. By the introduction of these innovations, the Japanese market may grow at an annual growth rate of 1 %.

The producers of sanitary napkins in Japan are Procter & Gamble, Unicharm, Kao, Shiseido and Daio Seishi. The top three, Procter & Gamble, Unicharm, and Kao, occupy more than an 85 % share.

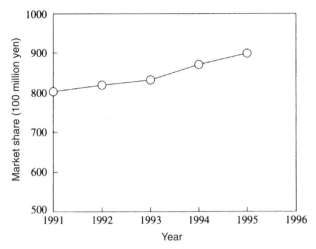

Fig. 2 Market share of sanitary napkins in Japan.

4 STRUCTURE OF SANITARY NAPKINS

Functions that are required of sanitary napkins are: (i) high absorbency/no leakage, especially no side leakage; (ii) no chafing; (iii) comfort; (iv) good fit to the body contour; and (v) no humidity that leads to skin rash. In order to improve on these functions, there have been many technological developments over the 35-year history of sanitary napkins. Currently, sanitary napkins consist of a top sheet, a diffusion layer, an absorption layer, a back panel, and adhesive tape to prevent shifting. In addition, flaps or 3D side gathering has been added. Figure 3 depicts a cross-sectional view of the general structure of sanitary napkins.

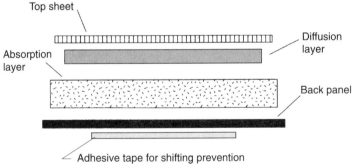

Fig. 3 Cross-sectional view of a sanitary napkin.

Nonwoven cloth was once the material of choice for the top sheet. In 1986, a new material, mesh sheet, was introduced and gained rapid popularity with young users. At present, the mesh sheet has completely replaced the nonwoven cloth. Based on numbers, the mesh sheet now occupies more than 60 % of the market. The mesh sheet is a polyethylene sheet with funnel-like holes. The polymer surface is treated by a surfactant to make it hydrophilic, thus blood does not adhere and its dryness and cleanliness are popular features among users. In particular, a dry mesh sheet, a 3D polyethylene film with holes, exhibits no obvious sign after use, making it an attractive feature for the consumers.

The role of the diffusion layer is to quickly accept the bodily fluid from the top sheet and transfer it to the absorption layer. As the diffusion layer, tissue, high-loft nonwoven cloth, or pulp laminate is used.

Fluff pulp or crepe pulp was used as the absorbent prior to 1978. Today, materials with a superabsorbent polymer dominate the choice of the material. There are basically two types of absorption layers—either a laminate in which a superabsorbent polymer is sandwiched by a tissue or nonwoven cloth, or a mixture of a superabsorbent polymer and pulp wrapped by a tissue around a nonwoven cloth. In addition, there are variations of these two of absorption layers. For example, a transport layer is introduced between the absorption core and the back panel. Furthermore, a small-sized laminate with a high polymer concentration can be placed on top of the absorption core. The amount of a superabsorbent polymer used for a sanitary napkin is approximately 0.6 to 1.5 g.

The back sheet is a water-nonpermeable sheet which makes contact with the clothing. It is mainly made of a polyethylene film. A laminate of polyethylene film and nonwoven cloth is also used as a cloth like back panel.

As shown in Table 2, there are many kinds of sanitary napkins depending on the style of side characteristics, size of product, and

Table 2 Types of sanitary napkins.

Classification	Product type		
Size	Regular, about 20 mm	Long, 26–30 mm	Extralong, ≈ 33 mm
Thickness	Thick, about 10 mm	Thin, 4–6 mm	Ultrathin, ≈ 2 mm
Side characteristics	No flap	With flaps	
Top sheet	Nonwoven cloth	Mesh sheet type	
Wrapping style	Usual wrapping	One-step wrap	
Side gathering	No 3D side gathering	With 3D side gathering	

thickness. The side characteristics can be classified as 3D side gathering or flap-types. The lengths are classified as regular (20 mm), long (26–30 mm), and extralong (33 mm); with regard to thickness, these are thick, thin, and ultrathin products. The thick napkin uses a mixed core consisting of a superabsorbent polymer and pulp, its total thickness is approximately 10 mm. The ultrathin napkin has a laminated absorption core and its thickness is at most 2 mm. The thin napkin has a thickness of 4 to 6 mm, and is the intermediate-sized napkin. Many consumers use different products depending on regular use, nighttime use or extended use.

Approximately 10 % of the users have experienced some type of leakage problems resulting in soiled clothes. This indicates that leakage protection, the most fundamental function of sanitary napkins, is not yet sufficient. Thus it is necessary to improve the absorbency of the absorption core to improve the protection function. A superabsorbent polymer that has a high absorbency towards blood, the so-called menses-specific superabsorbent polymer, is essential for a high-performance core and its development specific is strongly desired.

5 DEVELOPMENT OF SUPERABSORBENT POLYMERS FOR SANITARY NAPKINS

5.1 Absorbency for Blood

Superabsorbent polymers were first used for sanitary napkins prior to their use in disposable diapers for children. However, the polymers currently used are not specifically for absorbing blood but also for absorbing urine. Because menstrual blood contains large amounts of proteins, blood cells and other system waste, the viscosity and surface tension is higher than that of urine [6]. Current superabsorbent polymers have problems of low absorbency, slow absorption rate, and they readily cause gel blocking. For example, as shown in Table 3, the majority of commercial superabsorbent polymers exhibit approximately 40 to 45 g/g absorbency towards urine, while it is about 20 to 25 g/g towards blood. The absorption rate is about 0.2 to 0.5 g/g/s for urine whereas it is 0.05 to 0.1 g/g/s for blood. Also, due to the gel blocking, the blood absorbency is not necessarily good. Gel blocking is the blockage of the flow of the fluid when a superabsorbent polymer is swollen. External pressure causes the swollen particles to deform, adhere and eventually close the gap between particles, thus

Table 3 Absorptivity of commercial superabsorbent polymers.

	Blood	Urine	Distilled water
Absorptivity (g/g)*	20–25	40–45	450–600
Absorption rate (g/g/sec.)	0.05–0.1	0.2–0.5	0.5–0.8
Gel blocking	observed	none	none

*Centrifugal (1,500 rpm) absorptivity

blocking the flow of the fluid [7]. Adsorption of proteins in the blood further strengthens the adhesion of particles and accelerates the gel blocking.

For superabsorbent polymers used for sanitary napkins, characteristics such as high absorbency, fast absorption rate, and no gel blocking are required. In order to improve absorbency towards blood the following approaches have been investigated: (i) a method to treat the surface of ordinary superabsorbent polymers; (ii) blending with inorganic or organic materials; (iii) use of superabsorbent polymers with different compositions; (iv) increased porosity; and (v) use of fibrous superabsorbent polymers. Details of these approaches will be described in the following.

5.2 Improvement of Superabsorbent Polymers

Absorbency of superabsorbent polymers to blood is determined by the following four processes: (i) wetting and wicking of blood on the surface of a superabsorbent polymer; (ii) diffusion of blood into the matrix when the blood makes contact with the polymer; (iii) swelling of the polymer; and (iv) transport of the blood between particles by capillary effect. Changes of the chemical compositions and physical structures can accelerate these four processes and improve absorbency.

5.2.1 Changes in Chemical Compositions

Chemical compositions of superabsorbent polymers strongly influence the aforementioned processes (i), (ii), and (iii). For example, there is a close correlation between the polymer structure, type and concentration of ionic groups, and salt concentration with the absorbency. Polyethyleneoxide (PEO)-type superabsorbent polymers compare favorably with poly(acrylic acid)-type polymers in the absorbency of electrolytes. However, because of the slow absorption rate, the application range would be limited.

The degree of neutralization of carboxylic acid and control of the counter ion are investigated on a current poly(acrylic acid)-type superabsorbent polymer to improve absorbency towards blood [8]. The maximum absorbency toward urine is obtained when the degree of neutralization of carboxylic acid is approximately 70 %. By contrast, this value drops to about 30 to 40 % for blood. The polymer with potassium or lithium as a counterion rather than sodium is reported to have higher absorbency towards blood.

5.2.2   Surface Treatments

Gel blocking may be accelerated by the proteins that adsorb onto gel particles. Hence, by improving the dispersion and wettability of superabsorbent polymers in blood, it is expected that the absorption and absorption efficiency will increase. For example, coating an ordinary superabsorbent polymer with poly(ethylene glycol)(PEG) or its derivatives has been investigated[9,10].

The desirable molecular weight of PEG is 300 to 2000 and the amount added is 0.5 to 3 % with respect to the superabsorbent polymer. With this approach, wettability and wicking of the superabsorbent polymer with blood has been improved.

5.2.3   Blends with Additives

Blending a water-soluble organic or inorganic compound with a superabsorbent polymer has been studied [11]. Organic compounds that can be used are acetic acid salts, formic acid salts, adipic acid salts, citric acid and urea, whereas for inorganic salts, hydrochloric acid salts, sulfuric acid salts, and phosphoric acid salts can be used. These additives remain between superabsorbent polymer particles to inhibit contact of the particles, to control gel blocking in blood, and to improve absorbency.

5.2.4   Porous Superabsorbent Polymers

Because porous superabsorbent polymers possess large surface areas as well as channels and capillaries for liquid diffusion, high blood-absorption rates can be expected. As shown in Fig. 4, the lower the bulk density (the property that relates to porosity) of superabsorbent polymers, the higher the blood-absorption rate. Thus, various porous superabsorbent polymers have been developed and their absorbencies are under investigation.

It is possible to prepare a porous superabsorbent polymer with extremely low bulk density without losing absorbency if a superabsorbent polymer that is swollen with water by 5 to 20 times is freeze dried [12,13]. A porous superabsorbent polymer prepared by adding 0.1 to 2 % of a cationic polymer such as a PEG or amine-epichiorohydrin adduct shows 5

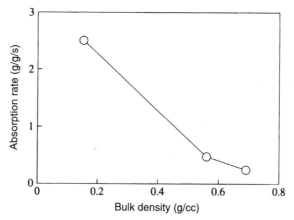

Fig. 4 Correlation between the bulk density of a superabsorbent polymer and absorption rate of blood.

to 10 times high blood-absorption rate than that without the additive. When acrylic acid polymerizes, a porous superabsorbent polymer can be prepared in the presence of a blowing agent. As a blowing agent, an inorganic compound such as carbonate salt, or a low-boiling-temperature organic solvent such as 1,1,2-trichlorofluoroethane (Freon 1,1,2) or a hydrocarbon solvent may be used [14–16]. The bulk density of the obtained porous polymer is as low as 0.01 to 0.05 g/cc.

Another method of preparing porous superabsorbent polymers is by agglomeration [17]. This method involves the formation of agglomerates of a fine superabsorbent powder in the presence of a water-soluble polymer. Such agglomerates are porous and are expected to show high absorbencies.

5.2.5 Fibrous Superabsorbent Polymers

Fibrous superabsorbent polymers have larger surface areas than particulate ones. Furthermore, they have excellent diffusion and permeation of liquid by capillary effect and thus high absorbencies are expected. These fibers can be manufactured with ordinary structural fibers such as cellulose fibers or synthetic fibers by the airlaid technique, and the potential for further development is high.

Representative fibrous superabsorbent polymers are Fibersorb and Fiberdri fibers manufactured by Camelot Superabsorbent of Canada, Ltd., Oasis fiber by Allied Colloid Ltd. (USA), and Lansdale F fiber by Toyobo

[18–20]. Both Fibersorb and Fiberdri are fibers made of copolymers of isobutylene and maleic acid. They have a fineness of 10 deniers and their fiber length is from several millimeters to several tens millimeters. Oasis fiber is based on poly(acrylic acid) and is prepared by spinning a bulk-polymerized linear polymer followed by crosslinking. Lanseal F is prepared by heating acrylamide fibers in an alkaline solution to cause hydrolysis and crosslinking. This is a double-layered fiber consisting of an outer layer of acrylic acid salt and an acrylamide core.

A nonwoven cloth made of a superabsorbent polymer and a synthetic fiber has been shown to have an excellent absorbency towards blood [21].

5.2.6   Quasifibrous Superabsorbent Polymers

Quasifibrous superabsorbent polymers are a composite of ordinary fibers and superabsorbent polymer powder that is adhered to the fiber by a binder. Various quasifibrous superabsorbent polymers are being investigated. For example, a composite made of a cellulose fiber and a superabsorbent polymer has already been developed [22]. Polycarboxylic acids, polyamines or polydiols can be used as a binder. The interaction between the cellulose fiber and the binder or the superabsorbent polymer and the binder is hydrogen bonding or covalent bonding. Quasifibrous superabsorbent polymers do not exhibit gel blocking and a capillary effect caused by the fiber networks, thus providing favorable conditions for blood absorption.

5.2.7   Superabsorbent Polymer Sheet

Superabsorbent polymer sheets have also been developed. If a particulate superabsorbent polymer is crosslinked between particles by a cationic polymer, for example, an amine-epichlorohydrin adduct, a superabsorbent polymer sheet can be prepared [23]. A particulate superabsorbent polymer, cellulose acetate and a plasticizer are dispersed in an appropriate solvent followed by casting onto a nonwoven cloth substrate. Upon irradiation by ultrasound, a superabsorbent sheet can also be prepared [24]. However, such material has poor compatibility with blood and lacks capillary channels. Thus, further improvement is necessary. Accordingly, various new superabsorbent polymers have been developed. Unfortunately, these polymers show insufficient property improvement or are too costly to be commercially available.

6 FUTURE DIRECTIONS

The sanitary napkin market grew steadily in the past several years despite the worldwide recession, and it is expected that the market will continue to grow by 6 to 7 %. Important properties of future sanitary napkins are: (i) improved leakage protection; (ii) greater comfort and convenience; and (iii) thinner and more compact size.

To improve leakage protection, improvement in the absorbency of the absorption core and in the fitting of the product to the body contour are important. The following three points are important directions for the development of the absorption core. First, high-performance superabsorbent polymers must be developed. Development of menses-specific superabsorbent polymers that have high absorbency, high absorption rate, and lack of gel blocking in the presence of blood will be more important. Development of high performance per cost fibrous or sheet superabsorbent polymers will further be accelerated in the future. The merits of fibrous or sheet superabsorbent polymers are their ability to be used at a much higher concentration than the particulate one, good core integrity, high absorbency, and good flexibility in manufacturing processes. Second, the development of high-performance acquisition and transport materials is important. A new material that can transfer bodily fluid from the top sheet surface to the core with little reverse permeation is indispensable to prevent leakage. Recently, high loft nonwoven cloths that possess a good balance of hydrophilic and hydrophobic properties are attracting attention. Finally, core designs to offer a good fit to the body contour and proper absorption profile will be more important. Napkins with higher absorbency near the center are already available on the market.

Development of napkins with not only leakage protection but also comfort and convenience will be important. New top sheets with no reverse permeation, that do not stick or chafe, and ensure cleanliness and dryness will be developed. The market share of ultrathin napkins ensuring good fit and allowing easy movement will continue to grow in the future. Furthermore, napkins will be designed that are based on human engineering. These napkins will stretch and adjust to the movements of the user. Although there is some desire for flushable products, there are technical and environmental difficulties. However, in order to respond to the needs of consumers, it is fairly certain that there will be some effort in this direction.

REFERENCES

1 Suzuki, T. (1994). *Nonwoven Cloth Report*, p. 4.
2 Uggla, E. (1995). *Nonwovens World*, Spring, 71.
3 Lee, H.J. (1994). *Nonwovens World*, April, 53.
4 Dunn, T.C. (1994). *Nonwovens World*, Summer, 47.
5 (1995). *Kokusai Shogyo* **5**:38.
6 Nagorski, H. (1992). *The New Nonwovens World*, Fall, 101.
7 Osada, Y. and ??? (1995). *Development of and the Latest Technology of Functional Polymer Gels*, CMC Press, p. 111.
8 (1993). US Pat., 5,241,009.
9 (1993). US Pat., 4,190,563.
10 WO 95/22355.
11 (1992). US Pat., Re 33,839.
12 WO 95/22357.
13 (1986). US Pat. 4,624,868; Tokkyo Kaiho 304127 (1988); Tokkyo Kaiho 304128 (1988); and Tokkyo Kaiho 16143 (1992).
14 (1994). US Pat. 5,314,420; US Pat. 5,118,719 (1992); US Pat. 5,154,713 (1992).
15 (1994). US Pat. 5,338,766.
16 (1993). US Pat. 5,236,965.
17 (1986). Tokkyo Kaiho 97333; and US Pat. 5,122,544 (1992).
18 Cheyne, I. (1994). *The New Nonwovens World*, Winter, 103.
19 (1991). US Pat., 4,997,714.
20 Takahashi, T. (1989). *Kobunshi* **38**:1074.
21 (1994). Tokkyo Kaiho 207358 (1994).
22 WO 95/19191.
23 (1992). US Pat., 5,149,334 (1992); and US Pat., 5,321,561 (1994).
24 Euripides, J.M. and Ryans, W.T. (1995). *Nonwovens World*, Spring, 90.

CHAPTER 2

Daily Commodities

Chapter contents

Section 1
Cosmetics

FUMIAKI MATSUZAKI

1 SKIN CARE COSMETICS

1.1 Introduction

General skin care cosmetics are designed to maintain cleanliness of the skin as well as to provide moisture and oil effectively to skin that is affected by external factors [1, 2]. In order to maintain or to have healthy skin various skin care cosmetics that satisfactorily address these efforts have been developed. They include cleansers, creams, and emulsions. Various polymers are added to these products for their aesthetic appearance, good tactile quality during use, stabilization of the base materials, and for the appearance of the intended effect.

Table 1 lists polymers used for skin care cosmetics. As can be seen from the table, the polymers used for skin care exhibit generally pseudoplastic or thixotropic flow and possess properties to form gels readily [3]. The roles these polymer gels play are extremely significant from the viewpoints of ease of use and functionality. Here, viscosity control, stabilization of emulsion, and maintenance of moisture are the functions of polymers in skin care cosmetics.

38

Table 1 Polymers used for skin care cosmetics.

Classification			Classification	Items
Organic materials	Natural		Polysaccharides	Guar gum, locust bean gum, carrageenan, garactan, Arabia gum, tragacarth gum, pectin, mannan, starch
			Microbe-type	Dextrin, caldron, hyaluronic acid
			Animal-type	Gelatin, casein, collagen
	Half synthetic		Cellulose-type	Methyl cellulose, ethyl cellulose, hydroxyethyl cellulose, hydroxypropyl cellulose, carboxymethyl cellulose, ethylhydroxyethyl cellulose, methylhydroxypropyl cellulose
			Starch-type	Soluble starch, carboxymethyl starch, methyl starch
			Arginic acid-type	Arginic acid
	Synthetic		Vinyl-type	Poly(vinyl alcohol), poly(vinyl pyrrolidone), poly(vinyl methylether), carboxy vinyl polymers, sodium salt of poly(acrylic acid)
			Others	Poly(ethylene oxide), ethylene oxide-propylene oxide copolymers
Inorganic materials			Clay minerals	Bentonite, labonite
Oil-soluble polymers	Organic materials	Half synthetic	Cellulose-type	Ethyl cellulose, ethylhydroxyethyl cellulose, methylhydroxypropyl cellulose
		Synthetic	Vinyl-type	Poly(vinyl alcohol), poly(vinyl pyrrolidone), poly(vinyl methylether), carboxyvinyl polymers

1.2   Viscosity-Increase Functions

1.2.1   Hydrophilic Gelation Agents

Hydrophilic gelation agents can be classified into nonion, anion and cation type based on their charges. The hydrophilic gelation agents for skin care must be safe to the skin and at the same time have the ability to increase viscosity. For these reasons, the nonionic or anionic type is generally used. Figure 1 shows the viscosity of nonionic and anionic gelation agents, including carrageenan, carboxymethyl cellulose, and sodium arginic acid.

Other important requirements for the skin care hydrophilic gelation agents are the aesthetics and the feel of the gel to the skin. Carboxyvinyl polymers, which are crosslinked poly(acrylic acid), satisfy all these requirements. Carboxyvinyl polymers form transparent gels in water or aqueous ethanol, and there is neither a fiber-forming property nor tackiness. There are various commercial carboxyvinyl polymers and the viscosity of some of them are shown in Fig. 2 [4]. As seen in the figure, the differences in viscosity depend on the molecular weight and the degree of crosslinking. These materials show high viscosity enhancement near the neutral pH due to the repulsion of the dissociated carboxylic groups. Figure 3 depicts the relationship between the pH of carboxyvinyl polymer

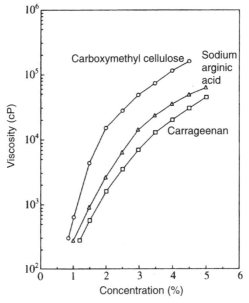

Fig. 1 Viscosity of various water-soluble gelation agents.

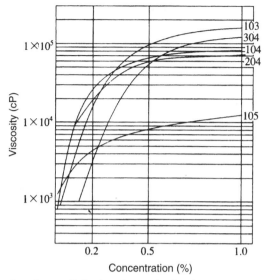

Fig. 2 Viscosity of crosslinked carboxyvinyl polymers at pH = 7 ± 0.2 (cross-linked carboxyvinyl polymers: Highvis Wako 103, 304, 104, 204, and 105, Wako Pure Chemicals).

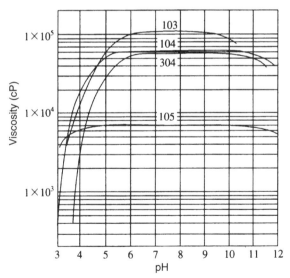

Fig. 3 Viscosity of crosslinked carboxyvinyl polymers at pH = 7 ± 0.2 (cross-linked carboxyvinyl polymers: Highvis Wako 103, 104, 304, and 105, Wako Pure Chemicals).

aqueous solutions and the viscosity [4]. It is known that a general shortcoming of polyelectrolyte gel is the extreme reduction of viscosity in the presence of a salt. Similarly, carboxyvinyl polymers also are influenced by a salt as shown in Fig. 4. Such a property relates to the problems when a cosmetic is to be removed from a bottle. If the cosmetic makes contact with a fiber on which a large amount of salt exists, the viscosity reduces at the cosmetic/finger interface, resulting in difficulty in removal from a bottle by fingers. To solve this problem, it is necessary to use more polymer or make simultaneous use of a nonionic polymer. However, a simpler and more fundamental solution is to develop a polymer that is influenced little by a salt. A carboxyvinyl polymer will form large balls if a large amount of polymer is added to the solvent at once.

Recently, a new carboxyvinyl polymer that has addressed this problem has become commercially available (trade name: Carbopol Ultrez 10®). This polymer shows a low viscosity even at a high concentration and exhibits no large aggregated balls. However, after neutralization, the viscosity enhancement effect is the same as the traditional materials. Figures 5 and 6 show viscosity enhancement at initial dispersion and after neutralization, respectively [5].

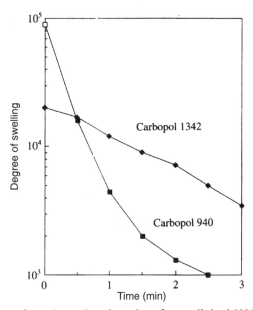

Fig. 4 Influence of a salt on the viscosity of crosslinked (1% concentration) carboxyvinyl polymers [5].

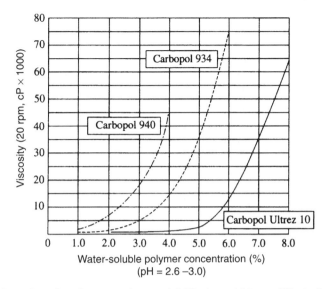

Fig. 5 Viscosity of an improved material (Carbopol Ultrez 10) at dispersion.

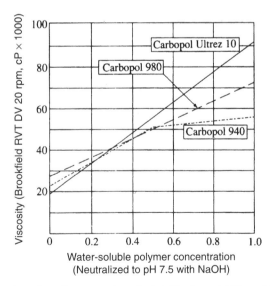

Fig. 6 Viscosity of an improved material (Carbopol Ultrez 10) after neutralization.

A new hydrophilic gelation agent, acrylic acid-based polymer/water-swelling clay composite that was recently developed will be described in the following [6]. In general, acrylic acid-type polymer and water-swelling clay are often used as gelation agent for cosmetics. However, it is known from experience that if both are mixed in water, a heterogeneous gel is formed and the original functions are lost. Hence, simultaneous mixing has been avoided. On the other hand, there have been several reports in the interaction between clay and polymers from the viewpoint of soil improvement [7–9]. Based on reported soil improvements, we attempted the formation of a composite. As an example, results on a composite made of an acrylic acid polymer, carboxyvinyl polymer (Highvis Wako 104; Wako Pure Chemicals) and water-swelling clay (Laponite XLG; Laporte Co.) will be presented. First, Highvis Wako 104 and Laponite XLG aqueous solutions at the same concentration are prepared. These two solutions are mixed at various compositions ϕ and homogeneous gels are formed.

This composite gel is precipitated by acetone, dried and crushed; composite powders with various ϕ are obtained. If the obtained powder is again dispersed in pure water, the material exhibits a higher gel strength prior to drying. Figure 7 shows the gel strength before and after a drying treatment. The gel strength depends on the mixing ratio ϕ and it does have an optimum ratio. Upon comparison of the gel strength before and after the drying, the gel strength increased by a factor of two upon drying. This trend was observed on other acrylic acid-type polymer/water-swelling clay composites. Furthermore, it was also observed in any drying method,

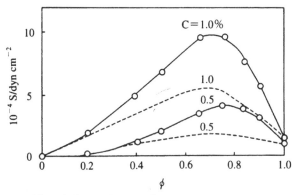

Fig. 7 Compositional dependence of the gel strength of Highvis/Laponite mixture gel (dotted line) and its regenerated gel.

such as the freeze-drying method, hot air drying, *k* vacuum drying, and spray drying. At present, there is no logical explanation for this.

However, it is thought that the molecular distances were shortened by drying and that the hydrogen or ionic bonding strengthened. Figure 8 shows the time dependence of the degree of swelling of an acrylic acid-type polymer/water-swelling clay composite powder in pure water along with the data of a hydrophilic polymer, Aquaric CA-W 4® (Japan Catalyst). As can be seen from the figure, both curves almost overlap. Thus, they have fundamentally the same absorption rate and absorbtivity. Although not shown here, polymers of the carboxylic group and water-swelling clays generally form stable composites. While there are some differences, they exhibit water absorption and gelation capability. Accordingly, mixing hydrophilic gelation agents, which were not desirable for simultaneous use, can sometimes result in a new gelation agent.

1.2.2 Oleophilic Gelation Agents
The feel of oleophilic gels for other than cleansing and sunscreen is undesirable for general skin care cosmetics. Thus, there is much less need

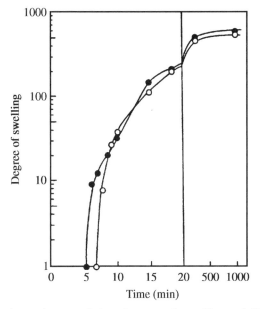

Fig. 8 Time dependence of the degree of swelling of Highvis/Loponite composite powder prepared at the optimum condition (○) and hydrophilic polymer, Aquaric (●), in water.

for oleophilic gels than for hydrophilic gels. Additionally, there are few oleophilic gels that show similar effective viscosity enhancement to oils as hydrophilic gelation agents. This also contributes to the lack of oleophilic gelation agents. Among these, organic clays are oleophilic gelation agents that are not sticky. The organic clay is the clay from which exchangeable cations between silicate layers are exchanged by quaternary amine. The Na^+ ions of one of the water-swelling clays, montmorillonite, can be cation exchanged with relatively large amine molecules. This is called organophilic bentonite (trade name: Benton®) and does not swell in water but does exhibit swelling and gelation in nonaqueous solvents [10, 11]. This Na^+-type montmorillonite can be exchanged not only with quaternary amines but also with a nonionic surfactant that possesses ethylene oxide (EO) chains. Yamaguchi *et al.* found that Na^+-type montmorillonite that was exchanged with an EO chain-containing nonionic surfactant can gel liquid paraffin that is often used in skin care cosmetics [12]. Figure 9 shows the viscosity of montmorillonite-added oleophilic gels in the presence of a nonionic surfactant. The viscosity changes depending on the type and amount of the nonionic surfactants. From the result of x-ray diffraction, it was found that a monomolecular layer of the nonionic surfactant is chelated into the clay galleries into which the oil penetrates. If a nonionic surfactant is chelated into the quaternary amine-exchanged organic clay, the gel strength changes depending on the surfactant. The stronger the chelating ability of a nonionic surfactant, the stronger the oleophilic gels and the more effective they are as water repellants. This water-repelling ability is useful as the base material for sunscreen agents, because it is not washed away by perspiration or water (waterproof).

1.3   Emulsification

To effectively provide water and oil to the skin, O(oil)/W(water)-type or W/O-type emulsified cosmetics are important in skin care. An emulsion of liquids that are not miscible is not in thermodynamic equilibrium and thus is separated after a prolonged period. The separation processes can be divided into creaming, aggregation, and homogenization.

1.3.1   Stabilization of O/W-type Suspension by Water-soluble Polymers

Figure 10 shows particle-size distribution of liquid paraffin (10%) emulsion using a carboxyvinyl polymer (0.5%). In general, several

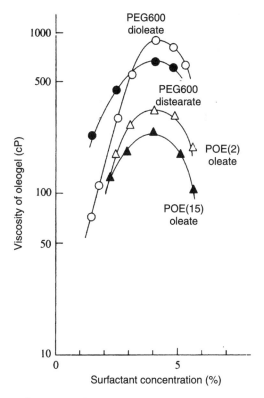

Fig. 9 Viscosity of montmorillonite oleogel under the existance of various nonionic surfactants.

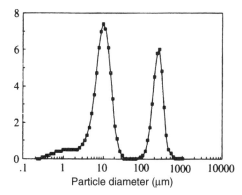

Fig. 10 Particle-size distribution of a liquid paraffin (10%) by using a cross-linked carboxyvinyl polymer (0.5%).

micrometer-sized emulsified particles are obtained when a surfactant is used, whereas the diameter of emulsified particles by a water-soluble polymer is large and heterogeneous. This is due in part to the small ability of carboxyvinyl polymers to lower the interfacial tension as compared with surfactants. Nonetheless, the emulsified particle diameters do not change as a function of time. Hence, there are different mechanisms in polymer emulsification than that by surfactants.

The generally considered mechanisms of emulsion stabilization are creaming inhibition by the reduction of the floating rate of the emulsified particles, steric stabilization, electric repulsive stabilization, and the inhibition of aggregation by a nonadsorbing, nonionic polymer [13, 14]. Various water-soluble polymers have been used as emulsion stabilizers. However, attempts have been made to use polymers with hydrophobic groups that have similar amphoteric structures as surfactants. These polymers include pulronic-type block copolymers, acrylic acid-alkyl methacrylic acid copolymers (trademark: PEMULEN®) [15], and lipohetero polysaccharides (Emulsan). Additionally, use of proteins [16] or proteins with small molecular weight surfactants [17] are also being studied.

1.3.2 *Emulsification of W/O-type by Oleophilic Polymers*

Yamaguchi *et al.* found that oleophilic gels made by the aforementioned surfactant/organic clay composites can include a large amount of water in a stable manner [18]. After mixing a liquid paraffin and surfactant/organic-clay composite to prepare an oleophilic gel, the gel can incorporate water by three to four times the amount of the initial oleophilic gel to become W/O-type emulsion. This emulsion is very stable in a wide temperature range without using a W/O-type emulsifier. It can also emulsify at room temperature.

1.4 Moisture Holding

One of the functions of skin care cosmetics is to soften the skin to allow it to retain moisture. Therefore, moisturizing agents in addition to water are added to skin care cosmetics.

Among moisturizing agents are frequently used low-molecular-weight compounds such as polyol-type, pyrrolidone carboxylic acid and lactic acid, and polysaccharides such as hyaluronic acid. Generally speaking, low-molecular-weight organic acid salts exhibit high moisture absorption and moisturizing ability. However, they are also easily affect by external conditions such as relative humidity. By contrast, although there

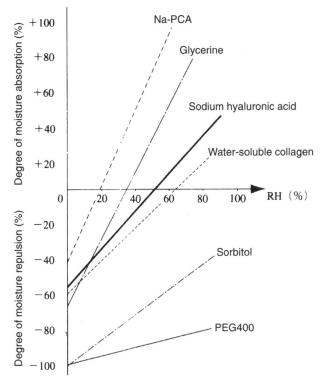

Fig. 11 Moisture-absorption properties of various moisturizing agents [19].

are differences, polymeric moisturizing agents are not significantly influenced by external conditions. In particular, sodium hyaluronic acid exhibits little change in moisture up-take as shown in Fig. 11 and is considered an ideal moisturizing agent among polymeric materials [19]. It is also known that hyaluronic acid maintains its skin-softening ability for a long period of time when it is used with a low-molecular weight moisturizing agent, glycerine [20].

1.5 Conclusion

The role of polymeric gels in skin care cosmetics has been discussed. In recent years, skin care products without perfume and preservatives and with natural ingredients have been favored. Higher functional cosmetics to whiten or prevent wrinkling are also desired. In such situations, considerations for additional safety and high functionalization such as biocompatibility and biodegradability have become important subjects.

2   HAIR CARE COSMETICS

YASUNARI NAKAMA

2.1   Classification of Hair Care Gels

The worldwide consumption of hair care cosmetics is increasing every year. Along with this, the growth of gel products for hair care has also been remarkable. At present, it is reported that 38% of the women and 18% of the men in the United States and Europe use hair care gel products daily.

There are three types of gels used for hair care products: a styling gel; a conditioning gel; and a pomade [21]. The styling gel is used to obtain a desired hairstyle. On the other hand, the conditioning gel is used to adjust the condition of hair [22]. Thus, conditioning gels add flexibility and body, make combing easy, and prevent hair from being affected by static. Variations of styling gels are spraying gels and gel mists, which are also used to achieve a certain hairstyle or to maintain a hairstyle.

Pomades are mainly made of natural ingredients, and contain perfumed oils that are made by absorbing perfumes from plants into beef fat or lard. They were initially used for hair styling, but at present are used to add a shine to hair in addition to styling [23]. Hence, they have both styling and conditioning functions, there is an excellent monograph on pomades by Goode [24] and thus a detailed description of them will not be described here.

The most popular of the gel products are those for styling. Thus, styling gels and conditioning gels will be explained here in more detail.

2.2   Styling Gels

The unique aspects of styling gels in comparison to other hair care gels are that they are absorbed onto the hair, form a desired hairstyle, and hold the style for a long period. Such a styling effect is achieved mostly by hair-fixative polymers, which are applied to the hair by a gel.

A good hair care gel must satisfy the following conditions [21]. It is:

(i)   colorless and transparent;
(ii) exhibits a shear thinning rheology, which is important for removing the gel from a container and spreading it through the hair; and

(iii)   shows stability towards shear, heat and light and is nontoxic to humans.

To be a further desirable styling gel it must also have excellent holding power for a fixed hairstyle and not feel sticky upon use. The viscosity of styling gels is usually in the range of 45,000 to 65,000 cP and even low viscosity spraying gels have 7000 to 10,000 cP.

A styling gel has at least the following components.

- Water: adjusted in such a way as to total 100%
- Alcohol: 0–6%
- Hair fixative polymer: 1–5%
- Gelation agent: 0.25–1.0%
- Neutralizing agent: 0.5–1.5%
- Nonionic surfactant: 0.5–2%
- Perfume: 0.1–0.4%
- Antioxidant: 0–1%
- UV stabilizer: <0.1%
- Chelating agent: <0.1%
- Others: 0–01%

Alcohol is used as a co-solvent when other components do not dissolve in water sufficiently.

Requirements for hair-fixative polymers are [25, 26]

(i)   that they be either water-soluble or able to dissolve in water with a small amount of alcohol, leading to the necessity of having hydrophilic groups in the molecule;
(ii)   that they have sufficiently strong fixative strength for an extended period of time, that is, that the polymer have a constant adhesive strength and excellent film-forming capability; and
(iii) that they to be safe to humans and be odorless.

Table 1 summarizes often-used hair-fixative polymers. Among these, the first synthetic polymer to be developed as a hair-fixative polymer was poly(vinyl pyrrolidone) [27]. This polymer is widely used because it has excellent water solubility and film-forming ability and in addition it is nontoxic. Although it is still widely used today, it has the shortcomings of being too hygroscopic and it becomes sticky under high humidity. In order to solve these problems, various copolymers have been developed (see Table 1).

Table 1 Listing of hair-fixative polymers.

Chemical names (INCI Name)	Commercial products	Manufacturers
Poly(vinyl pyrrolidone)	PVP K-30, 60, 90	ISP
	Luviskol K-30, 60,. 90	BASF
Copolymers of hydroxyethyl cellulose and dimethyl ammonium chloride (Polyquaternium-4)	Celguat H100, L200	National Starch
Copolymers of vinyl pyrrolidone and quaternized dimethylaminoethyl methlacrylate (Polyquaternium-11)	Gafquat 734, 755N LuviquatPQ-11	ISP BASF
Copolymers of methylvinyl ether and maleic acid ethyl ester	Gantrez ES-225	ISP
Copolymers of methylvinyl ether and maleic acid butyl ester	Gantrez ES-425 Gantrez A-425	ISP
Copolymers of octylacrylamide (or acrylate) and butylaminoethyl methacrylate	Amphomer Amphomer LV-71 Lovocryl-47	National Starch
Copolymers of acrylamide and dimethyldiarylammonium chloride (Polyquaternium-7)	Merquat 550	Calgon
Copolymers of acrylamide, dimethylacrylammonium chloride, and acrylic acid (Polyquaternium-39)	Merquat Plus 3330	Calgon
Quaternized polyurethane-amine (Polyquaternium-39)	Merquat 295	Calgon
Copolymers of vinyl acetate and crotonic acid	Resyn 28-1310	National Starch
Copolymers of vinyl pyrrolidone, crotonate, vinyl neodecanoic acid	Resyn 28-2930	National Starch
Copolymers of vinyl pyrrolidone and acrylic acid ester	PVP/VA E-735 PVP/VA E-635 PVP/VA S-630 PVP/VA W-735 Luviskol VA 64	ISP BASF
Copolymers of vinyl pyrrolidone and acrylic acid ester	Luviflex VBM 35	BASF
Copolymers of vinyl pyrrolidone and dimethylaminoethyl methacrylate	Copolymers 845, 937, 958	ISP
Hydroxypropyltrimonium guar chloride	Jaquar C135, C16S, C17	Rhone-Polunc
Copolymers of vinyl carprolactum, vinyl pyrrolidone and methacrylic acid ester	Gaffix VC-713 H2OLD EP-1	ISP

Table 1   *(Continued)*

Chemical names (INCI Name)	Commercial products	Manufacturers
Hydroxylpropyltrimethlammonium-2-chloride (Polyquaternium-10)	Polymer JR	Aermchol
Copolymers of vinyl pyrrolidone and methylvinyl imidazole chloride (Polyquaternium-16)	Polymer JR	Luviquat FC 370, 550, 950, HM552 BASF
Copolymers of vinyl pyrrolidone and methylvinyl dimethylaminoethyl methacrylate (Polyquaternium-28)	Gafquat HS-100	ISP
Mixtures of poly(vinyl pyrrolidone) and dimethicone	PVP/Si-10	ISPR
Mixtures of Polyquaternium-28 and dimethicone	Gafquat Hsi	ISP

Table 2 summarizes gelation agents currently used. The most effective and the most widely used is carbomer. Carbomer is a crosslinked poly(acrylic acid). It assumes a relaxed conformation in an acidic solution. Upon neutralization with an appropriate base, repulsion occurs among negatively charged groups and the main chain is extended (see Fig. 1). In addition, due to the physical crosslinks between polymers, networks are formed. Only 0.5% of neutralized carbomer readily leads to a viscosity of 50,000 cP [21]. Hence, it is an extremely effective gelation agent. However, such a negatively charged carbomer and positively charged hair-fixative polymer may interact ionically and form an insoluble

Table 2   Gelation agents used for styling gels.

Chemical names (INCI Names)	Trade name	Manufacturer
Crosslinked poly(acrylic acid) (carbomer)	Carbopol 940 Carbopol 980 Carbopol ETD Carbopol Ultrez 10 Acritamer 940 Acritamer 941	BF Goodrich RITA
Hydroxyethyl cellulose	Cellosize HEC QP Cellosize HEC WP	Amerchol
Crosslinked copolymers of methylvinyl ether-maleic acid ethyl ester	Stabileze 06 Stabileze QM	ISP

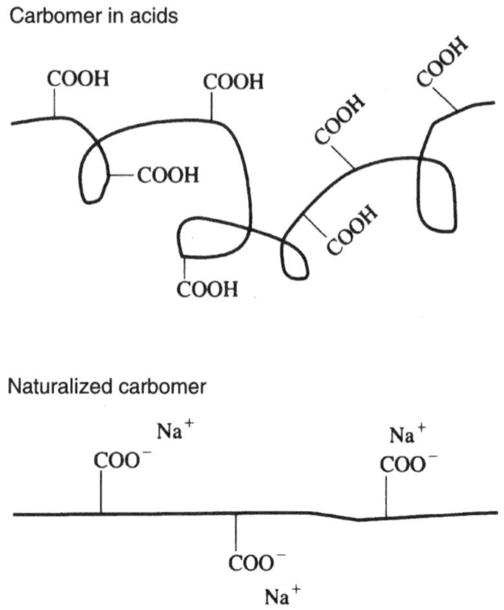

Carbomer in acids

Naturalized carbomer

Fig. 1 Viscosity enhancement mechanism of carbomer.

complex. Therefore, it is necessary to pay attention to its compatibility with other hair-fixative polymers when carbomer is to be used.

Unlike carbomer, hydroxyethyl cellulose is nonionic. Although it is not as effective as carbomer as a gelation agent, it does not interact with negativity or positively charged hair-fixative polymers. Therefore, this polymer exhibits excellent compatibility with other components in the gel.

Stabileze are recently commercialized gellation agents and are crosslinked copolymers of anhydrous maleic acid and methylvinyl ether. It has the same gelation mechanism as carbomer but it has excellent film-forming ability. Neutralizing agents for carbomer include the often-used compounds triethanolamine, sodium hydroxide, potassium hydroxide, diethanolamine, amino-methylpropanol, aminoethylpropanediol, and sodium hydroxymethylglycidic acid.

Listed in the following is the actual recipe used for a commercial styling gel.

- Water: 86.5%
- Carbomer 940: 1.0%
- Sodium hydroxide: 1.4%

- PVP/VA copolymer: 10%
- Quaternary sodium salt of EDTA: 0.1%
- PEG-60 glyceric acid almond: 1.0%
- Pigments, fragrance, and antioxidant: adjusted to become 100%

This gel product is manufactured by the following processes.

1) Add water into a mixing vessel with a stirrer and begin stirring. Add carbomer and stir until it dissolves completely.
2) Neutralize by sodium hydroxide.
3) Dissolve a PVP/VA copolymer in water placed in another container and add this solution slowly to the carbomer solution.
4) Add quaternary sodium salt of EDTA and mix well.
5) Add the remaining additives and stir until homogenization.

2.3   Conditioning Gels

Conditioning gels basically consist of the same components as styling gels except for the use or a conditioning polymer rather than a hair-fixative polymer. In addition, as a conditioning accelerator, a protein derivative or quaternary ammonium compound is often used at 0.2 to 5% [21]. The previously mentioned gelation agent is also needed here.

Conditioning polymers are usually positively charged water-soluble polymers. Because they are cationic, they have better compatibility with hair than hair-fixative polymers. However, they show rather weak film-forming capabilities [21]. Often-used conditioning polymers are Polyquaternium-4,7,10,11 and 24, hydroxypropyltrimonium guarchloride, and dimethicone copolyol. This dimethicone copolyol is a silicone modified to hydrophilic by copolymerizing with ethylene glycol. This copolymer is used for its lubricating ability.

The following is an example of a conditioning gel [21]

Component A

- Water: 95.20%
- Carbomer 940:   0.5%
- Trimethanolamine: 0.5%

Component B

- Polyquaternium-11:   1.00%
- Glycerin: 1.00%

- Propyleneglycol: 1.00%
- Diazolidinyl urea: 0.30%
- Methylparaben: 0.11%
- Propylparaben: 0.03%
- Polysorbitol 20:   0.20%
- Benzophenone-4:   0.10%
- Disodium salt of EDTA: 0.05%
- Pigments, fragrances: adjusted to become 100%

Because a cationic polymer and carbomer form an insoluble complex, mixing under pH 4.5 must be avoided. Conditioning gels are prepared by the following procedures.

1) Preparation of component A: Carbomer is dispersed into half the amount of water, to which half the amount of triethanolamine is added.
2) Preparation of Component B: Added to the remaining amount of water are Polyquaternium-11, glycerin, and the rest of triethanolamine, which are then dissolved into water completely. Then diazolidinyl urea, methylparaben and propylparaben are dissolved into propylene glycol and this solution is added to the aqueous solution already prepared. Added to this solution are Polysorbate 20, pigments, and fragrances. Finally, benzophenone-4 and disodium salt of EDTA are added.
3) Components A and B are mixed and stirred until homogenization.

Thus far, hair care gels have been explained by dividing them into styling gels and conditioning gels. However, multifunctional products are currently available on the market. Moreover, a styling gel also offers some level of conditioning, and also some hair-fixation ability. Hair care is indispensable to daily health and aesthetics. Therefore, new products that are easy to use and are highly functional are in demand.

3   COSMETICS FOR CLEANING

The word, gel, is ill-defined with regard to cosmetics. Generally, those materials that have high viscosity, and that are homogeneous and transparent/translucent are called gels. Gels used for cleansing cosmetics also are used to clean other cosmetic products such as those used for make-up, hair care, and body care. However, the components included are different depending on the type (hard or soft) and where the cosmetics are applied. Thus surfactants rather than polymers are the main components. Nonetheless, polymers that exhibit film-formation and gelation effects have important roles in the products, and their application and the amount used are increasing. Here, technical details of gels commercialized for cleansing products that utilize interaction between a polymer and a surfactant will be described.

3.1 Face Cleanser

A face cleanser is used to clean oxidative degradation products from the skin, physiological by-products such as perspiration residue, dirt from air and make-up foundation cosmetics. There are many types of face cleansers designed for such factors as the age of the user, skin quality, type of dirt, and personal preferences. Other distinctions can be made based on the type of surfactant, for example, soap and cleansing foams, the type of solvent, for example, a cleansing oil in addition to the peal-off type used for pack for the surfactant type, carboxyvinyl polymers and sodium acrylic acid are used as gelation agents for cleansing foams.

Recently, a liquid crystal face cleanser that does not use a polymer has been introduced [28]. This material is made of a lyotropic liquid crystal formed by a nonionic surfactant/polyol/water system, which makes use of the phase-transition behavior caused by the amount of oil and water. A scrubbing agent such as polyethylene powder or sodium poly(acrylic acid) can be added to improve cleaning power and leave a fresh feeling. Pack is used for several hours by forming a coating on the face to provide nutrients and water to the skin as well as to remove dirt from the skin. Poly(vinyl alcohol) or carboxymethyl cellulose is used as the film former. A typical recipe is listed in Table 1 [29].

Table 1 Components used for pack.

Poly(vinyl alcohol)	15.0%
Carboxymethyl cellulose	5.0
1,3-Butylene glycol	5.0
Ethanol	12.0
PQE oleilalcohol ether	0.5
Fragrance	Appropriate amount
Preservative	Appropriate amount
Buffer	Appropriate amount
Purified water	62.5

3.2 Body Cleansers

Body cleansers are used to clean dirt and perspiration from daily life and a typical example is soap. As stated in a book written around the first century AD by Brinius, soap is the oldest cleansing material. It is made of an alkali metal salt of fatty acid such as those from beef and palms. In recent years, change of the use environment such as taking a shower or cleaning tools such as a sponge or nylon towel has led to the introduction of liquid or gel products. One such is a shower gel that uses a synthetic surfactant [30]. A shower gel gives slippery feeling during rinsing, and thus it is not favored by Japanese who do not like the feeling of a residue on the skin. However, shower gels are widely used in the United States and Europe. Its content is the gel that is obtained by mixing a specific ratio of an anionic surfactant and amphoteric surfactant in an aqueous solution. Figure 1 shows a ternary phase diagram of hexadecyldimethylammonio-propanesulfonate (HDPS)/sodium dodecylsulfate (SDS)/water systems [31]. In the region where relatively high HDPS exists, gelation of the system takes place and viscoelastic properties are observed. The extrapolated region of this component, which showed viscoelasticity at high concentrations, is hexagonal liquid crystal. Hence, this phenomenon is considered to be caused by the precursor of the hexagonal liquid crystal, a rod-like micelle. In the region of gelation, it is known that stimulation to

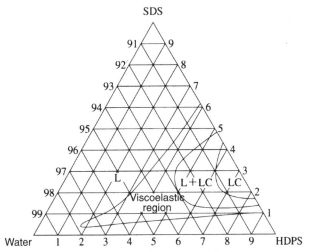

Fig. 1 Ternary phase diagram of dilute HDPS/SDS/water systems.

Table 2 A recipe for shower gels.

Triethanolamine lamylsulfate	20%
Palm fatty acid amide propylbetain	15
Glycol	3
Hydrolyzed protein	1
Fragrance	Appropriate amount
Preservative	Appropriate amount
Buffer	Appropriate amount
Purified water	61

the skin is reduced due to the static interaction of both surfactants [32], which results in a mild skin cleanser. Depending on the type of surfactant, the mixing ratio to gelation changes. In practice, convenience, stimulation properties, and economy are considered and the composition is determined. A typical recipe is shown in Table 2.

3.3 Hair Cleansers

Hair cleansers, shampoos, and hair rinses are used to clean and maintain cleanliness of the hair and scalp. Thus, shampoos that have a high cleansing ability have been developed using mainly soaps or synthetic surfactants. However, in recent years, because people wash their hair almost everyday, strong cleansing power is not as important as before. Shampoos that have little stimulation, ease of use, and are safe are now desired. Low stimulation of the scalp was achieved by a high safety surfactant, acrylmethyltaulin salt [33] or a proper mixing technology of surfactants [31]. For convenience, conditioning shampoos have been developed. The conditioning effect of a shampoo is to provide smoothness for the passage of fibers through the hair and to minimize entanglement of the hair [34]. Natural cationic polymers, such as cationic cellulose and cationic guar gum, and synthetic cationic polymers are used. Typical cationic polymers are shown in Table 3. If the main component is an anionic surfactant which shows a strong degreasing ability, the hair entangles during rinsing, resulting in hair loss, breakage of hair or surface damage [35]. Upon addition of a small amount of a cationic polymer, these problems can be avoided. Table 4 lists a general recipe for conditional shampoos [36]. The cationic polymer and anionic surfactant form a complex and provides the conditioning effect. In particular, Goddard *et al.* investigated these complexes in detail [37, 38].

Table 3   Chemical structures of cationic polymers.

Natural cationic polymers	Synthetic cationic polymers

Cationic cellulose

Copolymer of diarylammonium chloride and acrylamide

Poly(diaryldimethylammonium chloride)

Table 4 Recipe for a conditioning shampoo.

Lauryl poly(oxyethlene (3) sulfate) ester triethanolamine salt (30% aqueous solution)	19.0
Lauryl poly(oxyethylene (3) sulfate) ester sodium salt (30% aqueous solution)	20.0
Laurylsulfate ester sodium salt (30% aqueous solution)	5.0
Lauryldiethanolamide	3.0
Lauryldimethylaminoacetic betain (35% aqueous solution)	7.0
Cationic cellulose	0.2
Ethyleneglycoldistearic acid ester	2.0
Protein derivative	0.5
Fragrance	Appropriate amount
Preservative	Appropriate amount
Metallic ion chelating agent, pH buffer	Appropriate amount
Purified water	52.3

Figure 2 is a phase diagram of a cationic cellulose and an anionic surfactant using the results from lowering surface tension. The ordinate is the surface tension and the abscissa is the concentration of the anionic surfactant. When the amount of anionic surfactant that adsorbs onto the cationic polymer became a monomolecular equivalent, the largest amount of water-insoluble complexes was formed. Some combinations will produce gels insoluble to water [39]. Commercially available conditioning shampoos take advantage of the formation of the complexes. On appearance, the polymer is dissolved in a large amount of surfactant to show a

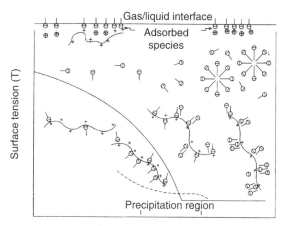

Fig. 2 A phase diagram of cationic cellulose/anionic surfactant systems.

homogeneous liquid. Upon rinsing, the shampoo is diluted and the complex precipitates to show conditioning effects by adsorbing onto the hair [40].

A hair rinse is used to add smoothness to the hair and to facilitate styling. It is an oil dispersed in a cationic surfactant/higher alcohol/water system. Figure 3 is a phase diagram of hexadecanol/octadecyltrimethyl-ammonium chloride/water ternary systems at 55 and 75°C investigated by thermal analysis. The G_2 phase that exists in a broad molar range exhibits a thixotropic gel that has a lamellae structure below its melting point (see Fig. 4) [41].

This gel adheres to the hair effectively and reduces the friction coefficient of the hair, provides an antistatic effect, and protects the hair from damage. Thus, it is used as a base material for hair rinse and hair treatment.

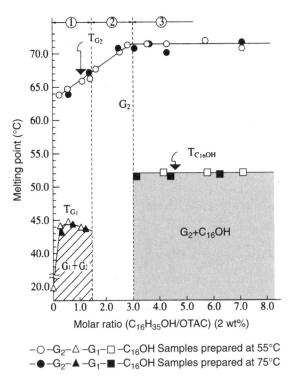

$-\bigcirc-G_2-\triangle-G_1-\square-C_{16}OH$ Samples prepared at 55°C
$-\bullet-G_2-\blacktriangle-G_1-\blacksquare-C_{16}OH$ Samples prepared at 75°C

Fig. 3 A phase diagram of $C_{16}H_{35}OH/OTAC/water$ ternary systems.

Short surface (0.415 nm)

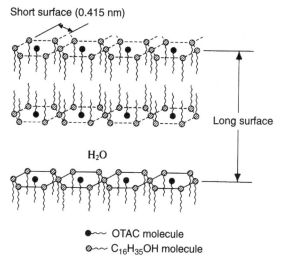

H₂O

●⌇ OTAC molecule
⊘⌇ C₁₆H₃₅OH molecule

Fig. 4 The gel structure at a molar ratio of $C_{16}H_{35}OH/OTAC$ at 3.

Recently, changes in lifestyle, such as the increased number of times the hair is washed and washing the hair prior to leaving the house by the young generation, has led to the commercialization of a further-advanced concept of conditioning shampoo and rinse-in shampoo [42, 43]. (The latter is literally a shampoo in which a rinse is mixed.) The rinsing agent contains oil, anionic-cationic surfactant complex (ion pair) [44] or high-molecular-weight polydimethylsiloxane (silicone) derivatives. In addition, aminoalkyl modified silicone that is a dimethylsiloxane to which the aminoalkyl group is introduced, has been developed to improve forming characteristics, cleaning power, and adsorption to the hair [45, 46] (see Fig. 5 [46]). It provides a rinsing effect that differs in appearance from that of traditional systems.

Accordingly, the addition of polymers to hair cleansing systems is used to improve the ease of use rather than to increase the viscosity.

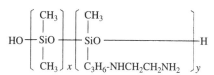

Fig. 5 Chemical structure of aminoalkyl modified silicones.

REFERENCES

1 (1979). Federation of Japan Cosmetic Engineers. *Cosmetic Science Guide Book*, pp. 15–55.
2 Mitsui, T. (1993). *New Cosmetic Science*, Nanzan-do, p. 323.
3 Nakamura, M. (ed). (1973). *Water-soluble Polymers*, Kagaku Kogyo Publ., pp. 214–215.
4 *Wako Pure Chemicals*, Highvis Wako pamphlet.
5 BF Goodrich, Brochure on Carbopol Ultrez 10.
6 International Report, Jpn., WO91/5736.
7 Emerson, W.W. (1955). *Nature*, **4479**: 461.
8 Imoto, M. (1963). *Kagaku to Kogyo* **16**: 442.
9 Michales, A.S. (1958). *Ind. Eng. Chem.* **46**: 1435.
10 Hauser, E.A. (1950). U.S. Pat. 2,531,427.
11 Jordan, J.W., Hook, B.J., and Finlayson, C.M. (1950). *J. Phys. Colloid Chem.* **54**: 294.
12 Yamaguchi, M. (1990). *Abura Kagaku*, **39**: 100.
13 Mitsui, T. (1993). *New Cosmetic Science*, Nanzan-do, pp. 185–190.
14 Ihara, F. and Fususawa, K. (1990). *Modern Colloid Chemistry*, Kodan Publ., p. 80.
15 Nikko Chemicals. (1995). *Technical Bulletin* (TDS-182), PEMULEN.
16 Mita, A. (1986). *Abura Kagaku* **35**: 389.
17 Horiuchi, T. (1995). *Handbook of Water-soluble Polymers*, NTS, p. 35.
18 Yamaguchi, M. (1991). *Abura Kagaku* **40**: 491.
19 Sotooka, N. (1985). *Hifu* **27**: 296.
20 Ozawa, T., Nishiyama, S., Horii, I., Kawasaki, K., Kumano, Y., and Nakayama, H. (1985). *Hifu* **27**: 276.
21 Lochhead, R.Y., Hemker, W.J., and Castaneda, J.Y. (1987). *Cosmet. Toiletries* **102**: 89.
22 (1988). *Science of Hair Care*, Kao Life Sciences Institute, Shokabo, pp. 79–86.
23 Kakihara, T. (1987). *Practical Knowledge of Cosmetics*, Toyo Keiza Shimpo-sha, pp. 111–114.
24 Goode, S.T. (1979). *Cosmet. Toiletries* **94**: 71.
25 Robbins, C.R. (1994). *Chemical and Physical Behavior of Human Hair*, New York: Springer-Verlag, pp. 263–297.
26 Zviak, C. (1986). *The Science of Hair Care*, New York: Marcel Dekker, pp. 149–182.
27 Johnson, S.C. (1984). *Cosmet. Toiletries* **99**: 77.
28 Suzuki, T., Nakamura, M., Sumita, H., and Shigeta, A. (1991). *J. Cosmetic Engineers, Jpn.* **25**: 193.
29 Mitsui, T. (1993). *New Cosmetic Science*, Nanzan-do, p. 467.
30 Abama, G. (1993). *Shower Gels, Cosmet. & Toiletries* **108**: 83.
31 Soul, D., Tiddy, G.J., Wheeler, B.A., Wheeler, P.A., and Willis, E. (1974). *J. Chem. Soc. Faraday Trans.*, *I*, **70**: 163.
32 Miyazawa, K., Ogawa, M., and Mitsui, T. (1984). *J. Cosmetic Engineers, Jpn.* **18**: 96.
33 Miyazawa, K., Tamura, U., Katsumura, Y., Uchikawa, K., Sakamoto, T., and Miyata, K. (1989). *Abura Kagaku* **38**: 297.
34 Fukuchi, Y., Ookoshi, M., and Muroya, K. (1988). *J. Cosmetic Engineers, Jpn.* **22**: 15.
35 Tamura, T. (1975). *Fragrance J., Jpn.* **11**: 36.
36 Mitsui, T. (1993). *New Cosmetic Science*, Nanzan-do, p. 421.
37 Goddard, E.G., Phillips, T.S., and Hannan, R.B. (1975). *J. Soc. Cosmet. Chem.* **26**: 461.
38 Goddard, E.G. and Hannan, R.B. (1977). *J. Am. Oil: Chem. Soc.* **54**: 560.

39 Leung, P.S. and Goddard, E.D. (1991). *Langmuir* **7**: 608.
40 Goddard, E.D. and Harris, W.C. (1987). *J. Soc. Chem.* **38**: 233.
41 Yamaguchi, M. and Noda, A. (1989). *J. Chem. Soc., Jpn.* **1**: 26.
42 Harusawa, F. and Nakama, Y. (1989). *Fragrance J., Jpn.* **10**: 11.
43 Harusawa, F., Nakama, Y., and Tanaka, M. (1991). *Shampoos, Cosmet. & Toiletries* **106**: 35.
44 Nakama, Y. and Yamaguchi, M. (1993). *Abura Kagaku* **42**: 366.
45 Tsubaki, Y. and Noda, I. (1990). *Fragrance J., Jpn.* **5**: 27.
46 Beppu, K. and Omiya, K. (1995). *Abura Kagaku* **44**: 45.

Section 2
Air Fresheners and Deodorizers

KEISUKE SAKUDA

1 INTRODUCTION

The market for air fresheners and deodorizers can be largely divided into those for rooms, toilets and cars. The market for toilet air fresheners was opened when Konoe Brothers introduced Airwik in 1952. However, an active development of the market occurred when gel air fresheners were developed in the 1970s. Because gel does not spill even if the bottle is dropped, it is easy to use and thus became widely popular. Today, hydrogel dominates the market with an 80% share with the rest of the market filled by oleophilic gel, liquid and spray-type air fresheners.

The market for room air fresheners opened at the same time as that for toilet air fresheners as there was no distinction as to use at that time. In the 1980s, many products were commercialized with different types of fragrances and for different applications (for pets, tobacco and entry ways of houses). It was around this time that the distinction between room air fresheners and toilet air fresheners became clear. Therefore, the market for air fresheners grew. Air fresheners for rooms have containers of various designs, indicating consumers preference not just for fragrances but also for designs for interiors.

The initial purpose of air fresheners for cars was to mask the tobacco odor. Gradually, its purpose has changed to establishing driving comfort or an expression of personal preferences in a limited space. Thus, various air fresheners with emphasis on design and color are available. Because their containers are easy to fill, liquid air fresheners are sold more. However, gels are also attractive due to their nonspilling characteristics. Even in the gel products, unlike for toilet purposes, there are transparent products or a product in which small toys (for example, a turtle toy swimming about) are included, emphasizing visual effects.

2 GEL AIR FRESHENERS AND DEODORIZERS

The most important function of an air freshener is to spread a fragrance effectively. In instances when there is an unpleasant toilet odor, an air freshener is used not just to enjoy the fragrance but also for deodorizing such odors. Most of the commercial products are manufactured with functions of this type in mind. It is important to vaporize the fragrance into the space at an appropriate concentration for prolonged periods of time. Thus, various products are sold with such mechanisms. Solid air fresheners and deodorizers using gelation agents occupy 80% of the market. The following items are important and must be considered when gel air fresheners and deodorizers are to be developed.

i)   There must be no change in fragrance with time and the strength of fragrance must last.
ii)   There must be no discoloration of pigments.
iii) There must be no phase separation of water.
iv)   A constant gel strength must be maintained.
v)   There must be no degradation of the gel with time.
vi)   The fresheners and deodorizers must be heat resistant.
vii) They must be able to recover even if they are frozen.
viii) There must be only a small amount of residue after use.
ix)   They must have an antimicrobial effect.

Those air fresheners and deodorants that meet the foregoing conditions can be divided into hydrophilic and oleophilic types.

2.1   Hydrophilic Types

The hydrophilic-type gels use water as the main component and is the dominant type for air fresheners and deodorants. Additionally, there is no danger of fire. Because they can be manufactured inexpensively, a large vaporization area can be made and the products with strong fragrance can be manufactured. The end point of the product can be readily identified because the gel shrinks when the water is completely evaporated. However, fragrances are oils, and hence various techniques are required to homogeneously disperse the fragrance and gel. The basic hydrogel preparation processes are given in what follows. However, readers are referred to other chapters on the gelation agents as they are mentioned only briefly here.

(a)   Carrageenan Gels

Carrageenan is a viscous material that is contained in the red-seaweed group. It is a mixture of various polysaccharides and is widely used for toilet air fresheners and deodorizers.

[Recipe]

Carrageenan base	2–3%
Stabilizer	1–5%
Preservative	0.1–0.2%
Surfactant	1–5%
Fragrance	5–8%
Pigment	an appropriate amount
Water	the required amount
Total	100 (wt%)

[Manufacturing Method]

The fragrance and surfactant are mixed in advance. The other components are mixed and heated to approximately 90°C, dissolved and cooled to 65°C, and then the mixture of the fragrance and surfactant is added. Stir well until homogenization and fill into a desired container.

The reason why there are ranges in each component is because the composition must be tailored to the aforementioned needs and situations. For example, a gel air freshener for a cold region will freeze if the stabilizer (for example, propylene glycol) is only 1%. Thus, the amount of propylene glycol must be increased to prevent freezing. In addition, the amount of surfactant depends greatly on the type of fragrances used. If the surfactant is insufficient, emulsification cannot be achieved, whereas excess surfactant adversely affects vaporization of the fragrances. Hence, great care must be paid to the composition of a fragrance and surfactant. In order to improve gel strength and to prevent phase separation of water, locust bean gum or potassium chloride is sometimes added.

(b) Agar Gels

Agar is a viscous material contained in seaweed. It is a mixture of various polysaccharides. The products that use agar are more expensive than those with carrageenan, therefore agar is seldom used today.

[Recipe]

Agar	1–2%
Stabilizer	1–5%
Preservative	0.1–0.2%
Surfactant	1–5%
Fragrance	5–8%
Pigment	an appropriate amount
Water	an appropriate amount
Total	100 (wt%)

[Manufacturing Method]

The fragrance and surfactant are mixed in advance. The other components are mixed and heated to approximately 90°C, dissolved and cooled to 65°C, and then the mixture of the fragrance and surfactant is added. Stir well until homogenization and fill into a desired container.

(c) Gellan Gum Gels

Consists of polysaccharides produced by a culture of microbes. The manufacturing method is patented by Kelco Co. [2] because it has excellent transparency and heat resistance, it is used for room and car applications where aesthetic aspects are important.

[Recipe] [3]

Gellan gum	0.3–1.0%
Stabilizer	5–10%
Calcium lactate	3–5%
UV stabilizer	0.1–0.2%

Preservative	0.1–0.2%
Fragrance	5–10%
Surfactant	5–10%
Ethanol	5–8%
Pigment	an appropriate amount
Water	remaining amount
Total	100.0 (wt%)

[Manufacturing Method]

Gellan gum and a stabilizer are dispersed in water and heated to approximately 95°C. After they dissolve, cool the solution to about 60°C. Add a surfactant and other mixed solution and stir to homogenization followed by filling into a container.

Because gels with excellent transparency can be prepared, various products have become commercially available. For example, gel in which beads of a small shell of a turtle are included, or a gel in which the color of the upper and bottom layers are differentiated. To prepare these special gels, it is necessary to adjust the recipe considering the density and gelation rate.

(d)   Collagen/gelatin Gels [4]

An elastic, transparent gel can be prepared. Few examples exist for commercialized products.

[Recipe]

Part A

Gelatin (molecular weight 55,000–1,000,000	4.5%
Sugar	0.5%
Water	75.0%

Part B

Fragrance	6.0%
Poly(oxyether nonylphenyul ether) (10)	4.0%
A copolymer of isobutylene-anhydrous	0.8%
Maleic acid (molecular weight 8,000)	
Ethanol	2.0%
Propylene glycol	0.2%
Water	7.0%
Total	100.0 (wt%)

[Manufacturing Method]

Mix components of Part A and heat them to 60 to 70°C to dissolve homogeneously. After cooling the solution to 40°C, add Part B, which was prepared in advance. Stir until homogenization and keep quiescent in a container at 25°C for 1 h.

(e) Aquagel [5, 6]

Aquagel is a gelation agent developed by Mitsubishi Petrochemicals. Its component is a urethane compound obtained by reacting a polyol, a polyisocyanate and imidazole (or 2-methyl imidazole). A transparent gel can be prepared at room temperature. However, the gel lacks thermal stability and is of limited applicability where temperature is likely to be high.

[Recipe]

Aquagel C-2020	12.0%
Fragrance	5.0%
Surfactant	5.0%
Sodium sulfite	0.2%
UV stabilizer	0.1%
Pigment	0.005%
Water	remaining amount
Total	100.0 (wt%)

[Manufacturing Method]

After the components other than Aquagel are homogeneously mixed, Aquagel is added and again homogeneously mixed. The solution is left until gelation. Gelation required for about 3 h at 10°C or about 1 h at 20°C (depending on the type of fragrance used, this gelation time varies significantly).

2.2 Oleophilic Gel Air Fresheners

A hydrocarbon such as limonene or isoparaffin is used as the main component in oleophilic gel air fresheners. These hydrocarbons have good compatibility with fragrances and the amount of fragrances used can be changed at will. However, use of limonene limits the fragrance to orange-type fragrance. On the other hand, isoparaffin, in addition to being flammable, has a characteristic solvent odor and therefore consideration must be given to masking such odor.

There are fewer commercial products than water-base products. In the following, the basic gel-preparation method and properties of gels will be described. Readers are again referred to other chapters for gelation agents.

(a) Soap Gels [7]

The main component is a stearic acid soap and so-called limonene gels. These gels are widely used for toilet air fresheners and deodorants.

[Recipe]

Sodium stearate	5–15%
Hydrocarbon (limonene and others)	40–90%
Alcohols	1–50%
Water	0.2–2.0%
Total	100.0 (wt%)

[Manufacturing Method]

Water, alcohol, hydrocarbon and sodium stearate are mixed, heated to 90 to 95°C and dissolved. After cooling to approximately 60 to 65°C, fragrance is added to the mixture, which is quickly stirred and filled into a container.

The reason for the range of the concentrations is that the composition must be adjusted in a manner similar to that for the hydrogels depending on needs. For example, if an air freshener is to be developed for an automotive application, it must take into account the possibility of the interior temperature reaching 70°C. Thus, it is necessary to develop a gel that does not melt at this temperature.

In this case, the amount of water is the important parameter because in an oleophilic gel water is not only important in dissolving sodium stearate, but also in controlling the melting temperature of the gel. Generally speaking, the lower the amount of water, the higher the melting temperature.

(b) Metallic Soap Gel [8]

Metallic soap gels are transparent elastic gels.

[Recipe]

Limonene	80–95%
Antioxidant	0.3–0.5%
Fragrance	5–10%
Aluminum 2-ethylhexanoate	5%
Aluminum stearate	5%
Total	100.0 (wt%)

[Manufacturing Method]

All the raw materials are mixed into a single container as nothing can be added later. The antioxidant is dissolved into the fragrance, to which is added aluminum stearate. After stirring thoroughly, the entire container is placed in an atmosphere of 150°C for 30 min while occasionally stirring. When the entire material becomes transparent, it is taken out of the container, and the bubbles will disappear after it is kept for one day.

(c)   Dibenzitidene Sorbitol Gel [9–11]

[Recipe]

Dibenzilidene sorbitol	1–5%
Polar solvent (3-methyl-3-methoxybutanol)	30–96%
Fragrance	3–69%
Total	100.0 (wt%)

[Manufacturing Method]

Dibenzilidene sorbitol and 3-methyl-3-methoxybutanol are mixed at 70 to 80°C to homogenization. After cooling to 60°C, the fragrance is mixed and stirred thoroughly, followed by filling in a container. The gel will be left cooled.

As oleophilic gelation agents [12], there are 12-hydroxystearic acid and amino acid type ((N-lauloyl-glutamate dibutylamide, etc.) However, they are seldom used for air fresheners and deodorizers.

3   VAPORIZATION OF FRAGRANCES FROM HYDROGELS [13]

A model fragrance as shown in Table 1 is evaporated from a pottery container, liquid, and gel. The composition of the evaporated gas as a function of time of the gel sample shows little variation in addition to stable evaporation. Despite the use of the same fragrance components and composition, the gel sample showed a quite different composition from the model fragrance A. That is, as soon as the fragrance was added to the gel, the balance of the fragrance became different from the balance intended by the fragrance designer. This is because various forces acting between the fragrance and surfactant change the activity coefficient of the fragrance and the evaporation balance becomes different from the compositional balance.

Table 1   Model recipe for a fragrance (fragrance A).

Component name	Functional Group	Quantity
Isoamylacetate	$\overset{\displaystyle O}{\overset{\|}{-C-O-}}$	10
cis-3-Hexane-1-ol	$\overset{\displaystyle O}{\overset{\|}{-C-O-}}$	20
cis-3-Hexenyl acetate	—OH	20
P-Cresol methyl ether	—O—	30
Benzaldehyde	$\overset{\displaystyle O}{\overset{\|}{-C-H}}$	50
Linalool	—OH	100
Benzyl acetate	$\overset{\displaystyle O}{\overset{\|}{-C-O-}}$	200
Anethole	—O—	300
2-Phenylethyl alcohol	—OH	270
Total		1000

Perfumers have long recognized that when a fragrance is added to a soap, shampoo or air freshener, its fragrance becomes different from that of the original blend. Hence, a perfumer adjusts the composition of the air freshener knowing what base material is used in order to achieve the final fragrance balance desired.

Here, a composition of fragrance B (Table 2) provides the same fragrance as intended for fragrance A except when fragrance A is added to a gel made for air fresheners.

By preparing an air freshener to which fragrance B is added, the initial gas-phase composition from the air freshener is in good agreement with that of fragrance A. In addition, the gel air freshener continues to release the same fragrance as the initial fragrance even after a prolonged period of time, and also maintains an evaporation balance.

Table 2 Model recipe for ragrance (fragrance B).

Component name	Functional Group	Quantity
Isoamylacetate	$\overset{\displaystyle O}{\underset{\displaystyle \|}{}}$ —C—O—	10
cis-3-Hexane-1-ol	$\overset{\displaystyle O}{\underset{\displaystyle \|}{}}$ —C—O—	15
cis-3-Hexenyl acetate	—OH	60
P-Cresol methyl ether	—O—	25
Benzaldehyde	$\overset{\displaystyle O}{\underset{\displaystyle \|}{}}$ —C—H	60
Linalool	—OH	70
Benzyl acetate	$\overset{\displaystyle O}{\underset{\displaystyle \|}{}}$ —C—O—	100
Anethole	—O—	120
2-Phenylethyl alcohol	—OH	540
Total		1000

4 FUTURE DEVELOPMENT OF GEL AIR FRESHENERS AND DEODORIZERS

In recent years, products that deodorize have become increasingly popular. Their original purpose has changed from masking odors by the inclusion of fragrances to deodorizing by combining fragrances and deodorizers (chemical deodorization). In the future, it is expected that this trend will become even stronger. Therefore, it is important to consider the following three parameters, specifically, more effective deodorizers, the base material and mechanism to be used in delivery systems to maximize their function, and appropriate fragrances for them in their entirety. It is important to develop new air fresheners and deodorizers based on these fundamental parameters.

REFERENCES

1 Horiuchi, T. (1994). *Book on Fragrances*, 182, 87.
2 Oohashi, S. (1986). *Food Chem., Jpn.* **12**: 61.
3 *Technical Bulletin*, Kelco Gel (Gellan Gum), Dai Nippon Pharmaceuticals.
4 (1989). Tokkyo Kaiho 297484.
5 (1984). Tokkyo Kaiho 52196.
6 *Technical Bulletin*. Aquagel, Mitsubishi Petrochemicals.
7 (1980). Tokkyo Kaiho 75493.
8 (1984). Tokkyo Kaiho 52196.
9 Mitsumata, T. (1979). *Fragrance J., Jpn.* **7**: 71.
10 (1984). Tokkyo Kaiho 77859.
11 *Technical Bulletin*. Gelol D, Shin Nippon Scientific.
12 Yamashita, T. (1992). *Yushi* **45**: 52.
13 Anma, K., Katoh, C., and Sakuda, K. (1995). *Fragrance J., Jpn.* **23**: 43.
14 (1972). Tokkyo Kokai, 25374.
15 (1972). Tokkyo Kokai, 25375.
16 (1973). Tokkyo Kokai, 56524.
17 (1976). Tokkyo Kokai, 10415.

Section 3
Disposable Portable Heaters

AKIO USUI

1 INTRODUCTION

A disposable portable heater first appeared on the market around 1976. During the first ten years the market for these heaters expanded to five billion. The introduction of a disposable heater that can be attached to clothing helped the expansion of the current market size to 10 billion heaters.

This type of portable heater consists of a porous film that prevents the movement of heating components by keeping them in a sheet form on one side. On the other side of the sheet is an adhesive by which the disposable heater is adhered to the clothing. This allows close contact to the body and a reduction of thickness, which results in energy saving due to low-temperature operation, preservation of resources and the reduction of waste. Improved ease of use and safety by reduced operating temperature were well received by consumers, who also were attracted by the widened options for placement of the heaters on the body.

The moisture-supplying agent that is used to produce heat shifted from traditional wood particles and vermiculite to superabsorbent polymers, which allowed the disposable portable heater to be thinner. The main reasons for the adoption of superabsorbent polymers are:

i)   a large amount of water can be maintained with a small volume of the material resulting in a thinner product; and

ii) it is easy to inhibit the movement of the sheet-formed heating material and also easy to maintain a constant thickness.

2   FUTURE TRENDS FOR DISPOSABLE PORTABLE HEATERS

There is much room for improvement in the current safety design, production technology, and quality control with regard to disposable portable heaters. In particular, there has not been sufficient progress in the development and production methods of the heating element. In addition, in the manufacturing process there are the problems of poor storage and metering, difficulty in manufacturing heating elements, of uniform thickness, high scrap rates, low productivity, and dust and noise pollution. As concerns product quality, it is desirable to improve safety with regard to low-temperature burning, unevenness of temperature distribution, and flexibility.

The establishment of the PL law now requires a warning label to be attached to such items with regard to their use. However, establishment of fundamental safety is strongly desired.

The heating is achieved by the chemical reaction of iron powder, water, and oxygen (air). The temperature can be controlled by controlling the amount of any materials that relate to the reaction. Current disposable portable heaters adjust temperature by controlling the amount of air supply through an air-permeable film. However, there is a danger in that the temperature may exceed the safety limit due to the change of thermal conductivity caused by the change of surrounding temperature. The exotherm is caused by the hydroxylation of the iron surface. This reaction is inhibited either by an extreme lack or an excess amount (water acts as a barrier to oxygen permeation beyond a certain thickness) of water. Hence, it is possible to develop a safer heating element by controlling the amount of air permeated and free water.

As a possible solution to these problems, we will evaluate the properties of hydrogels.

(1)   Use of Viscous Nature of Gels

Using the coating technology of printing, a uniform film of micrometer thickness can be placed on the packaging material. This can be done in a dust-free environment at a high production rate.

(2)   Use of Super-Controlled Release Properties of Gels

By using the super-controlled release properties of gels, the water-release rate can be controlled, resulting in controlled heat production. This will lead to the development of even safer heating elements.

(3)   Use of Thermoresponsive Nature of Gels

It is possible to control the heating reaction by releasing an excess amount of water and covering the iron powder above a certain thickness when the temperature of the heating element reaches the upper limit set for the product. When the temperature drops to the safety limit, the gel absorbs water and the heating reaction starts again. This will allow the development of a self-controlled heating device.

The conditions for the use of thermoresponsive gels as functional heating elements are that:

i)   they maintain high water absorption even at salt concentration around 8%;

ii)   they have high water absorptivity and can release more than a certain
amount of free water in a short period;

iii) the thermal-response temperature can be controlled freely within the temperature range of 35 to 50°C;

iv)   the error of thermal response be less than 2°C;

v)   they possess good reproducibility; and

vi)   there is no change in performance during storage for a prolonged period of time.

At the present time, fundamental research is in progress in the area of upper- and lower-limit temperature control, reproducibility, and long-term stability. Further development of gels is desired.

Potential applications of functional heating elements are:

i) as highly safe medical heating devices if they can be developed so that the difference in the upper and lower temperature limit is small; and

ii) if the difference is large and the heating cycle is long, the device can be developed as a supplemental heating element for skin-absorbing drugs.

In particular, as thermoresponsive gets are becoming very important for future development, the development of a new heating element will help grow the market by combining them with other products.

Section 4
Sanitary Products for Pets

MITSUHARU TOMINAGA

1 INTRODUCTION

Recent years have witnessed a growing relationship between people and their pets. This has resulted in a steadily growing market for pet products. Such products, or more specifically, sanitary products, have become a necessity for pet owners and are widely used for pets that are now held indoors. However, the market growth is also due to the efforts by the makers of pet products to develop functional products or higher-performance products.

The development of sanitary products for pets influenced strongly the generation of pet products. Of particular significance to this growth of the market is the application of superabsorbent polymers (hereinafter abbreviated as SAP). In the following, sanitary products for pets will be described with respect to the application of SAP.

2   PET SHEETS AND SAP

2.1   Use of Pet Sheets

Usually, small pets such as small dogs, cats, and hamsters are kept indoors more often. For the indoor care of small dogs, a product called a pet sheet is used, and it is estimated that about 600 million sheets are consumed. Dogs can be trained to urinate on the pet sheet, thereby avoiding indiscriminate urination in other locations in a room. The pet sheet, which can be used several times, is then disposed of as common household rubbish, in the same way as disposable diapers.

2.2   Structure of Pet Sheets

The appearance and construction of the pet sheets are shown in Fig. 1. As can be seen from the figure, the structure is nearly identical to the flat-type disposable diaper used by humans.

2.3   Pet Sheets and SAP

The amount of SAP used for pet sheets is around 3 g per sheet. Hence, the total consumption of SAP for pet sheets in Japan is estimated to be approximately 2000 metric tons. (The readers are referred to Part III, Chapter 3 for the basic concept of SAP for pet sheets.) In using pet sheets a dog walks around the sheet and may deposit a heavily local load on it. It

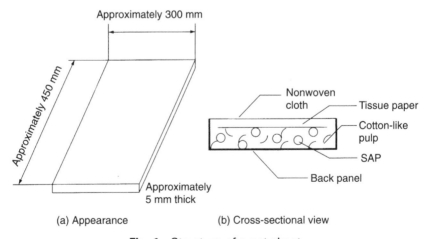

(a) Appearance   (b) Cross-sectional view

Fig. 1   Structure of a pet sheet.

is possible that the urine that is once absorbed by the sheet may be released under the load, soiling the dog's paws and then transferred to the floor as it walks around the house.

Thus, for pet sheets, the SAP must have a high water-maintenance capability (high gel strength). Attempts have been made to use a high water-absorption gel with a high gel-strength gel in a mixed or double-layered manner.

3   TOILET SAND FOR PETS AND SAP

3.1   What is Toilet Sand for Pets?

Toilet sand for pets is generally known as cat litter or pet sand. In the wild, cats dig a hole, urinate, and upon completion, cover the hole. Indoors, a container filled with litter or sand suffices. Traditionally, rather than sand, water-absorbing minerals that absorb urine better, such as zeolite and atapalgeit, have been used. Recently, crushed bentonite that clumps by absorbing urine several times its own weight has appeared on the market. Because of its easy disposal, this product helped the pet sand market to grow remarkably. Furthermore, other organic "sand," such as pulp, which absorbs much more urine than these mineral sands, is also becoming popular. According to a 1993 estimate, the consumption of zeolite mineral is approximately 15,000 t, bentonite mineral is about 90,000 t, and organic materials are 7000 t [1].

3.2   Use of Toilet Sand for Pets

Figure 2 shows the use of a pulp-type sand. A sand area is made by filling a tray to several tens of millimeters. Because cats customarily urinate in the same place, they are easy to train. Only the pet sand that absorbed the urine is removed and discarded. Even for the organic-type pet sand, there are products that form clumps in the same way as the crushed bentonite type. The used sand is discarded as domestic rubbish in a similar manner as disposable diapers for humans or pet sheets. There are products that can be discarded in a toilet as the result of improved dispersion of particles in water [2], and also products that contain a deodorant or antibacterial agent.

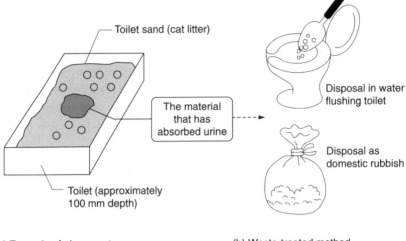

(a) Example of placement   (b) Waste-treated method

Fig. 2 An example of toilet-sand use.

3.3 Toilet Sands and SAP

Mainly 30 to 50% of the organic-type toilet sands contain SAP, and the concentration of SAP varies widely depending on the products. Thus it is difficult to estimate the annual domestic consumption, but a rough estimate is several hundred metric tons.

Table 1 Requirements for SAP for toilet sand.

Requirements	Effect in toilet sand
High absorbency (high absorption ratio) (possesses salt resistance)	Low amount of usage per each use (from the viewpoints of economy and environment)
High water retention (high gel strength) (does not contain fine particles)	Does not adhere to pets' paws and does not evaporate moisture (no odor)
Appropriate particle size distribution and shape	Good compatibility with pulp and no dust generation
Safety (appropriate size and viscosity) (not harmful and nontoxic)	Does not choke the throat if it is swallowed by mistake and also does not cause environmental pollution
Flammability (low ash content after burning)	Burned as domestic rubbish
Biodegradable	Minimal problem especially when it is flushed away
Low cost	Can be manufactured inexpensively

It is presumed that the product using SAP is not manufactured outside Japan.

The requirements for SAP used for toilet sand are listed in Table 1. However, as these requirements sometimes contradict, it is necessary to prepare an optimum product during the product design by deciding whether to use an appropriate property material or mixed use of materials. The recent trend in this material is that because it is a consumable, the economic factor plays an important role, and as expected, issues concerning environmental pollution cannot be ignored.

REFERENCES

1   Murai, Y. (1995). *Pet Buyer Monthly, Jpn.* 10.
2   (1994). Tokkyo Koho 85669.

Section 5
Photographic Films

TAKASHI NAOI

1 SILVER HALIDE PHOTOSENSITIVE MATERIALS

A photosensitive material using silver halide forms a cluster of several Ag atoms when light is absorbed by the silver halide in an emulsion layer. After this, the reducing agent in a developer penetrates into the emulsion layer and the reduction reaction of Ag^+ proceeds using the exposed AgX^* as a catalyst. The unexposed AgX is washed out as a complex in the fixer solution, resulting in the final photographic image. Hence, the emulsion layer must be a matrix made of network sizes through which only the reaction agent passes without losing the imaging element, that is, Ag particles; the pH of solutions varies. The developer is alkaline, the fixer is acidic, and water for washing is neutral. Hence, the photosensitive materials must absorb these solutions at all these pH and swell. Gelatin, a properly crosslinked amphoteric electrolyte, is chosen for this purpose. Photosensitive material is a good example of industrial products in which gelatin gel is effectively used.

2 GELATIN GELS

When a gelatin solution with an appropriate concentration is cooled, the viscosity and modulus increase and solidify to form a gel. Gelation of gelatin starts below 30°C and three-dimensional (3D) networks will be formed. The junction points of the 3D are crystallites made of three gelatin molecules [1, 2] (see Fig. 1).

Wherever each molecular chain approaches, crystallites are formed, leading to the formation of a 3D network. This structure explains well the main properties of the gel, such as the melting point, degree of optical rotation, modulus, and rubber-like elasticity. However, a few kinds of networks coexist in the photosensitive material depending on the manufacturing and development methods, the behavior of gelatin gels is very complex.

2.1 Physical Crosslinking of Gelatin Gels

The crystallites in a gelatin gel are often regarded as crosslink points. The crosslinking of crystallites can be readily removed by physical conditions, such as heat and moisture. Thus, the crosslinking by crystallites can move around, and it can be considered that the microstructure of the film is reversible until chemical crosslinking takes place.

2.2 Chemical Crosslinking of Gelatin Gels

Because photosensitive materials are developed near 40°C to reduce the development time, gelatin gels that are formed at low temperature will dissolve. Actual photosensitive materials are made insoluble by crosslinking gelatin molecules using the functional group of the gelatin.

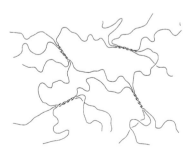

Fig. 1 Structure of a gelatin gel.

As a unique feature of the structure, aside from the crosslinking by crystallites, chemical crosslinking by covalent bonds also exists. The crosslinking agent reacts with lysine (primary amine) in the gelatin molecule and effectively turns them into a huge macromolecule with very small number of crosslinking.

In order to develop photosensitive materials rapidly, it is desirable to have large swelling in the developing bath and small swelling in the water-washing bath. For this purpose, a polyvalent metal such as aluminum is added to the fixative bath or a polymer with carboxylic acid is added to gelatin to control the swelling in the washing bath by ionically cross-linking between the gelatin and aluminum or gelation and carboxylic acid. Accordingly, a further characteristic of the structure is the existence of crosslinking by ionic bonds.

3   SWELLING OF UNCROSSLINKED GELATIN FILM

3.1   Swelling of Gelatin Film and Its Crystallinity

In general, the more crystalline the polymer, the lower the degree of swelling because of the impervious nature of the crystalline region and increased resistance toward deformation. However, as reported by Jopling [3], when the degree of crystallinity is higher, gelatin films swell more. He demonstrated that a gelatin film that was dried at 6°C swells 10 to 13 times whereas that dried at 40°C swells merely 3 to 5 times (see Fig. 2). Jopling thought that this was because the crosslink junction formed at higher temperature may be low in number but more perfect and stronger than that formed at the lower temperature. He further thought that the junctions are formed at the end of drying during which the polymer chains are more tightly packed. If these junctions are crystallites bonded by hydrogen bonding, these effects follow well the crystallization theory [4].

3.2   Swelling towards the Plane Direction of a Gelatin Film

The effect of crystallites on the swelling behavior of gelatin films appears stronger in the plane direction (horizontal swelling) rather than the thickness direction (vertical swelling) of the film. Kogure *et al.* [5] studied the correlation between the degree of horizontal swelling and the degree of optical rotation and found that:

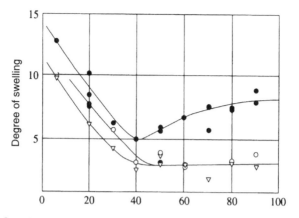

● : A film that was not humidity-conditioned
○ : A film that was conditioned at 20°C with 90% RH for 1 day
▽ : A film that was conditioned at 20°C with 90% RH for several days

Fig. 2 Swelling of gelatin films dried at various temperatures in distilled water at 20°C.

i) the greater the number of crystallites (higher optical rotation), the smaller the horizontal swelling (see Fig. 3); and
ii) the anisotropy of vertical and horizontal swelling agrees well with the degree of optical rotation (see Fig. 4).

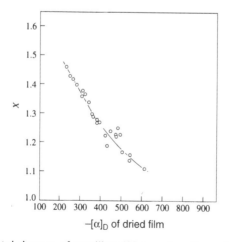

Fig. 3 Horizontal degree of swelling (*X*) as a function of the specific optical rotation of a dry gelatin film [5].

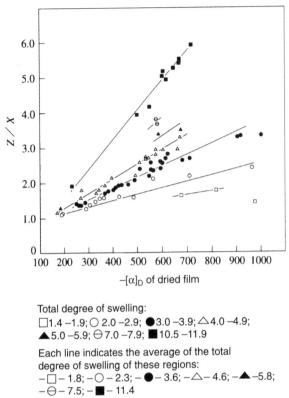

Total degree of swelling:
□ 1.4 –1.9; ○ 2.0 –2.9; ● 3.0 –3.9; △ 4.0 –4.9;
▲ 5.0 –5.9; ⊖ 7.0 –7.9; ■ 10.5 –11.9

Each line indicates the average of the total
degree of swelling of these regions:
– □ – 1.8; – ○ – 2.3; – ● – 3.6; – △ – 4.6; – ▲ –5.8;
– ⊖ – 7.5; – ■ – 11.4

Fig. 4   Anisotropy of the degree of swelling (Z/X) as a function of the specific optical rotation of a dry gelatin film [5].

These results indicate that the helical structure of the crystallites is oriented along the plane of the film and, further, the anisotropy of swelling can be used to measure the degree of orientation of molecules.

3.3   Other Parameters that Control Swelling

One of the characteristics of gelatin is its high degree of swelling. However, because it is an amphoteric electrolyte, the pH dependence of swelling varies depending on the number and type of ionic groups [6]. It exhibits the minimum degree of swelling at the isoelectric point where the number of dissociated groups becomes small. The gelatin treated with calcium carbonate and acid shows different behaviors. In particular, the

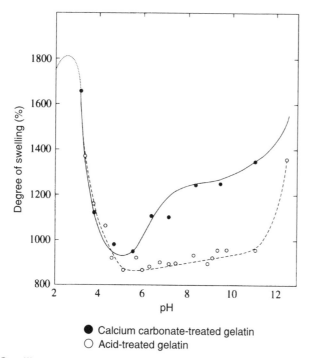

● Calcium carbonate-treated gelatin
○ Acid-treated gelatin

Fig. 5   Swelling curves of calcium carbonate- and acid-treated gelatin films.

acid-treated gelatin shows minimum swelling in a wide pH range from 5 to 10 (see Fig. 5).

4   SWELLING OF CROSSLINKED GELATIN FILMS

4.1   Rate of Swelling of Crosslinked Gelatin Films

It is important for gelatin, a binder of photosensitive material, to have high permeability of developing agent. This is simply the rate of swelling from the viewpoint of gelatin and also rate of permeation from the viewpoint of the developing agent. There have been only a few reports on the swelling behavior of gelatin [7–9]. Robinson [9] showed that the swelling behavior of gelatin film could be explained by Flory's theory of swelling. Although he ignored the effect of intramolecular crosslinking and crystallites on the swelling, the experimental results and the theoretical analysis generally agreed. As long as the χ parameter is known, it is possible to estimate the number of crosslink points for the modulus of the swollen film.

From the viewpoint of developing, the faster the rate of swelling, the more desirable. According to Libicky *et al.* [10], the swelling dynamics of a crosslinked gelatin gel can be expressed as:

$$\frac{dh(t)}{dt} = V_0 \left(1 - \frac{h(t)}{h(\infty)} \right)^p$$

where $h(t)$ is the amount of swelling at time t; $h(\infty)$ is the equilibrium swelling; V_0 is the rate of swelling at time 0; and p is a constant. It is considered that quadratic equation can approximate the experimental results well, but there is an opposing theory.

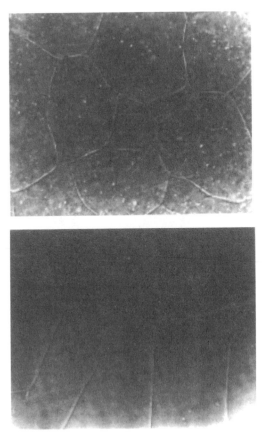

Fig. 6 The surface of a film on which reticulation appeared.

Claes *et al.* [11] measured the rate of swelling using various gelatin films. In the case of uncrosslinked gelatin film, the lower the temperature to form a film (higher crystallite concentration), the smaller the initial rate of swelling. The activation energy obtained from the temperature dependence of swelling is constant at 5.5 kcal/mol; however, that of the crosslinked gelatin film depends on the type of crosslinking agent and the degree of crosslinking.

When the crosslinking reaction takes place in a solution, the activation energy reduces as the degree of crosslinking increases. However, if the reaction takes place as the film dries, the activation energy is constant slightly above a certain value of an uncrosslinked film. It is thought that the crosslinking reaction inhibits the fold formation of collagen. The covalent bond by a crosslinking agent and network by the collagen fold complicate swelling behavior of gelatin gel.

4.2 Reticulation

The appearance of a wrinkle on the surface of the photosensitive material is called reticulation when a high-temperature development process is performed in an attempt to accelerate the rate of development using film with high degree of swelling. Gelatin film swells not only to the thickness direction, but also to the plane of the film. However, the swelling towards the plane direction is inhibited by the adhesion to the substrate and as a result stress is generated. To relax this stress, the wrinkle appears on the surface (see Figs. 6 and 7) and becomes the cause of undesirable loss of transparency for the photosensitive materials. [12].

A similar phenomenon can be observed when a polymer gel is placed in a container and swollen. Tanaka *et al.* analyzed this phenomenon using linear stability analysis [13]. For a gelatin gel, Amiya *et al.* reported that the same behavior as the other gels could be observed [14]. However, when the corresponding behavior on the photosensitive material was investigated, it behaved qualitatively similarly but its magnitude was smaller. It was hypothesized that this is due to the coexistence of collagen folds and crosslink structure.

4.3 Swelling of Gelatin Gels by Ionic Crosslinking

In order to shorten the development time, a gelatin film with large swelling in the development solution is prepared. This gelatin film is then

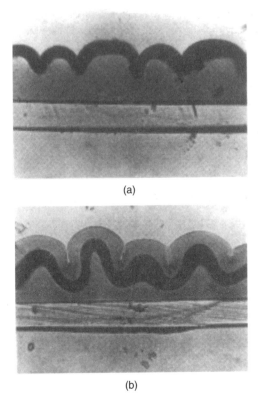

(a)

(b)

Fig. 7 The cross section of a film on which reticulation appeared.

coordination bonded with aluminum in the fixation solution. In the water-washing process, the swelling is minimized, and as a result the drying time is reduced.

It is possible to control swelling of gelatin film in the water-washing bath by blending pKa controlled, carboxyl-containing polymer into gelatin film. This is achieved because, in the fixation bath, the carboxyl-containing polymer and gelatin form a polyion complex. As a result, shrinkage takes place due to the dissociation of the carboxylic acid group.

4.4 Control of Development-Agent Permeation by Polymer Blends

In order to hasten the development time, there is a method other than crosslinking to blend a water-soluble polymer into gelatin and thus increase the rate of development-agent permeation.

Petrak [15] correlated the structure of surface film and permeation of methyl orange by preparing a film with voids formed by phase separation of a gelatin/dextrin/water ternary blend (see Fig. 8).

Some researchers also prepared films of gelatin/polyacrylamide/water ternary blend with differing degrees of phase separation and confirmed that the permeability of methyl orange drastically increased (see Figs. 9 and 10). If this technique to increase permeation of developing solution or swelling rate of gelatin film is used for medical films, there will be significant improvement in sensitometric properties [16, 17].

5 MECHANICAL PROPERTIES OF GELS

The modulus of gelatin gels from the viewpoint of scratch resistance of photosensitive materials has been studied for many years. The modulus of a gel is a measure of the gel hardness and rigidity and depends strongly on

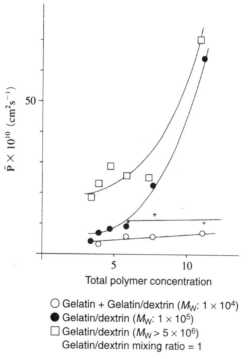

○ Gelatin + Gelatin/dextrin (M_W: 1×10^4)
● Gelatin/dextrin (M_W: 1×10^5)
□ Gelatin/dextrin ($M_W > 5 \times 10^6$)
Gelatin/dextrin mixing ratio = 1

Fig. 8 Dependence of permeation of various gelatin/dextrin blend films on the concentration of polymer in the casting solution.

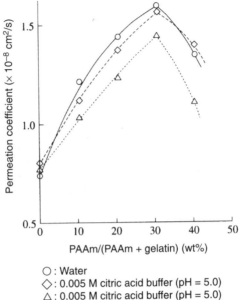

○ : Water
◇ : 0.005 M citric acid buffer (pH = 5.0)
△ : 0.005 M citric acid buffer (pH = 5.0)

Fig. 9   Dependence of permeation coefficient on the mixing ratio of poly-acrylamide/gelatin.

the degree of crosslinking. The measurement of modulus of a gelatin film is complex and difficult because:

i)   the modulus is reciprocally proportional to the temperature due to the existence of hydrogen bonding; and

ii)   below the melting temperature of the gel, only the modulus at quasiequilibrium can be obtained.

5.1   Relationships between the Modulus and Crystallites of Gelatin Gels

Todd [18] demonstrated that the following relationship exists between the modulus and the degree of optical rotation of a high molecular weight gelatin (see Fig. 11):

$$G^{1/2} \propto -[\alpha]_D$$

where the modulus of gelatin gels for a given concentration is determined by the number of crystallites created during gel formation. The modulus of gelatin also depends on the gelatin concentration. Ferry [19] found that the

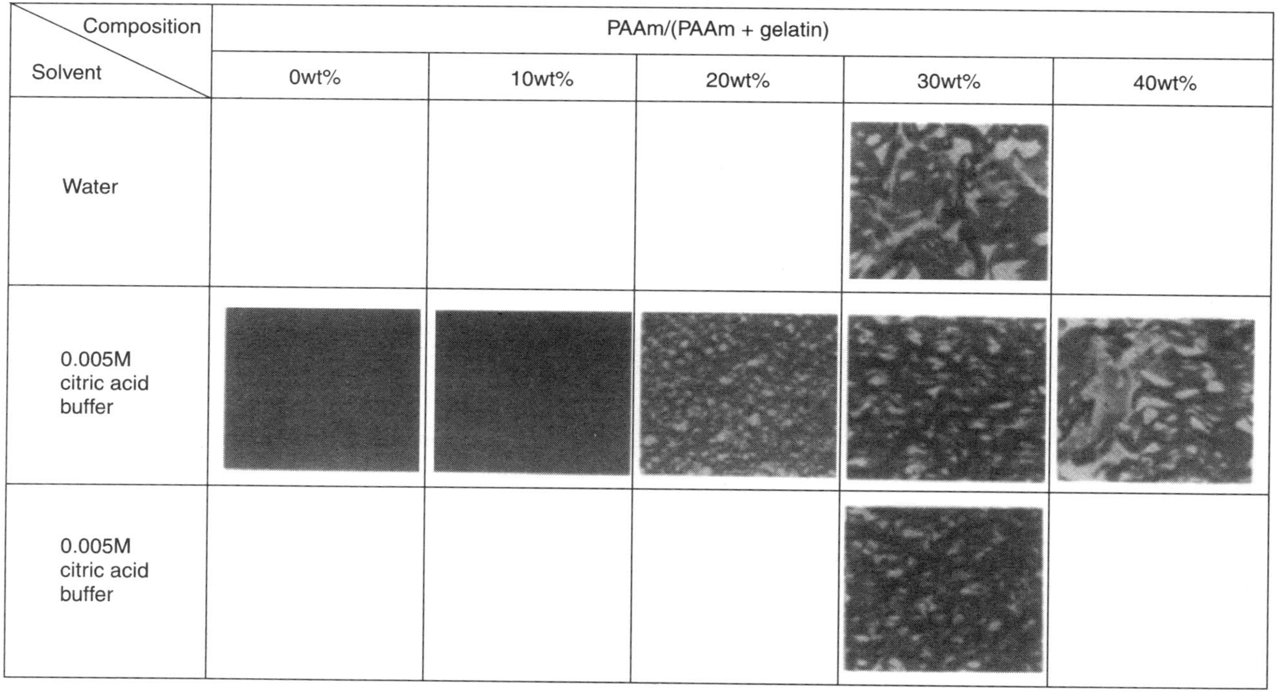

Fig. 10   Phase separation of gelatin/polyacrylamide (magnification: 300 times).

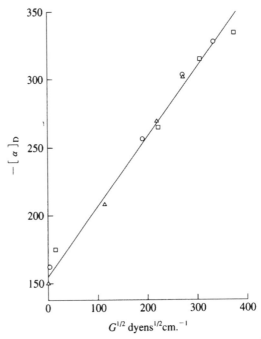

Fig. 11   The relationship between the specific degree of optical rotation and modulus of 5.5 % gelatin gel.

relationship between the modulus of gelatin gel G and concentration c is $G = Ac^n$ where A is the intrinsic to the type of gelatin, and n is a constant less than 2 (see Fig. 12).

5.2   Moduli of Gelatin Gels that Contain a Crosslinking Agent

In order to improve the efficiency of developing-agent permeation, photosensitive material should have a high degree of swelling. However, if the degree of swelling is increased, the gel strength decreases, which results in reduced scratch resistance and a decrease in the value of the commercial product. Taylor and Kragh [20] reported an approach to avoid these contradictory phenomena. They determined the moduli of gelatin films, which contain a crosslinking agent, with various degrees of swelling and found the following relationship (see Fig. 13):

$$(\text{modulus}) \times (\text{degree of swelling})^3 = \text{const.}$$

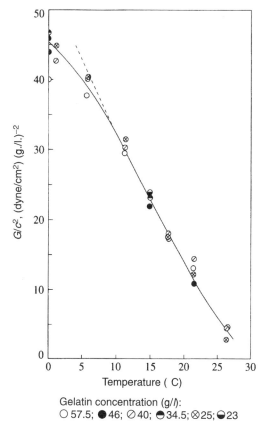

Fig. 12 Temperature dependence of modulus/concentration ratio (G/c^2) of gelatin gels at various concentrations.

From this relationship, it can be seen that the modulus of a gel is strongly influenced by temperature, pH, and ionic strength. Taylor and Kragh reported an interesting discovery. After swelling crosslinked and uncross-linked film in water for 10 min and subsequent immersion in 5% glutaraldehyde aqueous solution, the degree of swelling changes only slightly whereas the modulus greatly increases (see Fig. 14).

This implies that the relationship between the modulus and degree of swelling changes by changing the crosslinking method, in other words, the method that maintains good scratch resistance with increased degree of swelling is implied. This suggests an important approach to design photographic films that are to be rapidly developed.

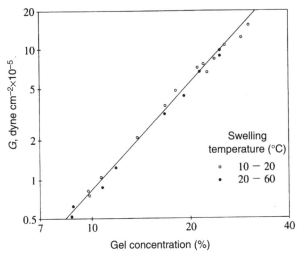

Fig. 13 Influence of swelling temperature on the relationship between the modulus and swelling of a crosslinked gelatin film.

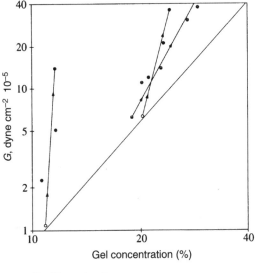

Swelling solvent
○ 1, Swelling in water
● 2, After swelling in water, the gel was crosslinked by glutaralldehyde
◑ 3, Swelling in glutaraldehyde aqueous solution

Fig. 14 Influence of crosslinking on the relationship between the modulus and swelling of a gelatin film under a swollen state [20].

6 CONCLUSIONS

The gels used for photographic materials are not as simple as other gels. Their complexity is caused by the process of manufacturing photosensitive materials, the existence of a few different kinds of networks because of the necessity of development methods, and the restriction of swelling to the plane direction during the development process. Over its long history of commercialization, extensive attempts have been made to replace photographic gelatin with more clearly defined synthetic gels [21–25]. Unfortunately, there are no materials that simultaneously equal the excellent properties of gelatin such as:

i)   excellent protective colloidal property;
ii) formation of an inclusion compound unique to photography; and
iii) sol-gel reversibility.

Most attempts have been at replacing part of the properties while maintaining the advantages of gelatin. Here, the natural and difficult-to-control gelatin has been described and its properties in an actually used environment were investigated. Photography cannot be discussed without mentioning the benefits of gelatin.

REFERENCES

1   Flory, P.J. *et al.* (1958). *J. Am. Chem. Soc.* **80**: 4835.
2   Boedtker, H. *et al.* (1954). *J. Phys. Chem.* **58**: 968.
3   Jopling, D.W. (1956). *J. Appl. Chem.* **6**: 79.
4   Mandelkern, L. (1964). *Crystallization of Polymers*, New York: McGraw-Hill.
5   Kogure, M. *et al.* (1976) *Photographic Gelatin II*, R.J. Cox, ed., London: Academic Press, p. 131.
6   Shepperd, J.E. *et al.* (1942). *J. Phys. Chem.* **46**: 158.
7   Chem, B.T. (1942). *Photogr. Sci. Eng.* **16**: 158.
8   Tanaka, T. *et al.* (1979). *J. Chem. Phys.* **70**: 1214.
9   Robinson, I.D. (1964). *Photogr. Sci. Eng.* **8**: 220.
10 Libicky, A. *et al.* (1976). *Photographic Gelatin II*, R.J. Cox, ed., London: Academic Press.
11 Claes, F.H. *et al.* (1978). *Photogr. Sci. Eng.* **22**: 28.
12 Tojo, E. *et al.* (1976). *Photographic Gelatin II*, R.J. Cox, ed., London: Academic Press, p.49.
13 Tanaka, T. *et al.* (1987). *Nature* **325–326**: 796.
14 Amiya, T. *et al.* (1987). *Macromolecules* **20**: 1162.
15 Petrak, K.L. (1984). *J. Appl. Polym. Sci.* **29**: 555.
16 Naoi, T. (1995). *Proc. 11th Polym. Gel Symp.*
17 (1986). Tokkyo Kaiho, 69061.

18 Todd, A. (1961). *Nature* **191**: 567.
19 Ferry, J.D. (1948). *J. Am. Chem. Soc.* **70**: 2244.
20 Taylor, D.J. *et al.* (1974). *J. Photogr. Sci.* **22**: 223.
21 (1968). Tokkyo Kokai, 14495.
22 (1972). Tokkyo Kokai, 25374.
23 (1972). Tokkyo Kokai, 25375.
24 (1973). Tokkyo Kokai, 56424.
25 (1976). Tokkyo Kokai, 10415.

Section 6
Domestic Oil-Treatment Agents

SHIN ICHIMURA AND TAKAMITSU TAMURA

1 INTRODUCTION

There are two types of waste oil-treatment agents for disposal of domestic waste oil used for cooking. One type solidifies the oil based on 12-hydroxystearic acid (12HS) and the other type uses pulp to absorb the oil. The former type occupies 90% of the market; for this, 12HS is added to the used hot oil above the melting point of 12HS at 80°C and the mixture is agitated. Upon solidification after cooling below 40°C, the solid is peeled off from the pan and discarded. This 12HS associates in an organic liquid and forms an organogel with several percent addition. This gel formation is based on the crystalline primary structure through hydrogen bonding or van der Waals forces. This primary structure aggregates to form oriented spiral fibers [1]. However, if 12HS alone is used to treat domestic waste oil, polar materials such as lecithin enters into the oil and interferes with the hydrogen-bond formation of 12HS molecules. As a result, the gel strength might become weaker than the unused oil and thus gel may not be formed [2]. As a result, the mechanism of gel-strength reduction was clarified and a waste oil-treatment agent was developed which does not interfere with gelation.

2 INHIBITION MECHANISMS OF GELATION

When oil is gelled by 12HS, the gel strength is reduced as the concentration of lecithin in the oil increases. Considering this phenomenon, model organic solvent systems such as cyclohexane and chloroform were used to study the structure of 12HS and lecithin using ^{13}C-NMR and FT-IR. Thus the reduction of hydrogen bonds among 12HS as the basic structure of the gelation in addition to the interaction between carboxylic acid and the polar group of lecithin [2] were confirmed. Further observation of the 12HS fibers by scanning electron microscopy (SEM) revealed the presence of fibrous networks when lecithin is not added (Fig. 1(a)). By contrast, as the lecithin concentration is increased, this fibrous structure changed into aggregates of particles (see Fig. 1(c) and (d)). Accordingly, the addition of lecithin caused the interaction

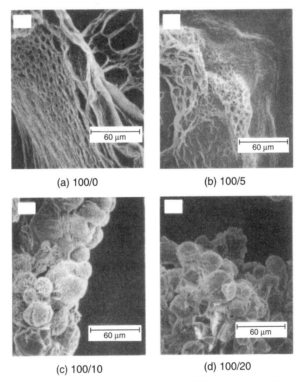

(a) 100/0   (b) 100/5

(c) 100/10   (d) 100/20

Fig. 1 SEM observation of 12HS (Co = 100 mM, solvent: cyclohexane) fibers at various 12HS/lecithin ratios.

between 12HS and lecithin, and the fibrous 12HS changed into aggregates of particles. This resulted in the disruption of network formation and the reduction of gel strength.

3 IMPROVEMENT OF GEL STRENGTH

3.1 Adsorption of Lecithin by Silica Gel

After melting 12HS at 90°C silica gel was dispersed uniformly. This mixture was added at 3.2 wt% (a standard concentration of waste oil-treatment agent) to a soy oil to which lecithin also was added. The gel strength of these samples was measured and are plotted in Fig. 2. Another mixture of oleic acid and silica gel was prepared and added to lecithin-added soy oil in a similar fashion. After centrifugation, the lecithin concentration in the supernatant solution was measured and also plotted in Fig. 2. By the addition of silica gel, lecithin is adsorbed onto the silica gel and the concentration of lecithin is reduced, thereby resulting in the increased gel strength.

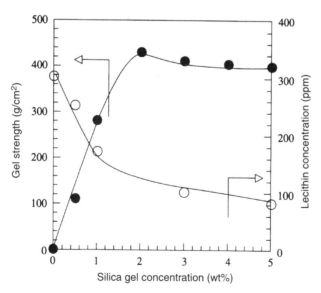

Fig. 2 Gel strength and lecithin concentration of lecithin-added oil.

3.2   Fiber Structure Changes by Addition of a Sodium Salt

It has been reported that the fibrous structure of 12HS changes from spiral to linear by the addition of a small amount of sodium salt [1]. When domestic waste oil was gelled by 12HS into which 0 to 1 wt% sodium stearate (hereinafter abbreviated as ST-Na) was added, a maximum was observed around 0.5 wt%. Here, SEM observation was made to verify the fiber structure of 12HS using model soy oil (see Fig. 3). If ST-Na is added

(a) 100/0   (b) 100/0.5

(c) 100/1.0

Fig. 3   SEM observation of 12HS ($Co = 3.2$ wt%, solvent: soy oil) fibers at various 12HS/ST-Na weight percentages.

by 1 wt%, the spiral structure of 12HS decreased and linear fat fibers appeared (see Fig. 3(c)). At 0.5 wt% ST-Na, a mixture of linear fat fibers and spiral fat fibers coexisted (see Fig. 3(b)). It was hypothesized that the maximum is caused by the good balance between coexisting linear and spiral fibers. Using these two techniques, development of a waste-treatment agent with higher gel strength was attempted.

REFERENCES

1   Tamura, T. (1995). *Hyomen* **33**: 73.
2   Ichikawa, M. and Tamura, T. (19). *Newsletter*, Div. Colloid and Surface Chem. (submitted).

CHAPTER 3

Foods and Packaging

Chapter contents

Section 1

Water-Absorption Sheet for Maintaining Freshness of Foods

TAKAO FUSHIMI

1 TECHNIQUES TO MAINTAIN FRESHNESS AND FUNCTIONAL PACKAGING MATERIALS

Along with the increased varieties of diet and nutrition have come stricter requirements on food. Thus maintenance and freshness have become important. A transportation system has been organized and various new and functional packaging materials have been developed and are being used. The relationship between these functional packaging materials and techniques to maintain food freshness is shown in Fig. 1. Fresh foods are divided into those that are alive even after harvesting, such as fruits and vegetables, and those that are not such as meat, fish and shellfish, and processed foods.

Fruits and vegetables continue to live even after harvesting. However, because there is no new supply of nutrition, aging occurs as a result of decomposition of its own nutrients. If an oxygen-deficient condition is created, there are material changes due to the generation of odor-causing compounds such as an aldehyde.

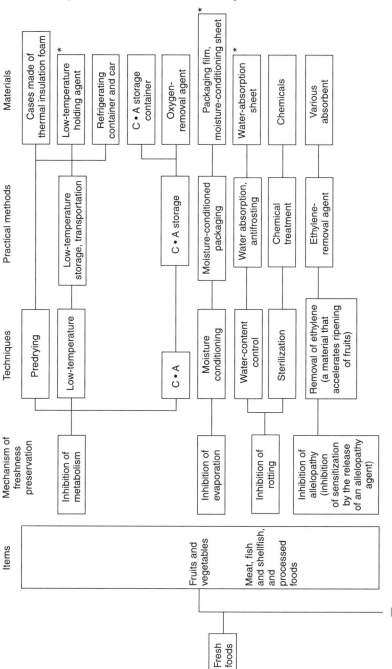

Fig. 1 Relationship between freshness-maintenance technique of fresh foods and materials from Nakagawa, K. Schematic diagram of freshness preservation for fresh foods [1].

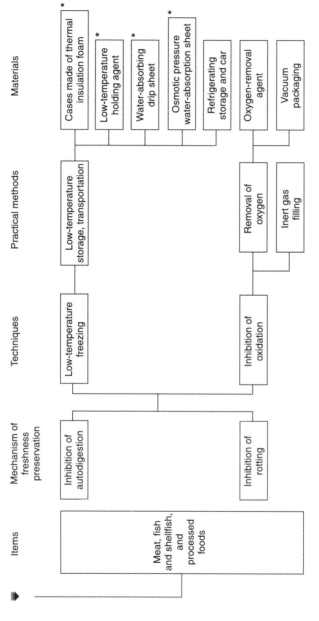

Fig. 1 Continued.

*The asterisk indicates those materials related to water-absorbing polymers

In order to maintain freshness of vegetables and fruits, well-balanced inhibition conditions must be established. Thus, a freshness-preservation technique was developed with controlled life activities by transporting fruits and vegetables at a low temperature. In addition, for those vegetables that have a high respiration rate and experience change in color such as broccoli and Shanghai green vegetable, a so-called controlled atmosphere (C.A) technique was developed to reduce the respiration rate by placing the vegetables in a low-oxygen high-carbon dioxide atmosphere. In practice, CA packaging is performed by:

i)   shipping foods in a packaging material that inhibits slightly the permeation of carbon dioxide;

ii) packaging foods in a cardboard box that is slightly hermetic in nature in the presence of a carbon-dioxide producing chemical; and

iii) shipping foods with an oxygen-absorbing agent.

To hold the freshness of fruits and vegetables, techniques such as the inhibition of metabolism, moisture conditioning to prevent evaporation of water (this is important for vegetables such as lettuce, inhibition of rotting by water absorption, and chemical treatment and ethylene removal to inhibit ripening (the absorption of ethylene gas is a growth and ripening accelerator produced by the plant itself) have been developed. To preserve the freshness of meat, fish and shellfish, techniques such as low-temperature storage (refrigeration, storage near the freezing temperature, and frozen food), and oxygen removal or inert gas filling to prevent oxidative, degradation and rotting by microbe growth have been developed. To achieve the foregoing, functional packaging materials as shown in Fig. 1 are used. Hydrogels, the subject of interest of this monograph, are also used as a water-absorbing sheet, a cold-temperature holding agent, and osmotic pressure water-absorbing sheet.

In the following, a water-absorbing sheet for freshness preservation, a multipurpose packaging material for freshness preservation, a low-temperature holding agent for food transportation, and a high osmotic pressure water-absorption sheet for food processing will be discussed. Here, a water-absorption sheet for the freshness preservation of foods will be described.

2 BASIC STRUCTURE AND FUNCTIONS OF WATER-ABSORPTION SHEET FOR FRESHNESS PRESERVATION OF FOODS

The basic structure of a water-absorption sheet for freshness preservation consists of superabsorbent polymers between two tissues, nonwoven cloth or thin polymer films as shown in Fig. 2. One side is made of thin tissue of nonwoven cloth that is water permeable. For freshness preservation of fruits and vegetables, it is important to prevent water loss by evaporation in addition to the inhibition of metabolism through respiration. When fruits and vegetables lose water, they lose shine and become wrinkled, resulting in loss of product value. Simply by wrapping with thick plastic film bags or airtight boxes leads to increased growth of microbes and eventual rotting due to the increased humidity. Therefore, a water-absorbent sheet is placed inside the package, and it absorbs moisture that causes frost. The moisture is fixed by the gelation of the superabsorbent polymer, thus when the environment becomes too dry the moisture is released to maintain proper humidity. This prevents wrinkling and weight reduction of fruits and vegetables. Hence, the water-absorption sheet for fruits and vegetables act not only to absorb water, but also act as a water-supplying source to prevent drying by evaporation from the leaves of the fruits and vegetables. In other words, this sheet functions as a moisture-controlling sheet.

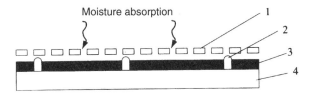

1 – Water-permeable sheet (nonwoven cloth, tissue, etc.)
2 – Adhesive (the adhesive is placed as a dot pattern.
It adheres to 1 and 4 while preventing debonding during water absorption and swelling of 3)
3 – Superabsorbent polymer layer (superabsorbent polymer powder is mixed with a binder and coated)
4 – Nonpermeable substrate sheet (a polypropylene or polyethylene film)

Fig. 2 Basic structure of water-absorption sheet used to preserve freshness in foods.

When mass-produced chicken is transported from factory to the market or tray-packaged materials such as fish, shellfish and cut meat are sold, it is important to use absorption sheets to prevent the loss of quality due to the dripping of water and blood.

Other products that utilize an absorption sheet include packaged materials of just-cooked rice or deep fried materials to prevent water condensation and reduction of the fresh taste. Absorption sheets are also used for keeping cut flowers fresh and to accelerate the growth of enoki mushrooms.

As the water-absorbing polymer for the absorption sheet, poly-(acrylic acid) is often used. In order to keep the total cost of food-packaging material low, a mass-produced expensive polymer is preferred.

In comparison with the absorption sheet using only a natural pulp sheet, the water absorption and preservation ability improves by the factor of several hundred % simply by exchanging only a small amount (approximately 10% by weight) of a superabsorbent polymer. In terms of cooling ability, the water-absorbed hydrogel is approximately equivalent to the same amount of water whereas the frozen hydrogel is about the same as a similar amount of ice.

3   IMPROVEMENT OF WATER-ABSORPTION SHEET FOR FRESHNESS PRESERVATION OF FOODS

As described earlier, the basic structure of water-absorption sheet for freshness preservation of foods is simple. However, sometimes there are problems, for example, leakage of flowing polymer gels, which absorbed water and swelled, thus leading to food contamination. Other problems might be the loss of shape or reduction of strength of the package that contains the flowable polymer gel, leading to food-packaging problems. One example to cope with such problems was the use of an absorption sheet that was completely sealed on all four sides of a package. To prepare this package, a superabsorbent polymer is first sprayed onto a substrate. This coated sheet is cut into a certain length and separated. An imperme-able sheet that is wider and longer than the coated sheet is placed on top of the coated sheet. The extra length is folded into the sheet to place the extra length or width beneath the coated sheet. This folded product is later sealed by heat to achieve a complete seal. The final product is prepared by cutting the unnecessary portion of this sheet without cutting the coated sheet [2].

Another example is to adhere films in sections which contain a superabsorbent polymer. Placed on the top surface is a water-permeable sheet but the backing sheet is made of a nonwater-permeable sheet. Between these sheets is a three-dimensional (3D) net made of an adhesive polymer. The woven points of the net are adhered to the top and backing sheets. Hence, the volume for swelling is sufficient. In addition, the strength as well as the shape is well maintained. If a metallic core fiber is included in the 3D net, box-like or L-shaped absorption sheets can be manufactured.

Although superabsorbent polymer powder is low cost and highly absorbing, it is easy to leak. Instead, nonwoven cloth made of a super-absorbent polymer-covered fiber can be used to obtain materials with good balance of water absorptivity and sheet strength [4, 5].

Yet, another approach is to disperse the superabsorbent polymer powder deep inside a polyurethane foam and fix it so that the powder will not leak. A superabsorbent polymer is thinly dispersed behind polyur-ethane foam with a connected pore structure. An adhesive layer is used to cover the powder layer to form an absorption sheet. If this sheet is adhered to a food tray, in addition to water absorptivity, additional functions such as impact strength and thermal insulation can be expected [6].

Many trays that incorporate a water-absorption sheet have been proposed. However, the substrate can be omitted and the structure can be simplified. The bottom layer sometimes has a ditch-like structure or many small nodule-like structures to prevent water from contacting the food [7].

4   DEVELOPMENT TREND OF WATER-ABSORPTION SHEETS FOR SPECIFIC APPLICATIONS

In addition to the improvement of the basic properties and structures of water-absorption sheets, the development of a water-absorption sheet for specific applications is also underway. In order to prevent the loss of freshness due to the formation of water condensation in a box of pizza or deep fried foods being delivered, a water-absorption sheet is being tested with the packaging materials. There is a carton that has improved thermal resistance, water absorptivity and oil absorptivity by adhering a water-absorption sheet made of rayon-type water-absorbing nonwoven cloth at the top and an oil-absorbing sheet at the bottom on the inside of the carton [8]. A superabsorbent polymer powder is adhered on one side of a paper on which a nonwoven polypropylene cloth is laminated. This material provides water and oil-absorbing layers and can be used to prepare a bag, box or a cardboard carton [9–11].

There are two methods in the application of packaging for micro-wavable frozen foods. The first method is a pre-soaked and water-swollen sheet that is included inside the package along with the frozen food to supply moisture during heating and thawing, thereby preventing dryness. This method will allow the preparation of moist and soft foods [12, 13].

The second method is to prevent the food from being soaked by an excessive amount of water during thawing so that the flavor of the food will not be lost [14–16]. In any case, the internal pressure will increase during heating, thus certain structures, such as easy failure of the heat-sealed portion, are built in to release the accumulated pressure.

In the following, the development trend for a specialized water-absorption sheet will be described. Figure 3 shows the container used by makers of commercially cooked rice that is then packed and delivered to

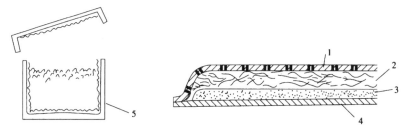

1 – Water-permeable thin film (a woven cloth or a film with holes is used. A heat-sealable hydrophobic nonwoven cloth, which is heat treated to minimize the appearance of surface fiber ends, is especially appropriate. Moisture passes through this surface sheet and prevents adhesion of the rice with the water-absorption sheet).

2 – Water passage structure (a nonwoven cloth, cotton, pulp or connected pore foam can be used. This structure maintains the passage of moisture from 1 to 3 while preventing the polymer gel 3 from leaking to the rice.

3 – Superabsorbent polymer layer (or hydrophilic polymer fiber layers).

4 – A container for cooked rice (water-absorption sheet is adhered to the top, bottom and sidewalls).

Fig. 3 Structure of water-absorption sheet and the placement of the sheet in a rice package.

restaurants, hotels, schools, and hospitals. The sheet acts to absorb a large amount of water that is released from freshly cooked rice to prevent water from depositing on the container wall and to maintain the flavor of the rice. The top, bottom, and sidewalls are all structured so that the enough moisture is absorbed while also preventing the adhesion of rice to the container [17]. As an example of CA packaging for fruits and vegetables by maintaining low-oxygen and high carbon-dioxide environment is a water-absorption sheet used for fresh herbal vegetables. A water-absorption sheet is packaged along with the vegetable in an polyolefin bag and sealed. This bag is kept below 5 °C, and can be stocked for as long as several months [18]. Other approaches include packaging a water-absorption sheet or a composite water-absorption sheet and deoxygenating agent with cut vegetables to prevent oxidation of the cut surface and drying of the vegetable. This method also prevents flavor reduction as well as rotting [19, 20].

Beyond the 1990s, the trend of food packaging will be to simplify the structure and reduce the cost (in part due to the enacted Anti-Pollution Law). It is unfortunate that these packages are not actively used in light of the development of high-performance packaging materials.

REFERENCES

1 Nakagawa, K. (1989). *Hoso Gijyustu* 28.
2 (1990). Tokkyo Kaiho 252558, Sealing method and device for water absorption sheet, Dai-Nippon Inc., Meter.
3 (1993). Tokkyo Kaiho 24141, Water-absorption sheet, Nippon Shokubai.
4 (1995). Tokkyo Kaiho 323498, Water-absorption sheet, Kanebo.
5 (1995). Tokkyo Kaiho 156972, Water-absorption sheet, Dai-Nippon Insatsu.
6 (1993). Tokkyo Kaiho 41769, Water-absorption and holding sheet, Sumitomo Seika.
7 (1994). Tokkyo Kaiho 239351, Water-absorption tray, Yama-ya, Eisa Chemicals.
8 (1994). Tokkyo Kaiho 183436, Carton, Dai-Nippon Insatsu.
9 (1993). Tokkyo Kaiho 269940, Packaging materials and containers for oil-based foods, Boko-ban Insatsu.
10 (1993). Tokkyo Kaiho 58668, Paper for fast food service, Ajinomoto.
11 (1993). Tokkyo Kaiho 65825, Food packaging container, Eiwa Chemical Industries.
12 (1994). Tokkyo Kaiho 153821, Foods in microwavable containers, Ajinomoto.
13 (1996). Tokkyo Kaiho 11947, Food packaging, Sanei Chemicals.
14 (1994). Tokkyo Kaiho 156546, Tray packaging for microwave cooking, Heisei Polymer.
15 (1993). Tokkyo Kaiho 16675, Packaging for microwavable foods, Fuji Seal Industries.
16 (1993). Tokkyo Kaiho 19171, Food packaging, Kuraray, Mizuno Sangyo.
17 (1990). Tokkyo Kaiho 49887, Internal sheet for rice packaging, Nippon Shokubai, Mitsuhashi.

18   (1995). Tokkyo Kaiho 137773, Packaging of flowering myoga and preservation method, Dai-Nippon Insatsu.

19   (1993). Tokkyo Kaiho 338675, Freshness preservation agents and method for vegetables, Nippon Soda.

20   (1994). Tokkyo Kaiho 10045, Bags for cut vegetables, Dai-Nippon Insatsu.

Section 2
Multifunctional Packaging Materials for Freshness Preservation

TAKAO FUSHIMI

1   WATER-ABSORBING, MULTIFUNCTIONAL PACKAGING MATERIALS

A sheet that absorbs dripping water or moisture from foods is not sufficient to preserve the freshness of foods. Additionally, as an approach to prevent health problems caused by bacteria on foods, an antibacterial function is often desired for food-related materials. Based on these needs, multifunctional packaging materials, with antibacterial and deodorant abilities, in addition to a water-absorption function for freshness preservation, are being developed. These new packaging materials will be introduced in what follows.

2   MULTIFUNCTIONAL PACKAGING MATERIALS WITH WATER-ABSORPTION AND ANTIBACTERIAL FUNCTIONS

A typical example of this class of materials is an antibacterial water-absorption sheet that consists of an inner water-absorption layer in which

121

an inorganic powder antibacterial agent is mixed as shown in Section 1, Fig. 2. This antibacterial agent is made of highly safe silver ion-replaced zeolite. Due to the antibacterial function of silver ion, which gradually dissolves, the growth of bacteria and fungi are suppressed inside the water-absorption sheet or contact point of the water-absorption sheet and food. Hence, the freshness of the foods is preserved for a longer period. If this sheet is placed at the bottom of a strawberry package, for example, the damage during transportation can be significantly suppressed [1].

The antibacterial water-absorption sheet described in the foregoing is a side or end-sealed superabsorbent polymer layer or powdery inorganic antibacterial agent mixed layer. Thus, it is somewhat disadvantageous in terms of manufacturing process, cost, and ease of use. As an improved type, the absorption polymer sodium poly(acrylic acid) is replaced by a pulp-sheet-type swelling material, a metallic salt of fibrous carboxymethyl cellulose. A swelling polyelectrolyte made of the same polymer but part of the polymer is a cationic water-soluble polymer, such as polyamine, that can also be used. A water-absorption sheet as shown in Fig. 1 is proposed.

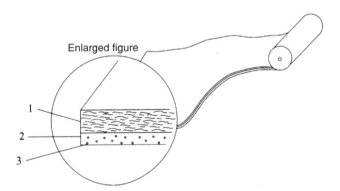

1 – Water-absorption layer (a swelling material made of fibrous carboxymethyl cellulose salt or the material made of the same polymer but part of the polymer is made of a cationic water soluble polymer, such as polyamine, that is forming polyelectrolyte complex).
2 – Nonwater-permeable film.
3 – Powdery inorganic antibacterial agent (silver ion-replaced zeolite or calcium phosphate-supported silver ion).

Fig. 1 Antibacterial water-absorption sheet that cuts easily.

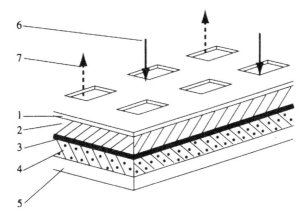

1 – Water-permeable sheet (polyolefin film with holes).
2 – Superabsorbent polymer.
3 – Allyl isothiocyanate permeable layer.
4 – Allyl isothiocyanate antibacterial agent layer.
5 – Water nonpermeable substrate sheet.
6 – Dripped water from the food.
7 – Antibacterial gas to the food (evaporated gas of
    allyl isothiocyanate).

Fig. 2 Water-absorption sheet that serves as a antibacterial sheet for the food by using an antibacterial gas [3, 4].

A nonwater-permeable substrate sheet in which a powdery inorganic antibacterial agent is mixed is laminated onto the water-absorption sheet. Because this sheet does not require a protective layer or side treatment, it can be cut freely. Hence, it can be used almost as a plastic wrapping material [2].

The aforementioned powdery inorganic antibacterial agent is easy to use and safe. However, the antibacterial effect can only reach the contact surface of the food. Instead, wasabi extract, which easily vaporizes and the gas surrounds the food, is being tested as an antibacterial agent. Wasabi extract consists of allyl isothiocyanate and is obtained either naturally or synthetically. As this material is permitted by the Department of Health and is currently used to prolong the freshness of foods, safety is not an issue. The basic structure of a sheet utilizing this material is shown in Fig. 2.

Because allyl isothiocyanate is an oily liquid it is used by mixing into the substrate, by mixing into the polymer as a solid powder of the inclusion compound with cyclodextrin, or by dispersing among fibers of a nonwoven cloth.

A longer-lasting antibacterial sheet by simultaneous use of this gaseous and aforementioned inorganic antibacterial agent has also been tested [3, 4]. In addition, a phosphonium salt monomer that is convenient for this purpose has been developed. This material does not leak out the antibacterial agent and thus a stable antibacterial function is expected (see Fig. 3) [5].

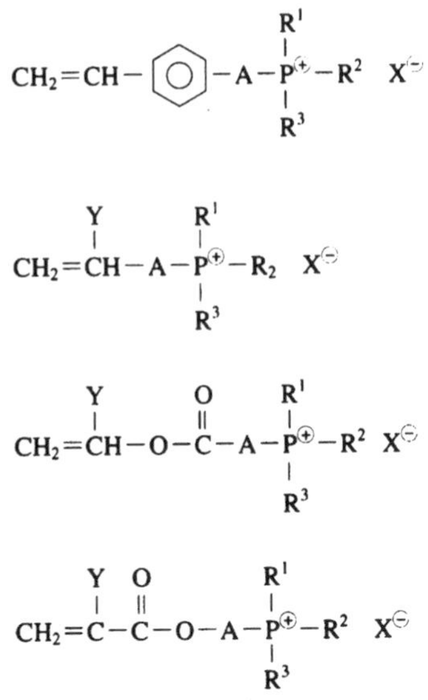

$$CH_2=CH-\overset{}{\underset{}{\bigcirc}}-A-\overset{R^1}{\underset{R^3}{P^{\oplus}}}-R^2 \quad X^{\ominus}$$

$$CH_2=CH-\overset{Y}{A}-\overset{R^1}{\underset{R^3}{P^{\oplus}}}-R_2 \quad X^{\ominus}$$

$$CH_2=CH-O-\overset{O}{\overset{||}{C}}-\overset{Y}{A}-\overset{R^1}{\underset{R^3}{P^{\oplus}}}-R^2 \quad X^{\ominus}$$

$$CH_2=\overset{Y}{C}-\overset{O}{\overset{||}{C}}-O-A-\overset{R^1}{\underset{R^3}{P^{\oplus}}}-R^2 \quad X^{\ominus}$$

(in chemical equations, R^1, R^2, and R^3 are, respectively, hydrogen atom, linear or branched alkyl chain with 1-18 carbons, allyl, aralkyl, aralkyl group substituted with hydroxyl or alkoxy group; X-expresses anion, A is alkylene, and Y is hydrogen atom, lower alkyl or allyl group). An antibacterial superabsorbent polymer is prepared by copolymerizing 0.001-20 wt% of one or two of the foregoing compounds, a small amount of a crosslinking agent, methylene-bis-acrylamide, and a vinyl monomer.

Fig. 3 Antibacterial superabsorbent polymer that does not require other antibacterial agents because a phosphonium salt monomer has been incorporated.

3 MULTIFUNCTIONAL PACKAGING MATERIALS WITH WATER ABSORPTION AND DEODORANT FUNCTIONS

The generation of odor-causing materials such as aldehyde, mercaptan, amine, ammonia, and ethylene is unavoidable when fruits and vegetables or fish are to be transported. Thus for freshness preservation, removal of these odors is also important. Various sheets in which a deodorant based on adsorption such as activated charcoal, zeolite or cristobalite have been mixed and tested [6, 7]. Additionally, deodorants, such as Fe-phthalocyanine complex, based on decomposition of odor-causing compounds, are also available [6, 7].

A deodorant sheet is placed on top of a tray that has many small nodules, and food is then wrapped with this sheet [8]. Other than the use of a deodorant, another method is to use aromatic compounds, such as hinokithiol and cinnamic acid, which are mixed to prevent food-related odors [9].

4 OTHER MULTIFUNCTIONAL PACKAGING MATERIALS

In addition to a superabsorbent polymer, a freshness-preservation agent that contains sodium carbonate, which produces carbon dioxide upon absorption of water, is used to preserve the freshness of sashimi. This material inhibits growth of bacteria and prevents rotting by using a carbon dioxide atmosphere [10]. The freshness preservatives obtained by combining a deoxygenating agent, such as activated iron oxide, and a superabsorbent polymer can prevent browning or the flavor change of foods [6]. Other packaging materials, such as freshening preservatives by combining more complex functions as described earlier or also combining with disposable trays, have been developed.

REFERENCES

1 (1994). Tokkyo Kaiho 14701, Freshness preservation method for strawberries, Dai-Nippon Insatsu.
2 (1995). Tokkyo Kaiho 156972, Water-absorption sheet, Dai-Nippon Insatsu.
3 (1995). Tokkyo Kaiho 123963, Antibacterial water-absorption sheet, Sekisui Chemicals.
4 (1995). Tokkyo Kaiho 17581, Freshness-preservation sheet and the container for its use, Dai-Nippon Insatsu.

5   (1996). Tokkyo Kaiho 92020, Antibacterial superabsorbent polymer, Nippon Kagaku Kogyo.

6   (1995). Tokkyo Kaiho 125762, Transportation sheet with freshness preservation, Nippon Tobacco Industries.

7   (1994). Tokkyo Kaiho 171675, Coolness-maintaining containers, Daiei Sangyo, SB Instruments.

8   (1994). Tokkyo Kaiho 1380, Transportation method of mushrooms, Itoh, S.

9   (1994). Tokkyo Kaiho 90661, Hinokithiol-containing freshness preservative, Sumitomo Seika.

10 (1993). Tokkyo Kaiho 95294, Water-absorbing packing materials, New Nippon Chemical Orment Industries.

Section 3
Coolants for Food Transportation

TAKAO FUSHIMI

1 BASIC STRUCTURE AND FUNCTION OF COOLANTS FOR FOOD TRANSPORTATION

In order to maintain fresh food without the loss of freshness, low temperature is needed. Therefore, when fresh foods are transported, foods are packed in thermally insulating styrofoam along with a coolant. Thus, the low temperature is maintained by the latent heat of melting. A polymer hydrogel coolant that contains a hydrogel in a flexible bag has better shape and cooling characteristics than ice or the more inexpensive dry ice. Thus, its consumption is steadily increasing.

2 IMPROVEMENT OF COOLANT SHAPES FOR FOOD TRANSPORTATION

Because polymer hydrogel coolants are more expensive than the traditional ice method, it is important to reduce the cost while improving the functionality and handling. Numerous improvements and innovations have been reported. Figure 1 presents examples of improved gel coolant packages. Figure 1(a) is a plate-shaped coolant that is wide with a

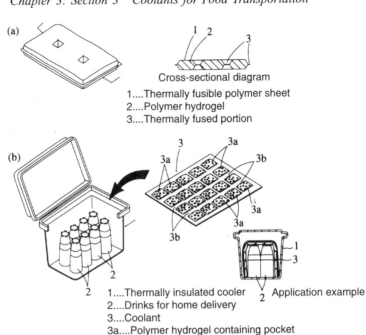

1....Thermally fusible polymer sheet
2....Polymer hydrogel
3....Thermally fused portion

1....Thermally insulated cooler
2....Drinks for home delivery
3....Coolant
3a....Polymer hydrogel containing pocket
3b....Thermally fused portion

3....Coolant
3a....Polymer hydrogel containing pocket
3b....Thermally fused portion
3c....Edge (handling portion)

Fig. 1 Improved shapes of polymer gel-containing bag used as coolant.

shape that is easy to maintain by adhering several places in the middle of the bag of melt adhesive polymer sheet [1]. Figure 1(b) shows small pockets of polymer hydrogels, which are organized to make the whole shape flexible in order to cover the food effectively [2]. Figure 1(c) is also similar to Fig. 1(b) except that it was designed to transport a large fish such as tuna without damaging its body using the small pockets as well as polyurethane foam on the surface [3].

3   IMMERSING SELF-ABSORBING COOLANTS

This coolant is kept as a lightweight and compact superabsorbent polymer bag. Prior to its use, it is immersed in cold water to absorb water and gel.

It is then frozen in a refrigerator until ready to be used. Figure 2(a) illustrates a typical example of an immersing self-absorbing coolant. At least one side of the bag is made of a water-permeable polymer sheet. The internal water-absorption sheet, which is made by sandwiching a super-absorbent polymer by pulp sheets, absorbs and holds water. Upon absorption of water, it gels and forms a gel, which becomes a usable coolant. The coolant, which absorbed half the capacity of the super-absorbent polymer and is frozen, can function as coolant as well as an additional water-absorbing material by absorbing the water when the ice melts. If part of the polymer-containing bag is made of nonwoven cloth or polymer sheet with holes, the aforementioned small-pocketed, wide flat sheet can also be used as an immersing self-absorbing coolant.

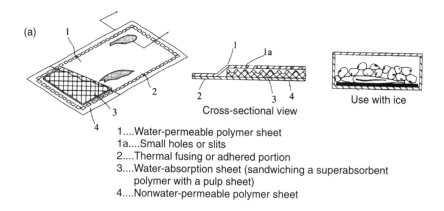

1....Water-permeable polymer sheet
1a....Small holes or slits
2....Thermal fusing or adhered portion
3....Water-absorption sheet (sandwiching a superabsorbent
polymer with a pulp sheet)
4....Nonwater-permeable polymer sheet

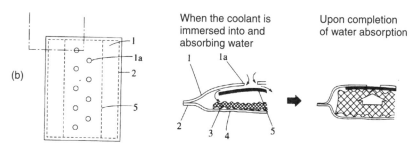

5....Nonwater-permeable polymer sheet for patching
the hole (1-4 are the same as above)

Fig. 2 Immersing self-absorbing coolant.

The hydrogel in this immersing self-absorbing coolant is not closed. Thus, when it is used under pressure, the polymer gel that became fluidic by absorbing water or water itself may be leaked, which could cause problems with delicate foods. Therefore, there are coolants in which the water-absorbing holes are designed to minimize this problem. This is done by placing a polymer sheet of slightly larger size than the hole inside of the polymer film that has holes. Upon immersion into and absorption of water, the gel swells and exerts internal pressure to seal the hole by pushing the polymer film from the inside [5].

4 IMPROVEMENT OF POLYMER HYDROGEL FOR COOLANTS

Efforts are underway to improve superabsorbent polymers used for coolants and water-absorption sheet materials. A water-absorption sheet material made of a polyurethane connected-pore foam from a special polyol and polyisocyanate can be used to disperse superabsorbent polymer powder. This water-absorption sheet has a high rate of water absorption, no volumetric changes, and excellent water-holding ability. Thus, it is ideal for the application of the aforementioned immersing self-absorbing coolant. In the presence of a superabsorbent polymer, urethane reaction takes place to form a polyurethane foam that includes the superabsorbent polymer powder [6].

It has been known that an aqueous solution of poly(vinyl alcohol) forms quasicrosslinking due to the hydrogen bonding between polymer chains when the solution is repeatedly frozen and thawed. This material is an elastomeric, high-modulus hydrogel with high water contents. If an inexpensive, soft and easy-to-handle poly(acrylic acid) superabsorbent polymer and poly(vinyl alcohol) are mixed, a hydrogel with good shape characteristics can be prepared. This material will not lose its shape even at room temperature, leak the gel, or become irregular in shape upon freezing. In addition, this coolant can be used repeatedly [7].

On the other hand, if a microhydrogel, which is made by absorbing water into microparticles of a superabsorbent polymer, is uniformly dispersed into a liquid silicone rubber matrix and the rubber is subsequently crosslinked, a soft and impact-absorbing coolant even at low temperature can be obtained [8].

5 APPLICATION EXAMPLES OF COOLANTS TAILORED FOR A PARTICULAR FOOD

Development of cold packaging with a coolant tailored for a particular food is being actively pursued. Figure 3 shows the packaging for transporting prawns from producer to consumers. Instead of the traditional use of ice and wood particles, prawns can be transported with minimum damage, and it is also easy to pack, remove and count the prawns. A long-lasting coolant to maintain the low body temperature of the prawns and to prevent the legs and other parts of the prawns from moving are the keys for successful application [9].

Figure 4 depicts the packing of desert cakes during freezing, cold storage and thawing. Freshly made cakes are stored in an aluminum container and covered by a water-absorption sheet in such a way as to direct the opening to the superabsorbent polymer layer. Under this condition, the container is placed in a fast refrigeration device to freeze, kept frozen, and removed and thawed when necessary. The cake can be placed into a refrigerator at 5°C and thaw in its container; there is no deformation, degradation and rotting of the cake. In this case, the water-absorption sheet functions as a moisture-conditioning sheet and stable, long-term coolant in addition to its water-absorption function [10].

Additionally, the combination of a cardboard box with one-sided water-repellant treatment and polymer hydrogel coolant is developed for

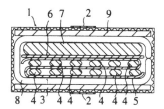

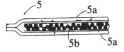

1....Carboard box
2....Polystyrene foam plate
3....Polystyrene foam plate
4....Prawns or crabs
5....Water-absorption sheet
6....Polythylene foam for thermal insulation
7....Coolant
8....Paper sheet
9....Polystyrene foam plate

5a....Nonwoven cloth water-permeable sheet
5b....Superabsorbent polymer

Fig. 3 Packaging for low-temperature transportation of live prawns and crabs [9].

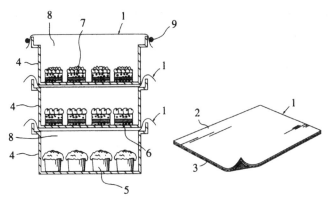

1....Water-absorption sheet that uses a superabsorbent polymer
   (in addition to its function as a water-absorbing sheet, it acts
   as moisture-conditioning and stable temperature coolant)
2....Nonwater-permeable polymer sheet
3....Superabsorbent polymer layer
4....Stackable containers made of aluminum
5, 6, 7....Desert cakes
8....Opening of the container
9....Fixation by a string or a rubber band

Fig. 4 Packing conditions during freezing, refrigeration storage, and thawing.

easy storage and disposal of fresh foods [11]. In the near future, the development of a specialized coolant that can function fully during transportation will become important.

REFERENCES

1 (1990). Tokkyo Kaiho 95837, Coolant, Nippon Byleen.
2 (1995). Tokkyo Kaiho 223677, Home delivery method of drinks in a bottle, Takuhei Sansho.
3 (1993). Tokkyo Kaiho 308896, Cold transportation method for large fish, Itoh, S.
4 (1990). Tokkyo Kaiho 273166, Coolant, Nippon Shokubai.
5 (1990). Tokkyo Kaiho 80489, Coolant, Nippon Byleen.
6 (1991). Tokkyo Kaiho 203921, Water-absorbing polyurethane foam and coolant, Sumitomo Seika.
7 (1990). Tokkyo Kaiho 232291, Coolant and its manufacturing method, Nippon Synthetic Chemicals.
8 (1990). Tokkyo Kaiho 36265, Water-containing silicone rubber elastic composite, Shin-etsu Chemicals.

9   1993). Tokkyo Kaiho 63, Packaging container for live prawns and crabs, Sekisui Chemicals.

10 (1994). Tokkyo Kaiho 153802, Freezing and thawing method of cakes, Sanyo Electronics.

11 (1994). Tokkyo Kaiho 42769, Shelf packaging for fresh foods, Rengo, Technos.

Section 4
Contact-Dehydration Sheet for Food Processing

TAKAO FUSHIMI

1 STRUCTURE AND FUNCTION OF CONTACT-DEHYDRATION SHEET FOR FOOD PROCESSING

A superabsorbent polymer that turns into a hydrogel by absorbing water has extremely high water-holding ability. However, it has weak absorptivity and will not dehydrate the water that is included in other materials. A food-processing material that dehydrates excess water from a food by combining the ability of the superabsorbent polymer to absorb and fix water and the strong osmotic pressure of sucrose has been developed. This amplified dehydrating material is widely used in food processing.

Figure 1 illustrates a contact-dehydrating sheet that shows a laminated structure with strong dehydrating ability [1]. This sheet absorbs excess water that was leaked or squeezed from the food. It also absorbs the free water that exists in fresh meat and fish. The hydrogels of superabsorbent polymers are usually employed because they gel by absorbing water and handling becomes easy. However, in this dehydration sheet, functionalizing the hydrogelation process itself by incorporating water

134

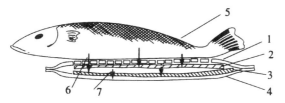

1. Semipermeable film
2. Concentrated sucrose solution
3. Superabsorbent polymer layer
4. Substrate sheet
5. The food to be dehydrated (example: fish)
6. The concentrated sucrose solution dehydrates
   excess free water in the fish
7. The superabsorbent polymer absorbs water
   from the concentrated sucrose solution

Fig. 1 Basic structure and function of contact-dehydrating sheet for food processing.

into the polymer is noteworthy. Furthermore, its unique function is to dehydrate foods for cooking without providing heat.

If a freshly caught fish is left for several hours in contact with this sheet, the water that is interacting with the flavor component will be left alone while only the free water will be dehydrated. Thus, overnight, dried fish can be obtained. Moreover, the free water, which is the cause of rotting and bad odor, is eliminated and the fish will be preserved for a longer period of time. Another benefit is the pre-drying of vegetables for pickling to preserve good flavor. Application to prawns that are used for frying will yield prawns that have appropriate taste. If it is used for materials for slow cooking, the material cooks faster and the flavor of the sauces penetrates the food better. Hence, this dehydrating material is now spreading widely to restaurants, maritime product producers, super-markets, and to gourmet consumers. In an area other than food processing, this dehydrating material is tested to condense or separate physiological polymer solutions such as sugar proteins [2, 3].

2 BENEFITS FOR COOKING AND APPLICATION EXAMPLES OF CONTACT-DEHYDRATING SHEET

The benefits of the dehydrating sheet for cooking are listed as follows.

Freshness preservation

It provides proper resistance when biting the food and delays the generation of bad odor
It prevents discoloration

Improved flavoring

It condenses the flavor
It reduces water and makes the meat firm from the core (see Fig. 2 (a))

Preservation of flavor

It removes generated odor
It preserves firmness and flavor of the fresh food even after freezing

(a) Excess water and odor of fresh food are eliminated. It can also be used prior to cooking so that the forementioned factors will not interfere with cooking.

Left in a refrigerator for 0.5 to 2h

(b) Shallow water fish and domestic beef are stored frozen (no frost will accumulate and the quality of the food will stabilize)

Store in freezer

(c) Frozen materials such as tuna, prawn, and scallop are thawed without damaging the quality (ice crystals will not be formed and cell damage of the food can be avoided)

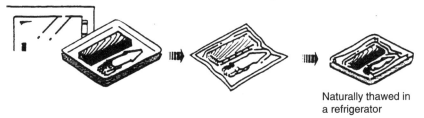

Naturally thawed in
a refrigerator

Fig. 2   Examples of contact-dehydrating sheet application.

Nonsalt dehydration

The meat can be made firm without the use of a salt

Improvement in cooking

The food can be baked faster and evenly
The cooking oil will not splash and the oil quality is preserved longer
The flavoring will penetrate better.
The lye can be minimized.

Several examples of cooking methods are depicted in Fig. 2. Recently, research is underway on an application that aims at the preservation of natural flavor as shown in Fig. 2 (b) and (c).

3 IMPROVEMENT TOWARDS LOW-COST CONTACT-DEHYDRATION SHEETS

It is necessary to enclose a concentrated sucrose solution that is placed on a semipermeable film when the contact-dehydration sheets in subsection 1 or shown in Fig. 1 are prepared. However, a concentrated sucrose solution has extremely high viscosity and handling is difficult. For example air bubbles remain when the solution was coated on the film. Thus, the processing difficulty increases the cost of the final product. In order to cope with this problem, preparation of a low-cost contact-dehydration sheet that is easy to manufacture and handle has been studied (see Fig. 3). Improvements include: i) various high osmotic pressure compounds have been tried; ii) antibacterial compounds such as silver ion ceramics powder, ethanol, egg white lisozyme, amino acids, or organic acids are mixed in order to prevent the water from becoming an environment for bacteria; and iii) hydrophilic alcohol moisture conditioners such as glycerin or propylene glycol is mixed so that excessive dehydration will not take place.

4 PACKAGE SYSTEMS THAT UTILIZED CONTACT-DEHYDRATION SHEET

Using a contact-dehydration sheet, a package system has been developed to transport blocks of meat such as beef and pork in a chilled condition without losing the freshness and appearance (see Fig. 4). While the excess water is actively absorbed by the contact dehydration sheet, thus prevent-

(a)

1, 3....Semipermeable poly(vinyl alcohol)
2........Water-absorbing polymer film (prepared by irradiating
and crosslinking the extruded poly(ethylene oxide) film)

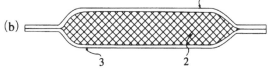

(b)

1, 3....Transparent semipermeable membrane (cellophane,
PVA film, slightly stretched nylon film)
2........Highly permeable material and microfibers (addition
of moisture holding material and anti-microbial agent will lead
to better results) (solbitol and other sugar aqueous solution for
food) (as for moisture holding material, glycerin and for
anti-microbial agent, ethanol and lizochime can be used).

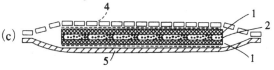

(c)

1....Water meapeable sheet with moisture controlling agent
(a pulp sheet with glycerin impregnated
2....Water absorbing polymer layer (sodium poly(acrylic acid))
4....Water permeable sheet (PP nonwoven sheet)
5....Water impermeable sheet (PE film)

Fig. 3 An example towards development of low cost, contact water absorbing sheet [4, 6].

ing the decay of the meat and shape, it also prevents physical damage by individual vacuum packaging or individual area formed by cardboard. Packaging methods similar to these include: preservation method for meat using a water-absorption sheet [9], frozen fish egg packages [9], and thermal shrinkage laminated package for meat using a contact-dehydration sheet made of thermally shrinkable poly(vinyl alcohol) [10].

In an actual use of a contact-dehydration sheet, often it is used with a packaging material such as polymer film. Hence, if a contact-dehydration sheet or water-absorption sheet is adhered inside of a film with a high gas barrier, it would be convenient for the transportation, processing and

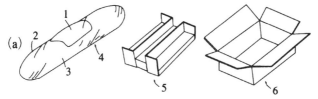

1....Contact water absorbing sheet (include highly permeable
    material and water absorbing polymer as shown in Fig. 1)
2....Synthetic resin packaging material (vacuum packaging)
3....Meat block (chilled)
4....Meat packaging material
5....Area separation material (six meat blocks can be separately stored)
6....External packaging box

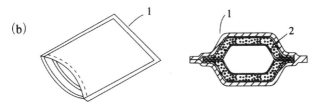

1....Packaging bag made of a film with high oxygen gas barrier
2....Contact water absorbing sheet or water absorbing sheet that
    are directly placed on the inner wall of the gas barrier film

Fig. 4 An example of a packaging system using contact water absorbing
sheet [7, 10].

preservation of fresh foods [11]. In the future, high performance and
multiple function packaging, which is suitable for a high-cost contact-
dehydration sheet, may be developed and is application further developed.

REFERENCES

1 (1983). Tokkyo Kokai 58124, Contact dehydration sheet using osmotic pressure,
   Showa Denko.
2 (1988). Tokkyo Kaiho 58086, Continuous dehydration method, Showa Denko.
3 (1987). Tokkyo Kaiho 97606, Concentration method of polymer aqueous solution
   and its device, Showa Denko.
4 (1993). Tokkyo Kaiho 227922, Water-absorption sheet, Mitsubishi Plastics.
5 (1995). Tokkyo Kaiho 185330, Water-absorption sheet, Kuraray.
6 (1990). Tokkyo Kaiho 39846, Moisture-conditioning sheet for meat, Showa Denko.
7 ( ). Toroku Jitsuyo Shin-an Registration 3014376, Shelving case for meat, Ito Ham.

8   (1991). Tokkyo Kaiho 19648, Preservation method for meat, Nippon Baileen, Fujino, K.

9   (1991). Tokkyo Kaiho 72842, Packaging for fish eggs, Mitsui Toatsu.

10   (1990). Tokkyo Kaiho 192941, Thermally shrinkable laminate packaging material for meat, Showa Denko.

11   (1996). Tokkyo Kaiho 53163, Absorption packaging and its manufacturing method, Showa Denko.

CHAPTER 4

Medicine and Medical Care

Chapter contents

Section 1
Gels for Cell Culture

KOICHI NAKASATO, MASAYUKI YAMATO AND
TOSHIHIKO HAYASHI

1 INTRODUCTION

Gels exhibit physically stable shapes and the ability to hold a solvent while allowing diffusion and permeation of solutes. If these two properties of gels are used to culture cells, the former property is what allows a gel to become the cellular growth matrix while the second of these properties provides diffusive growth factor. Gels are then able to provide both a base for adhesion and growth as well as the nutrients cells require. Multicellular animals have cells that are surrounded by a three-dimensional (3D) extracelluar matrix (ECM) that consists of proteins such as collagen, fibronectin or elastin and proteoglycan, which has glycosaminoglycan side chains (see Fig. 1) [1, 2]. The ECM is in a gelatinous state with high water content. Historically, cell culture work has been done in a petri dish. However, functional cell development also has been studied using a gel as a cell culture basis and planting cells in or on a gel. Collagen gel, agarose gel and matrigel, all of which are often used as cell culture bases, will be described here.

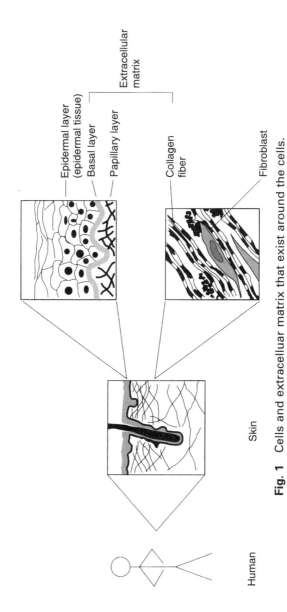

Fig. 1 Cells and extracelluar matrix that exist around the cells.

2 TYPE I COLLAGEN GEL

Collagen is a protein with a triple helix structure consisting of three polypeptide chains. For tropocollagen formation, each polypeptide chain must have gly-x-y (gly = glycine, x and y are other amino acids). Collagen has been identified in 19 molecular species and 33 genes. Collageneous proteins are considered a superfamily [2, 3]. Some collagens self-associate by adjusting solution temperature, pH, and ionic strength as a response to particular physiological conditions. As an example of this phenomenon, the most common protein, type I collagen, forms a gel by molecular association if protein concentration is increased (see Fig. 2).

Type I collagen can be prepared relatively easily in a large quantity and it is also commercially available. As a result, cell cultures with this collagen are common Indeed, this gel culture is widely used to study cancer cell transferability. The culture is made by first preparing a gel that contains a culture base and the cell is planted atop the gel. If cells are suspended in a collagen solution and the solution is gelled, a containing gel can be prepared [4].

This culture can be used to study cells that adhere three-dimensionally with fibrous collagen in the body, thus reproducing typical body conditions. In reality, *in vivo* studies smooth muscle cells [5], vein-like cavity formation of endothelial cells of fine blood veins [6], and secretion

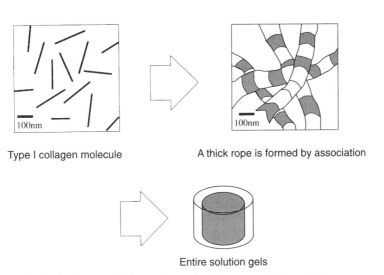

Type I collagen molecule   A thick rope is formed by association

Entire solution gels

Fig. 2 Reassociation of type I collagen and its gelation.

gland-like cavity formation of lacrimal cells [7] shows strong influence of the use of type I collagen substrate for their appearance of cell functions. Similarly, type I collagen and other extracelluar matrix play a role in the living organism.

In Type I three-dimensional collagen gel-cultured for fibroblasts or smooth muscle cell shrinks as the culture grows and the collagen fiber density approaches that of real skin (see Fig. 3) [8, 9]. Shrunken collagen gel is widely used as a model for real skin. Furthermore, artificial skin cultured from skin cells and artificial blood veins cultured from endothelial cells has already been developed. The growth of fibroblasts is significantly reduced as the gel shrinks, and then it ceases. The cells assume a long bipolar shape of several hundred micrometers [10]. Distance between the cells is sufficiently far that there is no direct contact with each other. Fibroblasts under this condition will not resume growth even after the addition of a cellular growth factor such as EGF. This differs from cellular growth that is inhibited as a result of platelet reduction or cellular adhesion. This phenomenon is called matrix contact interference (a result of the distance between cells). There is also no change in the number of EGF receptors or the bonding constant [11, 12].

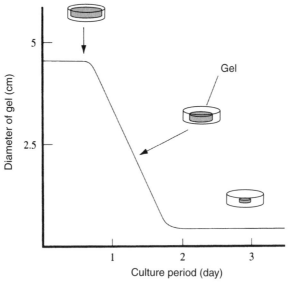

Fig. 3 Shrinkage of collagen gel by fibroblast.

Table 1 Changes in fibroblast in Type I collagen as a function of culture period.

Gel size	↓
Density of collagen fibrils in the gel	↑
Surface area of fibroblast	↑
Contact area of collagen and cell	↑
Growth of fibroblast and ability to synthesize DNA	↓
Synthesis of proteins	↓
Growth rate of fibroblasts	↓
Response to growth factor	↓
Migration and metastasis of cells	↓

→ No change, ↓ Reduction, ↑ Increase

Response to growth factors such as EGF, FGF, PDGF, and TGF-β, is either reduced or disappears in relation to 3D coverage of cells by collagen fibers [12,13].

The shape, growth and response ability of cells to cytokine in a 3D collagen gel culture differs significantly from those in a 2D culture (including a 2D culture on a collagen gel) [8]. The characteristics of a fibroblast in a 3D collagen gel culture are listed in Table 1 [11].

3 AGAROSE GELS

Agarose is a polysachharide that is the main component of agar and consists of a repeated structure of {D-galactosil-(b1- - -4)-3,6-anhydro-L-galactosil-(a1- - -3)}$_n$. Agarose does not readily dissolve in room-temperature water but it does so at $> 60\,°C$. When this solution is cooled to room temperature, viscosity increases gradually and the solution gels. If cells are added during this cooling time, a gel with cells inside can be prepared. It is not thought that a special receptor specific to agarose exists on the cell surfaces of multicellular animals. Instead, it is thought that the cells are either trapped in agarose networks or they adhere indirectly to agarose as a result of the ECM that the cell synthesizes.

Cells that are being cultured by agarose gels include chondrocyte (cartilage cells) and buoyant cells (blood cells, cancer cells, etc.). In 1986, Delbruck *et al.* cultured a human chondrocyte in an agarose gel [14]. They reported an increase in type II collagen, which is characteristic to chondrocyte, as well as glycosaminoglycan. They further reported that at the electron microscope level the cells resembled those in an *in vivo* condition. Chondrocyte is not uniform in cartilage. From the outer surface of cartilage to the inner ones, it begins gradually to change to osteoblastic cellular composition. Aydelotte and Kuettner sampled cartilage at various depths from the surface to the inside and cultured their samples in agarose gels [15]. They found that the shape and immune chromosome of each cell are in good agreement with the corresponding conditions in the body. Thus, it indicates that the environment provided by agarose gel is able to maintain the growth of chondrocytes.

When cells are cultured on an agarose substrate, it is thought that the cells do not directly and specifically adhere to the agarose, but it is possible that the properties of agarose do influence the cells.

4   MATRIGELS

Matrigel is a gelled associative compound made from the raw extract of basement membrane-like structural material from a mouse EHS tumor. A basement membrane is observed by electron microscopy as a membranous structure underlying epithelial cells, surroundings myocytes and neurocytes, and with a high electron density structure (see Fig. 4) [16, 17]. Thickness is on the order of several nanometers. Evidence from pathological examinations and cultured cells has demonstrated that normal functioning of epithelial and endothelial cells, requires adhesion with the basement membrane [18]. The main components of the basement

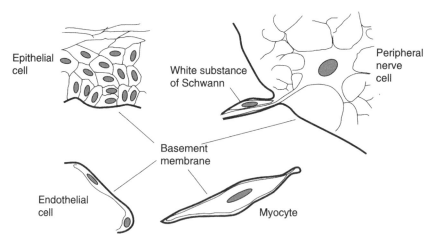

Fig. 4   Location of basement membrane.

membrane are type IV collagen, laminin, heparin (sulfuric acid ester) glycoproteins, nidogen (same as entactin) [19]. Matrigel, a gel consisting of these components, was reported by Kleinman *et al.* in 1985 [20]. Matrigel is regarded as a replacement material for the basement membrane.

When matrigel is used as a cell matrix, many epthelial cells exhibit and maintain their differentiation functions. Those functions include: formation of myoblasts [21]; mammary gland cells [22] and salivary gland cells [23]; ducts of cells; and formation of hepatic vessel lining cells [24]. In an investigation of cancerous metastases and infiltration, cancer cells were planted on the surface of matrigels. The distance the cells migrated and the number of infiltrated cells are measured as an index of cancerous cell metastasis. The cell response is not necessarily the same when separate components from the matrigel were used as the coated component of a culture dish in comparison with the response when matrigel itself was used [25]. This implies that the actual gel uniquely influences the cell.

Matrigel has been reported to contain various growth factors including TGF-β and FGF, in addition to mixed components of the epithelial basement membrane [26]. When the functions of various cells are used to clarify the biological functions of basement membranes, the existence of growth factors complicate the analysis. We reported that type IV collagen extracted from bovine lens capsules could form aggregates in a form of gel without the existence of laminin and heparin glycoproteins

Table 2 Comparison between Type IV collagen and matrigel.

	Type IV collagen gel	Matrigel
Starting materials	capsule of cow lens	EHS tumor
Extraction conditions	0.5 M acetic acid	2 M urea, 50 mM Tris-HCl (pH 7.3)
Components	Type IV collagen	Laminin 60 %, type IV collagen 30 %, heparin (sulfuric acid ester) proteoglycan 3 %, nidogen 5 %
Gelation conditions	2 M guanidine chloride, 10 mM dithiothreitol, 50 mM Tris-HCl (pH 8.1), 4 °C or 20 mM phosphate buffer (pH 7.3), 150 mM NaCl, 4 °C	150 mM NaCl, 50 mM Tris-HCl (pH 7.4), 35 °C

[27, 28]. This gel exhibits a meshwork construction similar to that of basement membranes when it is observed by electron microscopy [28]. It may be important to understand the structure and biological function of gels made only of type IV collagen in order to elucidate the structure and function of the basement membrane and its related organs. A comparison between the matrigel and type IV collagen is shown in Table 2 [27].

5 CONCLUSION

It is only recently that gels have been used as culture substrates for cells from high-vertebrate animals such as humans. Hence, little analysis has been done. At this point, it is important to clarify at the molecular level the various phenomena and mechanisms related to control of cellular functions. Studies on the functions of ECM molecules have begun from findings that the RGD arrangement in fibronectin is the cellular adhesion site. Other amino acid sequences on the various ECM structure molecules have now also been identified as cell adhesion sites. However, cell function formation and control that occur because the extracelluar environment is a gel cannot be sufficiently understood only from information on adhesion as influenced by the amino acid sequence. It is necessary to consider macroscopic properties (such as newly developed structure by molecular association, elasticity, and mechanical stimuli) in any evaluation of the effect of the matrigel on cell function. In fact, cell behavior apparently differs depending upon whether it is on either a type I collagen or coat a gel, or in a gel. Even in a gel, cell function development differs significantly depending upon the contact density of collagen fibers and cells. It may also be important to analyze the distribution of gels and the cell contact when analyzing functions. To understand the effects of gel on the control of cell functions, it is important to analyze the distribution of direct contact points between cells and the gel, the degree of contact concentration, and the existence or lack thereof of a specific adhesion structure. In other words, gels exhibit the plasticity of supramolocular structures. The higher-order structures involved in molecular associations, the packing of polymer chains, and conformation structures can provide signals to cells. Study of multicellular culture using gels will be important to our understanding of the formation and subsequent functional development of multicellular systems.

REFERENCES

1 Nakamura, K., Fuijyama, A., and Matsubara, K. (1995). *Molecular Biology of Cells* (Transl.), 3rd ed., Kyoiku Publ.

2 Tanaka, S. and Hayashi, T. (1992). in *Connective Tissue A, Biochemistry, Organs and Immunity*, T. Tsuji, ed., Bunei-do Shuppan, pp. 377–418.

3 Imamura, Y. and Hayashi, T. (1996). in *Collagen Family, Experimental Medicine, Terminology Library, and Cell Adhesion*, A. Miyazaka and I. Yahara, eds., Youdo Publ., pp. 52–59.

4 Yamato, M. and Hayashi, T. (1992). in *Culture in Collagen, Biomembrane and Transport*, vol. 1, *New Experimental Biochemistry Lecture Series 6* Soc. Biochem., Jpn., ed., Tokyo Kagaku Dojin, pp. 309–312.

5 Yamamoto, M., Nakamura, H., Yamato, M., Aoyagi, M., and Yamamoto, K. (1996). *Exp. Cell. Res.* **225**: 12.

6 Sato, Y., Okayama, K., Morimoto, A., Hamanaka, R., Hamaguchi, K., Shimada, T., Ono, M., Kohno, K., Sakata, T., and Kuwano, M. (1993). *Exp. Cell. Res.* **204**: 223.

7 Yoshino, K., Tseng, S.C.G., and Pflugfelder, S.C. (1995). *Exp. Cell. Res.* **220**: 138.

8 Nishiyama, T., Tsunenaga, M., Nakayama, Y., Adachi, E., and Hayashi, T. (1989). *Matrix* **9**: 193.

9 Yamato, M., Adachi, E., Yamamoto, K., and Hayashi, T. (1995). *J. Biochem.* **117**: 940.

10 Nishiyama, T., Tsunenaga, M., Akutsu, N., Norii, I., Nakayama, Y., Adachi., E., and Hayashi, T. (1993). *Matrix* **13**: 447.

11 Hayashi, T., Nishiyama, T., and Adachi, E. (1991). *Fundamental Investigation on the Creation of Biofunctional Materials*, T. Okamura and T. Tsuruta, eds., Kyoto: Kagaku Dojin, pp. 55–64.

12 Nishiyama, T., Akutsu, N., Horii, I., Makayama, Y., Ozawa, T., and Hayashi, T. (1991). *Matrix* **11**: 71.

13 Nishiyama, T., Horii, I., Nakayama, Y., Ozawa, T., and Hayashi, T. (1990). *Matrix* **10**: 412.

14 Delbruck, A., Dresow, B., Gurr, E., Reale, E., and Schroder, H. (1986). *Connect. Tissue Res.* **15**: 155.

15 Aydelotte, M.B. and Kuettner, K.E. (1988). *Connect. Tissue Res.* **18**:205.

16 Inoue, S. (1994). *Microsc. Res. Tech.* **28**: 29.

17 Yamasaki, Y., Makino, H., and Ota, Z. (1994). *Nephron* **66**: 189.

18 Watanabe, A. and Okazaki, K. (eds.) (1996). *Extracelluar Matrix—Application to Clinical Medicine*, Medical Review Co.

19 Timpl, R. (1989). *Rur. J. Biochem.* **180**: 487.

20 Kleinman, H.K., McGarvey, M.L., Hassel, J.R., Star, L.L., Cannon, F.B., Laurie, G.W., and Martin, G.R. (1986). *Biochem.* **25**: 312.

21 Maley, M.A., Davis, M.J., and Grounds, M.D. (1995). *Exp. Cell. Res.* **219**: 169.

22 Kim, N.D., Oberley, D.O., and Clifton, K.H. (1993). *Exp. Cell. Res.* **203**: 6.

23 Nogawa, H. and Takahashi, Y. (1991). *Development* **112**: 855.

24 Shakado, S., Sakisaka, S., Noguchi, K., Yoshitake, M., Harada, M., Mimura, Y., Sata, M., and Tanikawa, K. (1995). *Hepatology* **22**: 969.

25 Friedman, S.L., Roll, F.J., Boyles, J., Arenson, D.M., and Bissell, D.M. (1989). *J. Biol. Chem.* **264**: 10756.

26 Vukicevic, S., Kleinman H.K., Luyten, F.P., Robers, A.B., Roche, N.S., and Reddi, A.H. (1992). *Exp. Cell. Res.* **202**: 1.

27 Muraoka, M., Nakazato, K., and Hayashi, T. (1996). *J. Biochem.* **119**: 167.

28 Nakazato, K., Muraoka, M., Adachi, E., and Hayashi, T. (1996). *J. Biochem.* **120**: 889.

Section 2

Applications of Gel for Plastic Surgery: Artificial Breasts and Skin

NAOKI NEGISHI AND MIKIHIRO NOZAKI

1 HYDROGELS AND LYOGELS

Gels can be divided generally into hydrogels, where water is the solvent, and lyogels, which contain an organic solvent. Hydrogels are viscoelastic materials that are crosslinked hydrophilic polymers that are swollen by water (see Fig. 1). For medical uses, there are hydrogel dressings, soft contact lenses, and artificial lens. The following intermolecular cross-linked hydrophilic polymer gels have been developed: 1) crosslinked poly(N-vinyl pyrrolidone) (Nu-Gel^{TM}) for use as hydrogel dressings; 2) crosslinked poly(hydroxyethyl methacrylate) (Hydron^{TM}) for soft contact lenses; and 3) crosslinked hyaluronic acid (Hydron Gel^{TM}) for artificial lens. An artificial breast implant gel, crosslinked poly(N-vinyl pyrrolidone) (Bio-Oncotic Gel^{TM}) has also been evaluated.

Lyogels are viscoelastic materials that are crosslinked hydrophilic polymers that are swollen by an organic solvent (see Fig. 1). As representative examples, there are polyurethane and silicone gels [1]. Because lyogels use organic solvents, they are not usually used for food and medical applications. Currently, silicone gels are being used only for

154

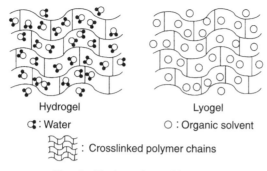

Hydrogel   Lyogel

🝔 : Water   ○ : Organic solvent

⋙ : Crosslinked polymer chains

Fig. 1 Hydrogels and lyogels.

external treatment. However, in the past, silicone gels were used as the breast implant material of choice. Problems with implant silicone gels will be discussed first here.

2 SILICONE GEL-FILLED BREAST PROSTHESIS

Silicone gels are viscoelastic materials made of a silicone sponge that is swollen by a silicone oligomer (see Fig. 2) [1]. They were originally developed as insulators for the electronics industry [2]. The silicone gel used as a breast prosthesis is a lyogel (see Fig. 3) [3–5]. Although silicone sponges and envelopes are solid, the solvent, an oily silicone, is naturally fluid. Thus, even if they have the same siloxane backbone, $-Si(CH_3)_2O-$, they cannot be considered to be the same [6–9]. Silicone fluid can create localized sensitivity reactions, but it can also cause either specific or overall immune responses by the body whether it remains *in situ* or migrates to other tissues. This may happen when peptides are stabilized in

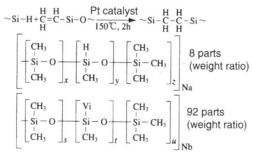

Fig. 2 Curing mechanism for silicone gels.

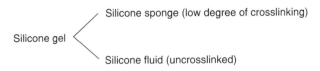

Silicone envelope (crosslinked)
Nondamaged: permeation and migration of the fluid
Damaged: leakage and migration of the gel
Fractured: sudden leakage and migration of the gel

Fig. 3 Composition, existence of crosslinking and envelope conditions of silicone gel-filled prosthesis and the behavior of the filler.

a water-in-oil (W/O) emulsion using a nonbiodegradable oil such as silicone [10, 11]

Nonbiodegradable oily silicone may cause abnormal reactions following prolonged sensitization [8, 9, 11]. When a peptide antigen is a quasi-antigen (similar cross-reacting antigen) that is similar to its own body proteins, there is the danger that an autoimmune disease can be triggered. However, compared to paraffin, because silicone fluid is highly viscous, the probability of autoimmune responses and/or diseases is expected to be small [9–11].

Because the nonbiodegradable oil is a fluid, it continues to leak from the capsule and migrate throughout the body after macrophages (Mϕ) can no longer clean it up. Following this migration in the body, local inflammation becomes widespread, even systemic. The most likely problem is prolonged local inflammation with a nonbiodegradable foreign fluid [12]. The problem then with a silicone gel is that it may trigger specific immune reactions.

The characteristics of silicone will be described here.

i)   The energy of the siloxane bond is high and not easily decomposed;
ii)   Uncrosslinked material is fluid even if it has high molecular weight, because the cohesive force is weak and mobility is high;
iii) Crosslinked silicone elastomer degrades by adsorption of fat [13, 14];
iv)   Uncrosslinked silicone fluid (oily silicone) forms an emulsion with the surfactant that exists in the body [15];
v)   Silicone fluid becomes an oily vehicle that stabilizes the peptide antigen as a W/O emulsion [8, 9];
vi)   It is a lymph-directive polymer that dissolves in fat. Rees *et al.* and Ben-Hur *et al.* reported from their pathological and histological

studies that silicone fluid migrates mainly into the reticuloendothelial system in the liver and is included in giant cells (such as Mϕ-type cells) [16, 17]. Furthermore, Kawakami *et al.* studied the distribution of radio isotope-labeled silicone fluid by autoradiography [7] and found that part of the silicone fluid migrates and is adsorbed by the skeletal system. A small amount was also discovered in digestive system excretions.

To date, the following side effects accompanying migration of silicone fluids and silicone gels have been reported [18–24]. These reports suggest that the problem might best be termed a foreign material diffusion syndrome as a consequence of silicone migration. In particular the influence was found in neutrophils, monocytes, and erythrocytes [18, 19], and fibrillation and capture [20–23] by various reticuloendothelial systems, including those of the lymphatic system nodes, the lungs, and the liver. Sanger *et al.* reported that the nervous system is subjected to contractile damage caused by the fibrillation that accompanies bodily foreign material reactions to silicone [24].

In general, if the foreign material is a solid rather than liquid, only a normal Mϕ reaction will be seen. That is, based on the size of the foreign material, Mϕ will induce a reaction, such as inclusion, formation of giant cells, and surrounding and encapsulation of the material [25]. If the embedded material in the body is solid, even if it is nonbiodegradable, it will remain localized due to the encapsulation effect. However, if it is a nonbiodegradable oil, it is able to leak and migrate [22, 23, 27–29] in the body as [26] the Mϕ cells become depleted. Because of this migration there is the possibility that local inflammation will occur in far-flung body organs. A similar problem occurs in artificial joints [30, 31] when friction caused by wear and tear creates a fine powder. The biggest problem is prolonged local inflammation by a nonbiodegradable foreign material that is liquid and thus can flow [32].

As already described, the solvent for the gel, silicone fluid, migrates throughout the body. However, the ratio between the crosslinked silicone (silicone sponge) and uncrosslinked silicone (silicone fluid) is unknown [4]. The data on silicone gel migration and ADME factors (A: absorption; D: distribution; M: metabolism; and E: excretion) are not at this time available. Although silicone fluid had to be classified as a new drug [33], the ADME of silicone, which serve to indicate safety within the body, is not publicly available as a result of industrial propriety [39]. Japanese

Table 1 Estimated order of foreign material migration in the body.

Mechanism of migration in the body
Initial migration in the body
  1. Mechanical effect
  2. Inflammation + emulsification
Later period migration in the body
  Movement towards lymph node and blood

interpret oily silicone as a nonbiological mineral oil. Oil is widely used to stabilize lipophilic drugs and to aid in fat solubility.

Biodegradable animal and plant oil must be used internally. Mineral oil is limited only to external uses because paraffin, as well as silicone oil do not decompose, and instead will migrate throughout the body (see Table 1) and accumulate. There are many reports on silicone fluid that indicate migration and accumulation in the body [7, 22, 23]. Migrated silicone fluid stimulates the Mϕ cells and as a result, severe fibrillation of surrounding systems takes place. Fibrillation by foreign material can cause contractile damage to nerves, which is a very serious problem [24]. As Mϕ cells disappear, the silicone fluid migrates again and repeatedly causes inflammation (see Table 2). While there is some opinion that silicone gel is safe because it is contained in a bag, there have been a publically issued dates from manufacturers in Japan regarding how long silicone envelopes remain intact [34]. Rather, the literature indicates that silicone bags tend to rupture 10 years after implantation [35, 36].

The problems caused by implanted silicone gels are listed here [12].

i)   Silicone gels are lyogels and their solvent, silicone fluid, migrates in the body.

Table 2 Example of foreign material inflammation by repeated stimuli.

Foreign material inflammation by the continuous activation of phagocytes
  Frictional material
    Powder formed by friction: repeated inflammation at the same location
  Migratory material
    Fluid: inflammation repeated at different location
Fibrosis proceeds by inflammation cycle

ii)   The migrated silicone fluid stimulates macrophage cells and, as a result, fibrillation of the surrounding system proceeds. This fibrillation can cause contractile damage to nerves.

iii) Following the disappearance of macrophage cells, silicone fluid migrates again. It is expected that inflammation and fibrillation of surrounding systems will be repeated.

iv)   The safe-if-used by date for silicone gel-filled prostheses is extremely insufficient and the risk/benefits balance is rather poor.

v)   In any examination of the foreign material migration syndrome, local inflammation should be tracked by using cytokine data and reduction of reticuloendothelial system functions (reduction of reticuloendothelial treatment capability) should be followed.

3 HYDROGEL-FILLED BREAST PROSTHESIS

Because hydrogels use water as their solvent, safer implant gels than silicone ones can be prepared if a hydrophilic biocompatible polymer is used. For breast implant gels, crosslinked poly(N-vinyl pyrrolidone) (Bio-Oncotic Gel^{TM}) [37] and crosslinked hyaluronic acid [38] have been evaluated. Although silicone gel has poor x-ray transmission, Bio-Oncotic Gel^{TM} has good x-ray transmission and does not interfere with a breast cancer diagnosis. Hyaluronic acid gels have good x-ray transmission. In addition, the membrane formation around the implant is thin and shrinkage of the membrane will not take place. Hyaluronic acid is already used as a wound dressing and safety data are plentiful. Even though the envelope that contains the gel can rupture, hyaluronic acid will not cause problems because it is biodegradable. Hyaluronic acid is the hoped-for material for future implant gels. Crosslinked hyaluronic acid (Hylan Gel^{TM}) is also being evaluated for use as an artificial vitreous [39].

4 SKIN SUBSTITUTES

Because the functions of the skin are complex, artificial skin can provide only a fraction of its functions.

In general, skin substitutes serve to reduce pain, prevent infection, accelerate epithelial cell formation, prevent bodily fluid from escaping or building up, and protect wounds. Thus, a nonstimulus profile, no water accumulation problems, and good adhesion characteristics are important design criteria. There are temporary skin substitutes used as wound dressings, durable skin substitutes for tissue reorganization templates, and permanent skin substitutes used in cultured skin [40].

4.1 Wound Dressings

Table 3 lists wound dressings of the temporary skin substitute type. Wound dressings can be classified into biological and synthetic dressings. Biological dressings include freeze-dried pigskin and reconstructed biomaterials such as collagen and chitinous membranes. Synthetic dressings include a collagen-coated nylon cloth (Biobrane^{TM}), which is a hydrocolloid that encourages epidermal growth [41, 42], hydrophilic dressing hydrogels [43, 44], and a hydrophobic poly(L-lysine) sponge that is good at retaining drugs. Table 4 lists classifications of wound dressings and artificial skin materials based on their applications. A dressing that maintains a moist environment and accelerates cure is

Table 3 Wound dressings.

Biological dressings
Protein type
    Freeze dried pigskin (Alloask D^{TM}): adhesive fiber (collagen)+elastic fiber (elastin)
    Collagen membrane (Meipack^{TM})

Polysaccharide type
    Chitin membrane (Beschitin-W^{TM})
    Alginate nonwoven cloth (Kaltostat^{TM}, Sorbsan^{TM})
    N-saccinilchitosan-aterocollagen sponge with gentamicin (Uresuc-C^{TM})
    Hyaluronic acid membrane (Hylumed Film^{TM})

Synthetic dressings
Polyamide, polyurethane type
    Collagen-coated nylon cloth (Biobrane^{TM})
    Polyurethane membrane (Biooclusive^{TM}, Op-Site^{TM}, Tegaderm^{TM})
    Polyurethane form (Allevyn^{TM}, Lyoform^{TM})
Polyaminoacid type
    Poly(L-lysine) hydrophobic sponge with AfSD (Xemex Epicuel^{TM})

Hydrocolloid type
    Polyisobutylene/pectin/gelatin/carboxymethyl cellulose composite (DuoDerm^{TM})
    Polyisopropylene/Caraya gun/carboxymethyl cellulose composite (J&J Ulcer Dressing^{TM})
    Poly(styrene-co-isoprene)/polycyclopentadiene.carboxymethyl cellulose (Comfeel Ulcus^{TM})

Hydrogel type
    Poly(ethylene oxide) hydrogel dressing (Vigilon^{TM})
    Poly(N-vinyl pyrrolidone) hydrogel dressing (Nue-Gel^{TM})

Table 4 Classification of wound dressing materials/artificial skin materials based on applications.

Dressing materials for temporary protection

Skin peeling wound (less than 12/1000 in), hallow second degree burn—wound dressing on skin papillary layer

Freeze dried pigskin (Alloask DMTM), collagen membrane (Meipack^{TM}), chitin membrane (Beschitin-W^{TM}), arginate nonwoven cloth^{TM} (Kaltostat^{TM}, Sorbsan^{TM}), polyurethane form (Allevyn^{TM}, Lyoform^{TM}), poly-L-lysine sponge (Xemex Epicuel^{TM}), N-succinilchitosan AC sponge (Uresuc-C^{TM}), hydrocolloid dressing (DuoDerm^{TM}, J&J Ulcer^{TM}, Comfeel Ulcus^{TM}), hydrogel dressing (Vigilon^{TM}, Nue-Gel^{TM})

Pressure relief dressings (Comfeel PRD^{TM})

Dressings for temporary closure of wounds

deep second degree burn—wound closing on skin papillary layer/third degree burn—wound closing on the muscle layer

Collagen-coated nylon cloth (Biobrane^{TM}), poly-L-lysine sponge (Xemex Epicuel^{TM})

Dressings for prevention of infection

Frasiomycin-containing gauze (Sofratulle^{TM}), Gentamycin sulfate containing N-succinilchitosan-aterocollagen sponge (Uresuc-C^{TM}), silver sulfur diazin containing poly-L-lysine sponge (Xemex Epicuel^{TM}), bacterial control biofilm-like arginate nonwoven cloth (Kaltostat^{TM}, Sorbsan^{TM})

Dressings for maintenance of moist environment

Moist wound healing—occlusive dressings

Polyurethane membrane (Biooclusive^{TM}, Op-Site^{TM}, Tegaderm^{TM}), polyurethane form (Allevyn^{TM}, Lyoform^{TM}), hydrocolloid dressing (DuoDerm^{TM}, J&J Ulcer^{TM}, Comfeel Ulcus^{TM}), hydrogel dressing (Vigilon^{TM}, Nue-Gel^{TM}), arginate nonwoven coth (Kaltostat^{TM}, Sorbsan^{TM})

Template and complex cultured skin substitute for permanent skin reconstruction

Loss of skin—template for skin reconstruction

Insoluble collagen/chondroitin-6-sulfate template (Integra^{TM}), bovine thermal-dehydration crosslinked fibrillar aterocollagen/thermally modified arterocollagen template (Terudermis^{TM}), pig chemically crosslinked aterocollagen template (Pelnac^{TM}), polygractin network sheet (Vicryl Mesh^{TM}), polygractin type composite cultured skin (Dermagraft^{TM})

called an occlusive dressing. There are also film dressings and absorbent dressings. The former is a bodily fluid cumulative wound dressing and the latter is a bodily fluid absorptive wound dressing. Film dressing is made of adhesive polyurethane film. A film in which moisture permeability is properly controlled is adhered to the wound. This creates a moist environment by accumulating the bodily fluid that escapes the wound and this then accelerates wound healing. Absorptive dressings are hydrophilic materials of hydrocolloids or hydrogel-type dressings. By absorbing bodily fluid on the surface of the wound, a moist environment is maintained and the cure is accelerated.

4.2 Tissue Reorganization Template

A tissue reorganization template is used as artificial skin and can be classified into polyester, collagen and polysaccharide types.

Yannas Burke Col/ChS Skin Template (Stage I)

The Yannas Burke type is made of a sponge sheet from a ionic complex of insoluble collagen fibrils (Col) and chondroitin-6-sulfate (ChS). It is crosslinked by glutaraldehyde and covered by a silicone elastomer. This template is called stage I [45–47]. It was commercialized by Marion Laboratories as the artificial skin Integra^{TM}. If this stage I skin by Yannas-Burke is used for wounds in which skin is missing, fibroblasts and capillary enter the sponge that faces the body. Collagen is also newly generated and a new, skin-like system is produced. The Col and ChS will both be replaced in ≈3–4 weeks by the body's own system. This sponge helps the body to heal and functions as though it were a mold. Hence, the use of the word template.

FAC/HAC Skin Templates

This is a template made of aterocollagen from which the antigenic determinant teropeptide is removed. Crosslinking is not formed by the usual chemical means, but by dehydration crosslinking by heating *in vacuo*, allowing the material to be less inflammatory. It is made of a fibrillar aterocollagen (FAC) and a heat-cured aterocollagen (HAC). Fibrillation of collagen improves biostability and heat-cured collagen improves cell compatibility. Hence, glycosaminoglycan is not added. In practice, the bottom layer that touches the wound uses a crosslinked FAC-HAC sponge for short-term thermal dehydration while a silicone membrane is used in the outer layer [48]. It was commercialized by Termo as a double-layer aterocollagen-type artificial skin with the trade-name Terudermis^{TM}. Similar double-layer aterocollagen-type artificial skin [49] is also sold by Gunze under the tradename Pelnac^{TM}.

5 SILICONE GEL SHEETING

Silicone gel sheeting is used to help prevent permanent hyperplastic scarring. Wounds that reach into the derma often form scar tissue if the wound covered a large area. This kind of wound is often accompanies by redness and itching. Scarred tissue does not stretch as much as normal skin does. On the other hand, keloid scar formation invades normal skin tissue. Keloid hyperplasia is often accompanied by severe itching. Radiation and chemical and pressure treatments have been used in an attempt to prevent both hyperplasia and keloid scan formation. In recent years, silicone gel sheeting has been used and it has attracted attention as a new treatment method for hyperplastic scarring [50–52]. The scar is covered by a silicone gel sheet, with the goal improvement of hyperplastic scarring. However, mechanism for improvement is not yet well understood. Some reports contend that simply maintaining moisture by use of an occlusive dressing will give a similar result. However, the advantages of the silicone gel sheet include its flexibility and ability to adhere, allowing easier application to the skin. If silicone gel sheeting is used daily for as long as possible, improvement after 6 months was reported to be 93–100 % better than the treatment without this gel [50]. Degree of improvement was evaluated based on the increased stretch capability of the hyperplastic tissue, diminution of color, and actual reduction of scarring. This approach was also effective in other wounds that were resistant to the pressure treatment method and there were no side effects [50]. Silicone gel sheets are not used for open wounds. The following skin protective sheeting materials are available today—Silastic™ Gel Sheeting, Dow Corning, and Cica-Care™ Gel Sheeting, T.J.Smith & Nephew Ltd.

REFERENCES

1   Matsuoka, H. (1987). *Manufacturing and Application of Functional Polymer Gels*, M. Irie, ed., Tokyo: CMC, pp. 52–66.
2   Nelson, M.E. (1962). US Pat. 3,020,260.
3   Cronin, T. (1966). US PAT. 3,293,663.
4   Brill, A.P. (1978). US Pat. 4,100,627.
5   Van Aken Redinger, P. and Compton, R.A. (1984). US Pat. 4,455,691.
6   Zivojnovic, R. (1990). *Silicone Oil in Vitreoretinal Surgery*, A. Yamanaka and K. Okubo, eds., Medical Aoi Shuppan, pp. 25–27.
7   Kawakami, T., Nakamura, C., Hasegawa, H., and Eda, S. (1987). *Dent. Mater.* **3**: 256.
8   Whitehouse, M.W., Orr, K.J., Beck, F.W.J., and Pearson, C.M. (1974). *Immunology* **27**: 311.
9   Wakamatsu, S., Negishi, N., and Hirayama, S. (1991). *Ijaku no Ayumi* **158**: 443.
10 Okabayashi, K. (1979). In *Immune Diseases* K. Okabayahi, ed., Tokyo: Bunko-do, pp. 165–223.
11 Nagishi, N., Nozaki, M., Isa, T. *et al.* (1992). *Nikkei Kaishi* **12**: 340.
12 Negishi, N., Nozaki, M., and Okano, M. (1995). *Seikei Geka* **38**: 411.
13 Carmen, R. and Mutha, S.C. (1972). *J. Biomed. Mater. Res.* **6**: 327.
14 Swanson, J.W. and Lebeau, J.E. (1974). *J. Biomed. Mater. Res.* **8**: 357.
15 Crisp, A., De Juan, Jr., E., and Tiedeman, J. (1987). *Arch. Ophthalmol.* **105**: 546.

16 Rees, T.D., Ballantyne, Jr., D.L., Seidman, I., and Hawthorne, G.A. (1967). *Plast. Reconstr. Surg.* **39**: 402.

17 Ben-Hur, N., Ballantyne, Jr., D.L., Rees, T.D., and Seidman, I. (1967). *Plast. Reconstr. Surg.* **39**: 423.

18 Andrews, J.M. (1966). *Plast. Reconstr. Surg.* **38**: 581.

19 Hawthorne, G.A., Ballantyne, Jr., D.L., Rees, T.D., and Seidman, I. (1970). *J. Reticuloend. Soc.* **7**: 587.

20 Ellenbogen, R., Ellenbogen, R., and Rubin, L. (1975). J. Am. Med. Assoc. **234**: 308.

21 Chastre, J., Basset, F., Viau, F., Dournovo, P., Bouchama, A., Akesbi, A., and Fibert, C. (1983). *N. Engl. J. Med.* **308**: 764.

22 Travis, W.D., Balogh, K., and Abraham, J.L. (1985). *Hum. Pathol.* **16**: 19.

23 Truong, L.D., Cartwright, Jr., J., Goodman, M.D., and Woznicki, D. (1988). *Am. J. Surg. Pathol.* **12**: 484.

24 Sanger, J.R., Matloub, H.S., Yousif, N.J., and Komorowski, R. (1992). *Plast. Reconstr. Surg.* **89**: 949.

25 Pasyk, K.A., Argenta, L.C., and Austad, E.D. (1987). *Clinics Plast. Surg.* **14**: 435.

26 Peacock, Jr., E.E. (1984). In *Wound Repair*, Philadelphia: W.B. Saunders, Co., pp. 1–14.

27 Harris, K.M., Ganott, M.A., Shestak, K., Loskin, W., and Tobon, H. (1991). *Radiology* **181**: 134.

28 Capozzi, A., du Bou, R., and Pennisi, V.R. (1978). *Plast. Reconstr. Surg.* **62**: 302.

29 Levine, R.A. and Collins, T.L. (1991). *Plast. Reconstr. Surg.* **87**: 1126.

30 Bommer, J., Gemsa, D., Kessler, J., and Ritz, E. (1985). *Nephron* **39**: 395.

31 Chiba, J., Rubash, H.E., Kim, K.J., and Iwaki, Y. (1994). *Clin. Orthop. Related Res.* **300**: 304.

32 Lossing, C. and Hansson, H.-A. (1993). *Plast. Reconstr. Surg.* **91**: 1277.

33 Ashley, F.L., Braley, S., Rees, T.D., Goulian, D., and Ballantyne, Jr., D.L. (1967). *Plast. Reconstr. Surg.* **39**: 411.

34 Rylee, R.T. (1991). *Dow Corning Materials News*, Mar/Apr., 4.

35 Van Rappard, J.H.A., Sonneveld, G.J., Van Twisk, R., and Borghouts, J.M.H.M. (1988). *Ann. Plast. Surg.* **21**: 566.

36 De Camara, D.L., Sheridan, J.M., and Kammer, B.A. (1993). *Plast. Reconstr. Surg.* **91**: 825.

37 Liang, J.H.E. and Sanders, R. (1993). *Br. J. Plast. Surg.* **46**: 240.

38 Lin, K., Bartlett, S.P., Matsuo, K., LiVolsi, V.A., Parry, C., Hass, B., and Whitaker, L.A. (1994). *Plast. Reconstr. Surg.* **94**: 306.

39 Dailey, W., Verstgraeten, T., Hartzer, M., and Blumenkranz, M. (1992). *in vivo. Invest. Ophthalmol. Vis. Sci.*, 33 (suppl.), 1314.

40 Negishi, N., Nozaki, M., and Tajima, M. (1992). In *Artificial Organs*, T. Agishi, H. Sakurai, and H. Takano, eds., Nakayama Shoten, pp. 263–272.

41 Hermans, M.H.E. and Hermans, R.P. (1986). *Burns* **12**: 214.

42 Nangia, A., Lam, F., and Hung, C.T. (1990). *Drug. Develop. Industr. Pharm.* **16**: 2109.

43 Foxjun, C.L., Modak, S., Stanford, J.W., and Bradshaw, W. (1980). *Burns* **7**: 295.

44 Mertz, P.M., Marchall, D.A., and Kuglar, M.A. (1986). *Arch. Dermatol.* **122**: 1133.

45 Yannas, I.V., Burke, J.F., Orgill, D.P., and Skrabut, E.M. (1982). *Science* **215**: 174.

46 Burke, J.F., Yannas, I.V., Quinby, Jr., W.C., Bondoc, C.C., and Jung, W.K. (1981). *Ann. Surg.* **194**: 413.

47 Murphy, G.F., Orgill, D.P., and Yannas, I.V., (1990). *Lab. Investigation* **63**: 305.

48 Koide, M., Osaki, K., Konishi, J. *et al.* (1993). *J. Biomed. Mater. Res.* **27**: 79.

49 Matsuda, K., Suzuki, S., Isshiki, N., and Ikada, Y. (1993). *Biomaterials* **14**: 1030.
50 Carney, S.A., Cason, C.G., Gowar, J.P. *et al.* (1994). *Burns* **20**: 163.
51 Gibbons, M., Zuker, R., Brown, M. *et al.* (1994). *J. Burn Care Rehabil.* **15**: 69.
52 Gold, M.H. (1994). *J. Am. Acad. Dermatol.* **30**: 506.

Section 3
Soft Contact Lenses

AIZO YAMAUCHI

1 INTRODUCTION

Well-known examples of polymer gel applications include disposable diapers made from superabsorbent polymers and soft contact lenses (SCL). Both of them utilize the function of gels fully and provide significant benefits to daily life. The SCL's value is extremely high. For example, if one contact lens weighs 20 mg and costs 10,000 yen, the price would be as high as 500 million yen/kilogram. Furthermore, the concentration of the polymeric raw materia is approximately one quarter. Some products even have various radii that allow lenses to be tailored to treat both astigmatisms and farsightedness. Contact lenses (CL), correct individual sight, are a unique and very expensive consumable product.

Currently, it is believed that 8–10 million people use CL in Japan. Of those, approximately 50 % are SCL users in comparison to the US, where SCL use is approximately 90 %. New users tend to choose easy-to-use SCL rather than hard contact lenses (HCL) and this ratio continues to increase. As economic development continues, the shift from eyeglasses to CL is occurring in various Asian countries and other developing countries.

166

2 HISTORY OF CONTACT LENSES

Attempts to place a lens directly on the cornea are actually old. Perhaps 100 years ago, a glass lens was used. However, active use started after World War II when poly(methyl methacrylate) (PMMA) lenses with excellent transparency were developed—these were called contact lenses [1]. The PMMA was first used for windshields during World War I. Since a fragment in the body caused by accident did not cause inflammation, it came to be regarded as a good biocompatible plastic quite by accident when fragments from a shattered windshield were embedded in the body it was seen that there was no inflammation. Because it has excellent characteristics, among them transparency, processability, and stability this material became favored for use in contact lenses. Manufactured in 1949 in Japan, and used in 1951 for the first time [1, 2], HCL of PMMA material has a long history. Because HCL are used on the cornea (a sensitive area), the wearer feels as though there is a foreign object in the eye. However, once users become accustomed to their lenses, they use them because lenses are easy to deal with and provide high optical accuracy.

Development of a material that resembles corneal material and thus provides comfort has always been the goal. A silicone rubber-type CL was developed for this purpose. However, as it was highly hydrophobic and harmed the cornea, it was withdrawn from the market. Subsequently, a hydroxyethyl methacrylate SCL was made [3]. It was commercialized by a U.S. Company, Bausch & Lomb, and use spread all over the world.

3 CLASSIFICATION AND COMPONENTS OF CONTACT LENSES

It is unnecessary to state that contact lenses are considered to be a medical device. As this is the case, very strict evaluation by every country where they are used is required regarding both safety and effectiveness. These are restrictions and standardizations placed on any medical device. The European Union (EU) in recent years established international standards (ISO) touching upon the material, product, performance, biosafety tests, and clinical trials.

Currently, contact lenses are controlled in Japan by the Ministry of Health, decree No. 302, *Standard on Vision Corrective Contact Lenses*. The materials used is controlled in a subsection which states that "the lens

Table 1 Classification and main components of contact lenses.

Major classification	Minor classification	Characteristics	Example of the polymeric main components that constitute lenses
HCL	PMMA-type HCL (oxygen nonpermeable HCL)	–	Homopolymer of methyl methacrylate (MMA)
	Oxygen-permeable HCL (RGPL)	Oxygen-permeable RGPL	Cellulose acetate butylate (CAB)
			Copolymers mainly from siloxanyl methacrylate (SiMA) and MAA
			Other
		Highly oxygen-permeable RGPL	Copolymers mainly from SiMA and fluoroalkyl methacrylate (FMA)
			Other
SCL	Nonwater-containing SCL	–	Silicone-type elastomers
			Acrylic-type elastomers
		Water content < 40 %	Homopolymers or copolymers of 2-hydroxyethyl methacrylate (HEMA)
	Water-containing SCL	Water content 40–60 %	Copolymers of HEMA, N-vinyl pyrrolidone (NVP), methacrylic acid (MA) and MAA
			Copolymers mainly from MMA and glycerol methacrylate (GMA)
			Other
		Water content > 60 %	Copolymers mainly from MMA, NVP, acrylamide (AAm)

Table 2   Representative monomers and polymers used as soft contact lens materials.

Chemical name	Chemical formula	Molecular weight (MW)
2-Hydroxyethyl methacrylate (HEMA)	$CH_2{=}C$ with CH_3; $C{-}O{-}CH_2CH_2{-}OH$; $\overset{\|\|}{O}$	(MW = 130.14)
2,3-Dihydroxypropyl methacrylate or glycerol methacrylate (GMA)	$CH_2{=}C$ with CH_3; $C{-}O{-}CH_2CHCH_2$; $\overset{\|\|}{O}$ $OHOH$	(MW = 160.16)
N-vinyl pyrrolidone (NVP)	$CH_2{=}CH$; N; $_2HC$ $C{=}O$; $_2HC{-}CH_2$	(MW = 111.14)
Methacrylic acid (MA) or sodium methacrylate	$CH_2{=}C$ with CH_3; $C{-}O{-}H$ (or sodium salt); $\overset{\|\|}{O}$	(MW = 86.09)
Acrylamide (AAm)	$CH_2{=}CH$; $C{-}NH_2$; $\overset{\|\|}{O}$	(MW = 71.1)
Poly(vinyl alcohol)	$\left[CH{-}CH_2{-}CH{-}CH_2 \right]_n$ with OH OH	

materials should not have voids, impurity, waviness, and discoloration, and should possess both chemical and physical stability." In a supplemental explanation, "the lens material for hard contact lens must be polyacrylic resins, in particular poly(methyl methacrylate). Due to safety considerations, glass is not allowed."

As already mentioned, because CL use began with HCL, the standard is based on HCL. However, it is also applied to SCL based on poly(hydroxyethyl methyacrylate)(PHEMA). It is also necessary to check on CL using *The Guidelines on Basic Biological Examination of Medical Devices and Materials*, [5] from the ministry of Health, June, 1995.

Accordingly, unless composition and components meet safety standards items that fall under the medical devices rubric cannot be offered for sale. However, CL will be described here only from the materials point of view. Table 1 lists overall classifications and the major components. Unless otherwise noted, the description refers to SCL.

All SCL basically consist of hydrophilic polymers that maintain their gel structure, with the ability to swell and not dissolve by intermolecular crosslinking. Therefore, natural polymers, such as agarose and gelatin, are potential candidates for its material. However, currently commercially available SCL are made only of copolymers of methacrylic acid derivatives, mainly PHEMA, and several other monomers because of stability and safety concerns. Lenses with a water content of 30–40 % are made mostly of PHEMA and those with water contents of 70–80 % are made of copolymers of N-vinyl pyrrolidone (NVP), acrylamide (AAm), HEMA, and methacrylic acid (MA). There are also poly(vinyl alcohol) derivatives. Table 2 shows the names and chemical structures of various SCL components [6].

4   MANUFACTURING METHOD OF SOFT CONTACT LENSES

There are basically two manufacturing methods for SCL [4] — the swelling method, in which a solid lens is first manufactured and then swollen with water, and the direct method, in which a monomer (s) that contains water or a solvent is polymerized in the lens mold. Whether it is a solid or a gel, the similarity of the manufacturing methods is shown in Fig. 1. Although frozen gel can be cut and polished (this has been done by the authors), it is not practical and is not currently used commercially. Polymers also can be injection or compression molded to make solid lenses and then later swollen. However, these methods are not actually used.

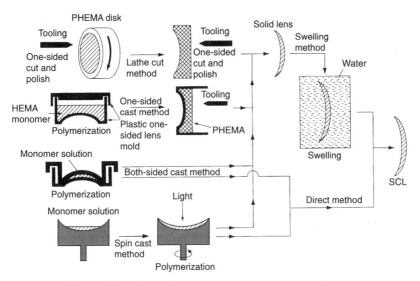

Fig. 1 Manufacturing method for soft contact lenses.

(1) Lathe Cut Method

A polymer in a rod or plate shape is cut and polished. The lens is then swollen by either water or saline solution. It is a fundamental manufacturing method for both HCL and SCL.

(2) One-Sided Cast Method

A monomer is polymerized in a mold with accurate curvature on one side and the other side is cut and polished to make a solid lens. It is later swollen to make SCL.

(3) Double-Sided Casting Method

Polymerization is performed in a mold having accurate curvatures on both sides and then later swollen to make a SCL, or a monomer that contains water or solvent is used to obtain a gelled SCL.

(4) Spin Cast Method

On a mold with certain curvature, the monomer is spin cast and polymerized. The lens is later swollen. Alternatively, a monomer that contains water or solvent can be spin cast and polymerized to directly obtain a gelled lens. If a solvent is used in either method (3) or method (4), the solvent is replaced with a saline solution or a SCL preservative solution.

The swelling method can be applied to all four of the methods listed. However, the direct method can be used only for methods (3) and (4). When the swelling method is used, removal of residual monomer is easily accomplished. The lens can be rim polished while it is a hard solid.

In particular, the lathe cut method (1) makes it possible to manufacture a lens with special curvature, this allows a high value-added lens specifically tailored to the user to be made. On the other hand, the direct method is suitable for mass production and is particularly suited for daily disposable lenses. Unfortunately, the current polymerization technique still has residual monomer and less molding precision than desired.

5   PROPERTIES OF SOFT CONTACT LENSES

Properties for SCL include safety, optical clarity, oxygen permeability, and ability to be easily handled and cleaned by the wearer.

(1)   Safety

Safety standards for CL manufacturing must include those described in both the forementioned *Standard on Vision Corrective Contact Lenses* and *The Guideline on Basic Biological Examination of Medical Devices and Materials*. In addition, they must meet the manufactured goods responsibility law (PL law) for medical devices from the standpoints of packaging, method of use, and display. In particular, boiling and chemical cleaning to prevent bacterial infections must be systematically performed. When a hydrogel SCL is boiled, decomposition, degradation, deformation and denatured adhered protein are potential problems. If a chemical treatment is used, it is necessary to perform an additional treatment to neutralize the chemicals. In addition when eyedrops are used with SCL in place, it is necessary for the lenses not to concentrate a specific component of the eyedrops, such as preservatives, with concomitantly, harm to the wearer's cornea.

(2)   Optical Properties

Optical correction properties are the fundamental function of all CL. It is extremely difficult to manufacture lenses of high optical clarity, refractive index, and curvature at high precision because SCL is a gelled material. In particular, the refractive index depends on the water content of the lens. Hence, along with oxygen permeability (to be discussed later), it is essential to understand the state of water in the gel. In recent years, interest in SCL for correction of both near- and farsightedness has become very intense and finding a high-performance technique to design and process lenses with nonsymmetric and nonlinear curvatures has increased in importance.

(3) Oxygen Permeability

Because the cornea lacks blood vessels, oxygen supply is provided by the eyes own moisture. If the amount of oxygen available is inadequate, eye damage results. Oxygen is supplied when contact lenses are worn by the use of a highly oxygen permeable polymer and the design of lenses to include an internal curvature which permits natural bodily fluid to bathe the cornea.

However, because they are soft and thus like the actual cornea, SCL will adhere to the cornea. Oxygen is mostly supplied through water transport in the lens material. Thus, oxygen supply depends heavily on the water content of SCL.

To use lenses continuously (continuous-wear lenses), it is necessary to design material in which very high water content is available so as to supply oxygen even during sleep. It is also necessary that the product be thin-walled because this improves oxygen permeability. Unfortunately, high water content and thin lenses create serious manufacturing difficulties. Also affected are deformation during sterilization, degradation, and reduction of optical properties. Thus, the right design for the intended purpose is very important. Oxygen permeability is usually shown by the coefficient of oxygen permeability/standard thickness. Using this quantity, water is said to be < 100 (10^{-11} $cm^3.cm.cm^{-2}.sec^{-1}.mmHg^{-1}$) and saline solution is reported to be 74 [7]. Figure 2 [9] shows the relationship [8–10] between water content and oxygen permeability. A lens with water content of < 30 % has extremely low oxygen permeability. It is thus hypothesized that the transport of oxygen depends on the free water movement throughout the gel. For continuous-wear SCL, it is desirable to use lenses of > 70 % water content.

(4) Cleanliness

Because SLC are always in contact with the wearer's own eye moisture, low molecular weight materials in these fluids penetrate the gel networks,

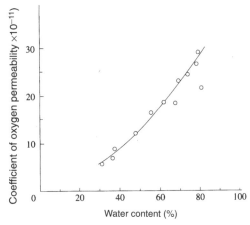

Fig. 2   Water contents and oxygen permeability of soft contact lens.

which of course means that both protein and fat adsorb onto the lens surface. Because soft contact lenses are made of a hydrophobic main chain with hydrophilic functional groups or side chains to improve hydrophilicity, they are considered an amphoteric or positive/negative charged polymer to which proteins and fats are attracted. Thus they must be washed during daily sterilization or a protein removal agent must be used. Coating or a surface treatment of the SCL surface has recently been evaluated. To date, however, there have been no satisfactory products.

(5)   Handling

Although SCL are more comfortable to wear, both strength and durability are inferior to that of HCL and handling can be problematic. The lens is transparent and has a refractive index that is close to water. Therefore, it is slightly colored, so that it is easier to find. The shape of high-water content lenses is difficult to maintain over time. Because it is difficult to recognize the inner from the outer surface of a lens, a mark may be imposed. There are any number of innovations to prevent mix up of the right from the left lens, including using a marked container for sterilizing and carrying of one's lenses. Disposable lenses eliminate the need for sterilization. They are easy both to use and carry. However, they are not yet widely used and when they are, it is only a temporary use.

6 FUTURE DEVELOPMENT

It is apparent that SCL will be the major vision correction device in comparison to eyeglasses and HCL. Near-future- and long-term future trends will be discussed in what follows.

Currently, SCL products satisfy certain product standards. Domestic and international standards for medical devices and materials will become even stricter than they are today. In particular, biosafety standards will be strengthened. Hence, in the near future, the use of historical PHEMA will be replaced following development of mass-production lenses as well as high value-added lenses to be used for astigmatism and farsightedness; manufacture of these newer lenses to have a nonlinear spherical surface will be pursued. At the same time, lenses for fashion and sports will be developed. In the long term, research and development on high-precision one pot synthesis for optical lens and high-performance hybridization with biomaterials will follow recent trends for sophisticated medical devices. Various subjects will be discussed here with these two items in mind.

6.1 Disposable Lenses

It can be easily foretold that the shift towards softer, and therefore more comfortable [11] materials is approaching.

One of the problems soft contact lenses have is the difficult, often complex, care they require. Sterilization using boiling water was required for SCL use in Japan. On the other hand, chemical sterilization (such as with hydrogen peroxide) has been used in Western countries. In Japan, the Ministry of Health allowed the hydrogen peroxide method, a cold sterilization method, to be used beginning in 1992 [12]. However, cleaning and sterilizing small, thin, fragile lenses remains quite cumbersome. The lens material degrades, weakens, deforms, hardens, and discolors as a result of repeated heat or chemical sterilization. Therefore, to reduce lens care frequent exchange type lens that combines several days of continuous use and then disposal of the lens after several weeks have begun to be used. Although these lenses do not require everyday cleaning, bothersome cleaning and sterilization still remain.

An extreme case of this frequent exchange type is the complete daily disposable CL (DDCL). From a medical point of view, these lenses eliminate concern regarding individual differences in care (e.g., amount of residual cleaning chemicals or an insufficiently cleaned lens). Although these seem to be ideal, there are problems inherent in manufacturing inexpensive lenses. Let us assume that the ordinary SCL, at 20,000 yen for each lens, can be used for two years. If the cost of care products is added, it will annually cost approximately 20,000 yen per eye. In order to offer the same cost per eye using daily disposable lens, 20,000 yen/365 days = 55 yen (currently 130 yen per lens). If cost can be brought down to 30,000 yen annually, the cost is still 80 yen per lens. If transportation and other costs are considered, another 20 yen per day must be added to our figure. Although the raw material cost is negligible, advanced manufacturing techniques that employ accurate optical precision remain necessary. Further, if 10 % of the current CL users in Japan uses DDCL, it is necessary to develop an annual mass production capability of 600 million lenses.

It is therefore very difficult to use the lathe cut method to satisfy this need. An appropriate response is to adopt the direct method using complete precision polymerization of an aqueous solution with a single step process in a mold. It is also necessary to develop an appropriate monomer, prepolymer, or photopolymerizable materials. At this point, optimization of design concept as shown in Fig. 3 becomes necessary. The raw material must be water-soluble monomers or polymers that can be aqueous-solution polymerized. During polymerization, 3D crosslinking must take place. Furthermore, the lens must be molded while it contains water and therefore dimensional accuracy will be guaranteed.

Although there are traditional gel processing methods, none of them are sufficient for an ultraprecision polymerization process capable of providing thin, exact water content lenses with the required optical properties (curvature). There are many problems involved in mass production, including the use of the mold itself as the lens case, establishment of a production technique for mass-produced lenses, ability to sterilize after polymerization, and transportation of the necessary number of lenses. There are many technologies not yet used in lens manufacturing that could accomplish these goals. Ultraprecision polymer processing has already been achieved by the electronics industry. Research and development of aqueous solution polymerization is also active. Accordingly, it is expected that, in the near future, consumption of daily disposable CL may increase substantially.

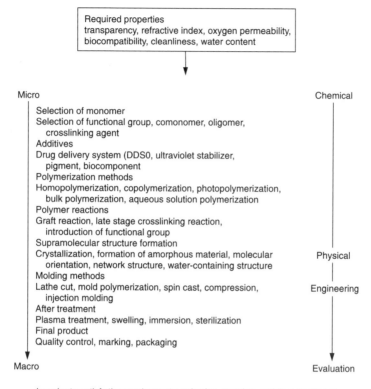

Fig. 3 Design concepts of contact lenses.

6.2 Hybridization

In recent years, environmental problems caused by plastic products have attracted significant attention. Interest is directed at the synthetic polymer materials made using petroleum chemistry and biodegradable plastics made by microbes. These newer materials have already been commercialized. In the area of medical device applications, a biodegradable thread is used and in the ophthalmology area, high molecular weight hyaluronic acid aqueous solution has been of use during cataract surgery. In the CL area, all currently commercialized hard and soft contact lenses are made of synthetic polymers. Only exceptionally, are PMMA-grafted polysaccharides used because of their good biocompatibility. Medical polymers have been in recent years made as a combination of tissues, cells, and other biopolymers such as collagen, chitin, chitosan and hyaluronic acid, and synthetic polymers. These hybrid materials exhibit excellent biocompatibility. Potential is high for use as the base materials for new functions such as artificial cells and tissues. In a similar direction, an artificial cornea using a hybrid material made of photopolymerized polymer and the epithelial cells of the cornea or the actual paranchyma cells from the cornea and CL material [13, 14] using photopolymerizable hyaluronic acid has already been reported.

These technologies are now possible because: i) the relationship between polymer gel structures and properties and functions have been analyzed on the molecular level; ii) understanding of the compatibility between biomaterials and synthetic materials, in particular, interpenetrating polymer networks of these materials, has improved; and iii) methods for evaluating bodily responses has improved vastly in recent years.

Although it might take some time due to the number of requirements (including legal safety issues) more biocompatible and contamination resistant hybrid materials suitable for mass production use will become available. Furthermore, going beyond the concept of CL as a vision corrective medical device, there may be an artificial cornea made of a futuristic intelligent material that can sense and recognize stimuli and respond to external factors.

REFERENCES

1   Mizutani, U. (1981). *Safe Contact Lenses*, Kyoiku Shuppan Center, pp. 26–28.
2   Nakajima, A. and Itoi, M. (eds.) (1984). *Contact Lens Preparation Manual*, Nan-ko-do, pp. 117–121.
3   Wichter, O. and Lim, D. (1960). *Nature* **185**: 117.
4   Ministry of Health, Division of Medical Device Development (Jpn.). (1995). Decree No. 99, *Guideline of Fundamental Biological Evaluation of Medical Device and Materials*, June 27.
5   Yamauchi, A., Kozaki, K., Saishin, M., and Kato, K. (1996). *Frontier of Clinical Application of Contact Lenses*, Kimpara Publ., pp. 151–164.
6   Yamauchi, A. and Yokoyama, Y. (1993). *Zairyo Kagaku (Mat. Sci., Jpn)* **30**: 15.
7   Sumioe, T., Kiyomatsu, Y., and Matsuyama, M. (1991). *Nichikore (J. Contact Lens Soc., Jpn.)* **33**: 191.
8   Nakabayashi, N., Nakabayashi, N., and Atsumi, K. (eds.) (1978). *Medical Polymers*, Kyoritsu Publ., pp. 20–25.
9   Yamauchi, A. (1983). *Nichikore (J. Contact Lens Soc., Jpn.)* **25**: 19.
10 Ikada, Y. (1988). *Biomaterials*, Nikkan Kogyo, pp. 135.

11 Inaba, A. (1996). *Nichikore* (*J. Contact Lens Soc., Jpn.*) **38**: S39.
12 Itoi, M. (1996). *Nichikore* (*J. Contact Lens Soc., Jpn.*) **38**: S25.
13 Nakao, H., Matsuda, T., and Saishin, M. (1993). *Ganki* (*Folia Ophthalmol. Japonica*) **44**: 247; **44**: 1107 (1993); **45**: 614 (1994).
14 Nakao, H. (1994). *Ganki* (*Folia Ophthalmol. Japonica*) **45**: 484.

Section 4
Absorbable Hydrogels for Medical Use
YOSHIHITO IKADA

1 INTRODUCTION

Various hydrogels have been used for medical purposes. Of particular, soft contact lenses occupy the major gel medical application niche. All disposable contact lenses are made of hydrogels. The next major application is to hemostatic agents. In these materials, gels are used as adhesives and sealants. However, use is low today, with the majority employed as hemostatic agents. In this subsection, biodegradable hydrogels will be described. As discussed in Part II, Chapter 2, Section 11, soft tissues in the body are all hydrogels.

2 HEMOSTATIC AGENTS, ADHESIVES, AND SEALANTS

The majority of blood coagulants use fibrin glue made of a two-liquid type. Its raw materials are fibrinogen, thrombin and the XIII factor protein from human blood. If aqueous solutions of these materials are placed on or in a bleeding area, the solution will gel and bleeding will cease. Gelation proceeds with the specific hydrolysis of fibrinogen into fibrin by thrombin, an enzyme. Fibrin then polymerizes to become fibrin aggregates and gels.

Table 1 Advantages and disadvantages of fibrin glue.

Advantages	Disadvantages
1. It can be directly coated onto the wound from a syringe needle	1. Possibility of infection because product is made of blood
2. Blood accelerates gelation	2. Adhesion with living tissue is weak
3. Healing of wounds is accelerated	3. Absorption is very rapid

This hydrogel is crosslinked by the XIII factor, which is a peptide bond synthesis enzyme, and it then becomes an even stronger gel. Although fibrin glue adheres to tissues during gelation, it is weak. Therefore, it is sometimes used as an adhesive or sealant when strong adhesion is not necessary.

Both the advantages and disadvantages of fibrin glue are listed in Table 1. As can be seen in this table, the major disadvantage of fibrin glue is the possibility that a viral infection could be transmitted because it is made of human blood. Perhaps because of this, it is not sold in either the U.S. or Canada.

In Japan, three fibrin glues are commercially available. These are: i) Beriplast P® (Hoechst Japan); ii) Tisseel® (Nippon Zoki); and iii) Bolheal® (Kaketsuken-Fujisawa Pharmaceuticals).

The commercial blood coagulants shown in Table 2 are also available.

Cyanoacrylate is also used as an adhesive, sealant and coating in medical applications. However, its use is insignificant compared with fibrin glue. Cyanoacrylate becomes hard upon polymerization and it is not a gel.

Table 2 Blood coagulants other than fibrin glue.

Abiten	Raw Materials	Manufacturer and sales
Novacol	Microfibrils from cow derma collagen	Alcon·Zeria Pharmaceuticals Co.
Helistat	Microfibrils from collagen	Date Scope·Takeda Phamaceuticals Co.
Oxycell	Oxidized cellulose	Marion Lagrator·Kenbu Seishudo
Surgecell	Crosslinked gelatin	Warner Lambert·Sankyo
Spongel	Chitin	J & J·J & J Medical
Gelform		Yamanouchi Pharmaceutical Co·Yamanouchi
Marintron		Pharmaceutical Co.
		Japan UpJohn·Sumitomo Pharmaceuticals Co.
		Unitika·Kyowa Pharmaceutical Co.

To replace fibrin glue, we continue to develop two-liquid type hemostatic agents that have no possibility of transmitting viral infections. Gelatin and poly(L-glutamic acid), bioabsorbable water-soluble polymers were chosen. As with fibrin glue, when these aqueous solutions are mixed, they gel quickly. Two gelatin methods were developed. In the first water-solution carbodiimide (WSC) is added to poly(L-glutamic acid) aqueous solution and then mixed with a gelatin aqueous solution [1–3]. Due to the effect of WSC, a peptide bond is formed between the carboxylic group of poly(L-glutamic acid) and the amine group of gelatin. This bond becomes a crosslink point and the mixture gels to obtain a hydrogel. This hydrogel adheres well with tissues and exhibits better hemostatic capability than fibrin glue. The adhesion strength is shown as a function of adhesion time in Fig. 1, in which mouse skin is glued together with this hydrogel. For comparison, the fibrin glue results are also shown in the same figure.

Another approach is to bond hydroxysuccinimide to poly(L-glutamic acid) and activate it. Then, this activated poly(L-glutamic acid) aqueous solution is mixed with a gelatin aqueous solution [4].

The reaction is schematically shown in Fig. 2. An example of gelation time is shown in Fig. 3. Activated poly(L-glutamic acid) is stable in the presence of water if it is under appropriate conditions. The gel obtained has good adhesion to tissue and exhibits excellent hemostatic capability. This gel is absorbed by the body. However, biosafety

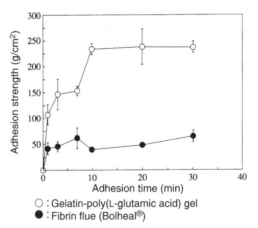

○ : Gelatin-poly(L-glutamic acid) gel
● : Fibrin flue (Bolheal®)

Fig. 1 Adhesion strength of mouse skins when they are glued by gelatin-poly(L-glutamic acid) gel crosslinked by 1-ethyl-3-(3-dimethylaminopropyl)-carbodiimide [3].

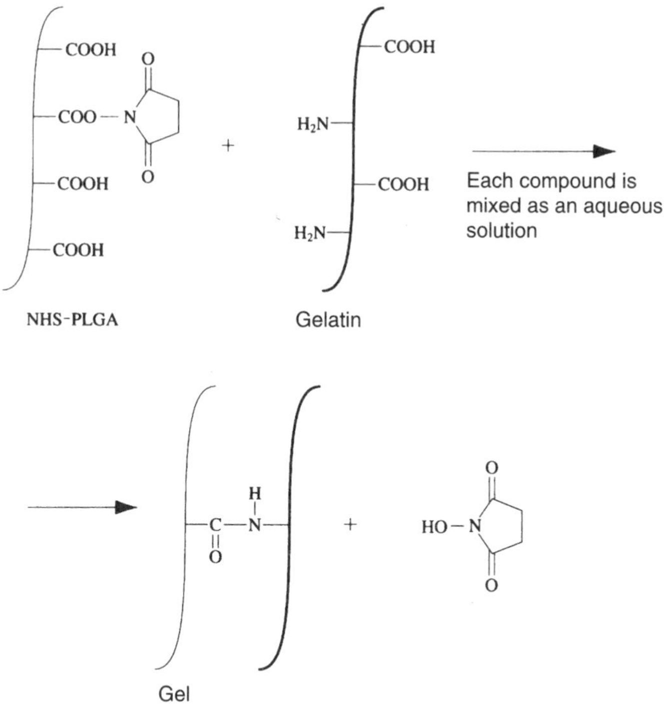

Fig. 2 Crosslinking reaction between poly(L-glutamic acid) (NHS-PLGA) activated by N-hydroxysuccinimide and gelatin.

characteristics must await the future research. The results obtained thus far indicate that hydroxysuccinimide is safer than WSC.

3 ADHESION PREVENTION

When the horn of the uterus, body tendons, and intestinal are damaged, tissue adhesion often results. (See Table 3 for data on adhesion preventative membranes for several types of surgery.) In order to prevent this phenomenon, a viscous polymer aqueous solution, silicone sheet, poly(4-fluoroethylene) sheet, amnion, regenerated collagen membrane, and an oxidized cellulose cloth have all been used. However, none of these methods provide satisfactory results. Unsatisfactory results probably occur as a result of blood clot formation, tissue damage from remaining materials, premature absorption by the body, lack of flexibility, and

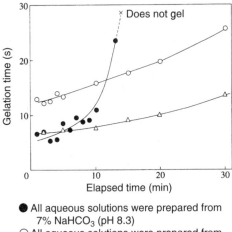

● All aqueous solutions were prepared from
   7% NaHCO$_3$ (pH 8.3)
○ All aqueous solutions were prepared from
   a phosphate buffer solution (pH 7.4)
△ Gelatin aqueous solution was prepared from
   7% NaHCO$_3$ and NHS-PLGA is from the
   phosphate buffer solution

Fig. 3 The relationship between elapsed time after aqueous solution preparation and the gelation time of N-hydroxysuccinimide activated poly(L-glutamic acid) (NHS-PLGA)(NHS-PLGA aqueous solution is 2 wt% and the gelatin aqueous solution is 20 wt%) [4].

Table 3 Operations and frequency of use of adhesion prevention membranes.

United States, 1991

	Annual occurrence	Frequency used	Number of cases for which membrane is used
Peritoneal adhesion prevention surgery	325,000	80%	260,000
Surgeries of muscles, tendons, fascia and cisterna	344,000	20%	68,800
Salpingostomy	423,000	40%	170,000
Hysterectomy	644,000	50%	322,000
Partial stomach removal, intestinal removal surgery	293,000	20%	60,000
Total	2,029,000		880,800

difficulty in fixation of the adhesion prevention material on the desired location.

Therefore, we started our research by preparing crosslinked hydrogel membranes from gelatin [5, 6]. Because this reproduced membrane has a high water content, the blood clot adheres very little and thus the contacted tissue is not damaged. By changing the degree of crosslinking, the rate of absorption can be controlled significantly. After cutting the periosteal tendon of a chicken's PIP joint, the flexor tendon was cut and then sawed. The gelatin membrane was wrapped around the tendon and observed after 4 weeks. As can be seen in Fig. 4, there is almost no appearance of mononuclear cells and little adhesion was observed. On the contrary, when a gelatin membrane was not used, severe adhesion was observed. Burns developed an adhesion prevention membrane Seprafilm™ from hyaluronic acid obtained from a microbe and carboxymethyl cellulose. Evaluation was completed recently [7]. This film disappeared in 7 days in the abdominal cavity and was completely excreted within 28 days. Upon application in the human abdominal cavity, 51 % of the patients showed that adhesion had been prevented. In comparison, when Seprafilm™ was not used, only 6 % patients avoided adhesions.

(a) Wrapped by acrosslinked   (b) No membrane used
gelatin membrane

Fig. 4 Tissue reaction four weeks after a gelatin membrane is embedded around a chicken flexor tendon (H.E. staining).

4 TISSUE ENGINEERING MATRICES AND SEPARATION MEMBRANES

The new technique to regenerate tissue *in vivo* or *in vitro* using cells and synthetic materials is called tissue engineering [8]. When regeneration takes place in the body, either cells are planted on the synthetic material prior to implantation or cells from the patient penetrate the embedded material. The materials used for tissue engineering function as the base for cell adhesion, growth, or as spacers for tissue growth. The former two are called the matrix and the latter is called an isolation membrane. Both absorptive and nonabsorptive materials in the body are used.

Absorptive hydrogels are used for both the matrix and isolation membrane types. The material used the most for the matrix is a porous collagen membrane, where it serves as the basis of skin tissue regeneration. For skin regeneration, covering the collagen membrane without implanting fibroblasts is sufficient to begin the regeneration of new skin [9]. However, in the case of epidermal regeneration, it is necessary to plant epidermal cells. Similarly, when a porous collagen hydrogel tube was sewn onto an area on an esophageal sample, esophageal soft tissue was regenerated [10]. In both cases, skin and esophagus, a silicone sheet was used as the protective membrane. This sheet is readily removed upon regeneration of the tissue. The collagen is absorbed by the body.

Gelatin and collagen are used as an isolation membrane in tissue engineering. For example, when a crosslinked gelatine tube was used as a connection guide for peripheral nerves to block invasion by external collagen tissue, nerves were regenerated along a 1-cm length defective portion of a sciatic nerve of a rat [11]. The gelatin hydrogel was completely absorbed by the body. A collagen membrane is also used following periodontal surgery to close up pockets in tissue and aid in healing and regeneration of tissue [12]. Tissue regeneration with such a polymeric membrane is called guided tissue regeneration (GTR) in the oral surgery field and for bone it is called guided bone regeneration (GBR). Bone regeneration that is done by combining an absorptive hydrogel and a cell growth factor is widely used [13, 14].

5 CONCLUSION

Synthetic hydrogels are used in the medical field where high strength is not required. Although hydrogels are also important in the area of drug delivery systems (DDS), it was omitted here because it is discussed in this Chapter in Section 9. When the absorptive materials must be especially strong, a glycol-lactic acid type polymer is mainly used. In place of biopolymers, synthetic absorptive polymers will be used in the future. Tissue engineering and DDS, where high gel strength is not required, offer great growth potential.

REFERENCES

1　Otani, Y., Tabata, Y., and Ikada, Y. (1996). *J. Adhesion* **59**: 197.
2　Otani, Y., Tabata, Y., and Ikada, Y. *Biomater.* (in press).
3　Otani, Y., Tabata, Y., and Ikada, Y. (1996). *J. Biomed. Mater. Res.* **31**: 157.
4　Iwata, H., Matsuda, S., Mitsuhashi, K., and Ikada, Y. (in preparation).
5　Ishida, H., Tomita, N., Tamai, S., Tomihata, K., and Ikada, Y. (1993). *Chubu Seisai-shi* **36**: 373.
6　Ishida, H., Tamai, S., Tomita, H., Tomihata, K., and Ikada, Y. (1993). *Nitte-kaishi* **10**: 215.
7　Burns, J.W. (1996). *Proc. 211th ACS National Meeting*, New Orleans, LA, March, Abstract, BTEC 004.
8　Ikada, Y. (1995). *Saibo Kogaku* **14**: 1455.
9　Suzuki, S., Matsuda, K., Maruguchi, T., Nishimura, Y. and Ikada, Y. (1995). *J. Plastic Surg.* **48**: 222.
10　Natsume., T., Ike, O., Okada, T., Takimoto, N., Shimizu, Y., and Ikada, Y. (1993). *J. Biomed. Mater. Res.* **27**: 869.
11　Li, Nakamura, T., Shimizu, Y., Tomihata, K., and Ikada, Y. (1993). *Jinko Zoki* **22**: 364.
12　Minabe, M., Kodama, T., Fushimi, H. Hori, T., Tatsumi, J., Kurihara, N., Ikeda, K., Hyon, S.H., and Ikada, Y. (1991). *Bull. Kanagawa Dent. College* **19**: 3.
13　Yamazaki, Y., Oida, S., Ishihara, K., and Nakabayashi, N. (1996). *J. Biomed. Mater. Res.* **30**: 1.
14　Yamada, K., Tabata, Y., Tamamoto, K., Miyamoto, S., Nagata, I., Kikuchi, H., and Ikada, Y. (1997). *J. Neurosurg.* **86**: 125.

Section 5
Bioadhesion Gels and Their Applications

RYOJI MACHIDA AND TSUNEJI NAGAI

1 WHAT IS BIOADHESION GEL?

The term bioadhesion indicates adhesion or absorption by the body. Therefore, bioadhesion gel (more correctly bioadhesive gel) should be called a tissue adhesive gel. However, the concept differs very little from the reality. When the word mucoadhesion is used, one must deduce that the general target is the mucosa, a tissue to which ordinary material does not easily adhere. The concept of gel also needs examination—materials in use today are not gels in the strictest sense. Those materials currently available are polymer matrices that absorb water and adhere to mucosa, but they do not facilitate maintaining of an original shape. They gradually decompose or dissolve. These materials include hydroxypropyl cellulose (HPC), cellulose derivatives such as sodium carboxymethyl cellulose, carboxyvinyl polymer type which is a poly(acrylic acid) (for example, Carbopole®), and plant-originated polysaccharides such as tamarind rubber.

In this subsection, discussion of actual application examples and research examples will focus on polymers that have mucoadhesive properties.

2   APPLICATION AREAS OF BIOADHESION GELS

The main application area for bioadhesion gels is mucosal tissue. Unlike the skin, which covers all external surfaces of the body, the morphology of mucosa membranes varies greatly depending on its location in the body. The alimentary canal, for example, has a mucous membrane that covers and protects a very large area that functions to digest and absorb nutrients. Because mucous glands function as secretors of mucous and thereby provide moisture to membranes, bioadhesion gels are perfectly suited to help with healing in this area because they can deal with the high fluid content in mucosal tissue.

The advantages of drug delivery to mucosal tissue rather than through the usual avenues of entry (i.e., by mouth and through feeding tubes) include:

i)   the first time passage effect can be avoided;
ii)   because overall permeability is better than through the skin, systemic drugs delivery is easily accomplished;
iii) selective administration can be performed relatively easily and precisely; and
iv)   drug delivery can be interrupted by removing the drug.

Characteristics of i) and ii) are why drug delivery via mucosal tissue rather than syringe injection of, for example, insulin or calcitonin is preferred. Characteristic iii), specific targeting, optimizes localized effects and helps prevent systemic side effects. Although it is difficult to interrupt drug delivery to the intestine (category iv), it is useful for other mucosal tissue if a side effect appears or a drug has been administered by mistake. To take advantage of these characteristics, the drugs must possess mucoadhesive properties. Although these characteristics are not applicable for oral drug administration, it is possible to design drugs that have a controlled rate of movement through the alimentary system. This will be discussed later.

3   ADHESION MECHANISMS OF BIOADHESION GELS

Bioadhesion gels generally employ the same mechanism as involves ointments and the tapes to hold them in place. However, the adhesion mechanism of ointments is different from the adsorption mechanism of bioadhesion gels. In the cases of the skin, the interface between the adhesive and the skin is apparent. The adhesive gradually deforms to the skin and then adheres. However in the case of bioadhesion gels, intermolecular interaction between the gel and the mucosal tissue plays an important role in the adhesion mechanism because the mucosal surfaces are bathed in mucous. Hence, as illustrated in Fig. 1, the molecular chains that are in contact with the surface of their mucosal tissue interpenetrate the mucous molecules and maintain adhesion mainly by simple hydrogen bonding [1].

Mortazavi reported that if urea (which disrupts hydrogen bonding) is added, the *in vitro* mucoadhesive force of the disks made of Carbopol 934P and poly(ethylene oxide) reduces. He then hypothesized that the mucoadhesives can remain inviolate for a prolonged period of time because the adhesive polymer and glycoprotein in the mucous physically entangle and then the polymer forms hydrogen bonds with the saccharide residues of oligosaccharides, and the network becomes stronger [2].

On the other hand, Chickering and Mathiowitz prepared hard microspheres made of both fumaric and sebacic acid and found that these microspheres strongly adhered to the mucosa of a rat intestine. Because such material cannot interpenetrate, it is hypothesized that hydrogen bonding is the cause of mucoadhesion [3]. Mortazavi and Smart measured the adhesive force of disks made of various mucoadhesive materials using a DiaStron rheometer. They concluded that the interaction between the adhesive and mucous is not an important factor because the disk adhered more strongly to poly(vinyl chloride) (used as a reference) than to the rat intestinal mucosa [4].

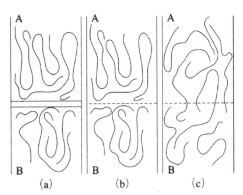

Fig. 1 Interpenetration between the gel polymer and mucous molecules during mucoadhesion.

As can be seen in the preceding discussion, the mucoadhesion phenomenon is complex and sometimes a single mechanism cannot explain the behavior of different materials. Even bioadhesion material are not necessarily gels when they are first used. Instead, they are dried and compressed materials (tablet) or a film. Upon administration of the material, there will be a rapid transfer of water from the mucosa. During this process, the polymer physically adheres to the mucosa. After this, intermolecular interaction probably maintains the mucoadhesive state.

4 APPLICATION EXAMPLES BASED ON LOCATION

4.1 Eyes

Because eyedrops and ophthalmic ointments must not contain any bacteria, it is obvious that the eyes are the most delicate organs in the body. Therefore, drugs that are administered to the eyes cannot contain any bacteria, and as well, particle size, osmotic pressure, and stimulus capability are all controlled. Thus, applying a drug that absorbs water and then swells would be difficult. Further, any materials that disturb vision cannot be used.

Mucoadhesive microspheres or nanospheres can be used to overcome these restrictions.

Durrani *et al.* [5] evaluated the residence time of microspheres, which are ^{111}In-labeled using poly(acrylic acid) (Carbopol 907) by the W/O method, in a rabbit eye. The clearance of the microspheres is bimodal, with fast disappearance of the initial stage followed by a slow disappearance. As shown in Fig. 2, when the microspheres are hydrated by pH 5.0 and administered, the clearance is slower than those hydrated at pH 7.4. This is in agreement with the *in vitro* results on adhesion characteristics. Because 25% of the administered material remained after the rapid clearance period, this can be a vehicle for the drugs that are not easily absorbed by the eye.

4.2 Nasal Cavity

Because the mucosal membranes in the nasal cavity have well-developed blood vessels, it is the preferred location for drug administration in order to avoid the initial passage effect and expect and the entire body effect, such as desmoprecin acetate for the cure of central diabetes insipidus,

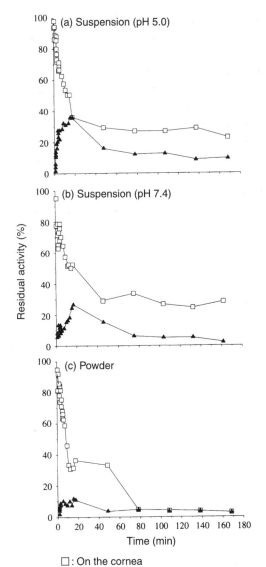

Fig. 2 Residence time of mucoadhesive microspheres in rabbit eye (*n* = 4) [5].

bucerelin acetate and nafarerin acetate for uterus intima disease. The nasal structure is complex. Surface area is large but the administered drug is sent quickly to the throat by ciliary movement. Thus, if a drug has poor absorption characteristics, a mucoadhesive that allows the drug to remain in the cavity will be advantageous. A drug that uses mucoadhesion has already been commercialized for use in treating allergic rhinitis. This mode of drug administration minimizes side effects on the body and exhibits a prolonged treatment effect. If a mixture of 50 mg of becrometazon propionic acid and 30 mg of hydroxypropyl cellulose is sprayed by a specialized small sprayer into the nasal cavity, administration of only 50 µg 2×day (100 µg total) was needed (see Fig. 3). [6] This is compared with traditional sprays that required 100 µg 4×day (400 µg total).

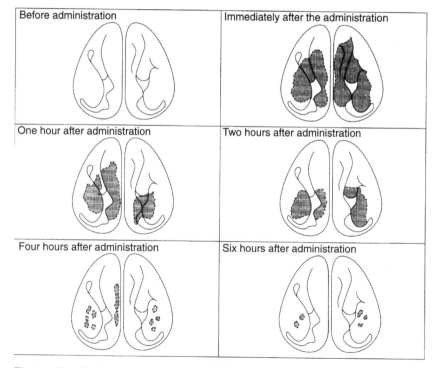

Fig. 3 Distribution and residence of HPC (30 mg) after spraying into the nasal cavity (a healthy male volunteer) [6].

Time (min) \ Drug	TL 102 M	Solution	Comparison of silver particle distribution
10	TL 102 M •••• ••••••••• CR ••••••• ••••••••• EP • • • • • • • CT ML	Solution ••••••• •••••••• CR • • • • • • EP • • • • CT • • • • ML • • •	The existence of both drugs can be detected on the oral mucosa surface. In TL 102 M, the distribution localizes near CR and EP. In solution, it distributes to CR, EP, CT and ML.
30	TL 102 M •••• ••••••••• ••••••••• CR ••••••••• ••••••• EP ••••••• ••••• CT • • • • • • • ML •	CR ••••••• ••••••••• EP • • • • • • • CT • • • • • ML • • •	In TL 102 M, the drug exists on the mucosa surface and the distribution spread to CT and ML. The distribution pattern of the solution does not change from the 10-min sample. However, the drug on the mucosa has already disappeared.
60	CR ••••••• ••••••••• ••••••••••• EP • • • • • • • • CT • • • • • ML • • • •	CR EP • • • • • • CT • • • • ML • • •	In TL 102 M, the distribution to CR and ML has increased from the 30-min sample. The distribution to EP and CT is somewhat reduced. In solution, there is no distribution to CR and distribution to each system is lower.
120	CR • • • EP • • • • CT • • • ML • •	CR EP • • CT • • ML •	In TL 102 M, slight distribution is observed in CR, whereas in the solution, there is no distribution to CR in the same manner as the 60-min sample. Both drugs reduced the distribution to each system. However, TL 102 M shows greater distribution than the solution.

CR: Keratinized epithelium; EP: multilayer flat epithelial; CT: Connective tissue; ML: Muscle layer, ●: Blackened silver particle; TL 102 M: Mucoadhesive drug; Solution: Solution of cottonseed oil

4.3 Oral Cavity

The oral cavity is related to important functions that include speech and taste. Hence, a drug capable of being absorbed continuously from the oral mucosa should not interfere with these functions and must resist picking up excess water. A double-layer tablet 7 mm wide and 1.1 mm thick is now commercially available for oral infections. The drug layer is 0.4 mm thick and HPC and Carbopol are added. These polymers exhibit mucoadhesive properties even if they are used individually. However, by mixing them, they form a durable gel through intermolecular interaction of the polymer molecules. This improved gel structure has excellent healing characteristics due to its ability to protect the infected area as well as deliver in a continuous dose the drug, triamcinonacetonido (0.025 mg/Tablet).

Because this tablet is not appropriate for a wide-ranging infection, a sprayer that also mixes 50 µg becrometazon propionic acid and 200 mg HPC can be used [7]. As shown in Fig. 4, the residence time of the main drug component is long in comparison with the solution because HPC adheres to the mucosa and forms a matrix.

An adhesive tablet (diameter 7.0 mm and thickness 2.1 mm) in which 2.5 mg/tablet of nitroglycerin is contained is necessarily intended to help control or ameliorate coronary angina. By adhering to periodontal tissue, it slowly releases nitroglycerin and the effect continues for 12 h. Similarly, there is a double-layered tablet that prevents heart attack. The fast releasing layer consists of D-mannitol and a low molecular weight poly(vinyl pyrrolidine) (PVP). The mucoadhesive sustained release layer is made of carboxyvinyl polymer and a high molecular weight PVP. By adsorbing on periodontal mucosa, satisfactory blood level contractions are sustained for as long as 10–12 h.

For use in the oral cavity, 3M commercializes a mucoadhesive tablet made of Carbopol 934P, polyisobutylene, and polyisoprene. Recently, Needleman and Smales studied the mucoadhesion properties of various materials as possible treatments for periodontal diseases. While other materials adhered for 1–5 h, chitosan and Eudispert adhered for as long as 4 days. Furthermore, chitosan exhibited good tissue moisturizing ability.

Fig. 4 Schematic distribution diagram of silver particles by microautoradiograph on the oral cavity mucosa upon administration of ^{3}H labeled becrometazon propionic acid (rat, $n = 8$) [7].

They concluded that hydration of the gel, control of swelling, and the moisturizing effect are major reasons to use long-term mucoadhesion [8].

4.4 Uterus and Vagina

The forementioned mucoadhesive drug used to treat oral cavity inflammations contained a mixture of HPC and Carbopol. However, this mixture was originally developed to help treat uterine cancer. Breomicin, carbocon or fluorouracil is mixed into a disk or a stick-shaped tablet and used to treat the cancer. Upon insertion in uterus, it adhered well and cancerous cells were eliminated [9–11].

Woolfson *et al.* recently reported on a mucoadhesive patch to which fluorouracil is mixed. The patch (diameter 26 mm) consists of a film made of a mixed solvent of ethanol : water = 3 : 7, 2% (w/w) Carbopol 981 and a glycerin plasticizer at 1% (w/w). A backing material made of a heat-treated PVC is applied to this film (see Fig. 5). They measured the adhesive strength for a human uterus, its drug delivery characteristics, and its tissue permeability. The adhesive strength increased as the film thickness increased to 0.09 mm as shown in Fig. 6 [12].

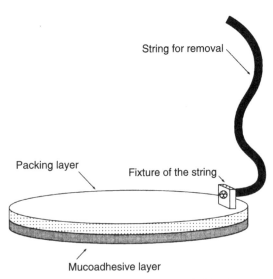

String for removal

Packing layer

Fixture of the string

Mucoadhesive layer

Fig. 5 The structure of an adhesive patch for the uterus [12].

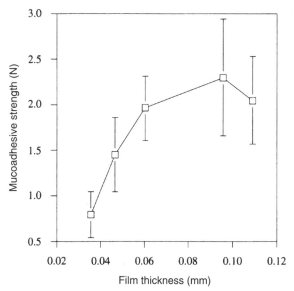

Fig. 6 The relationship between the thickness of the glycerine plasticized Carbopol 981 film and the mucoadhesive strength to the uterus tissue (average value ±SD, *n* = 5) [12].

4.5 Alimentary System

Bioadhesion gels are able to control the rate of movement through the alimentary system, which improves localizes effectiveness, and restricts the area the drugs affects and is meant to affect. For example, to the mixture of Carbopol 934 and HPC ferromagnetic powder was added and made it into granules in order to keep the drug at the cancer for a prolonged period of time. When the drug was administered, a strong magnet was used and this kept the granules near the cancer. The granules adhered to the cancer by mucoadhesive force and the drug was released continuously [13].

Akiyama *et al.* prepared poly(glycerol ester) (PGEF) microspheres (MS) on which Carbopol 934P (CP) was coated (CPC-MS) or MS in which CP is dispersed (CPD-MS). Mucoadhesion strength was compared with an MS without CP. In the *in vitro* experiment using the stomach and intestines of a rat, CPD-MS showed strong mucoadhesion capability because a new CPO particles are gradually and continually exposed from inside (see Fig. 7). When CPD-MS is administered to a starving rat, it took longer for alimentary passage in comparison to MS [14]. See Table 1 for MRT data.

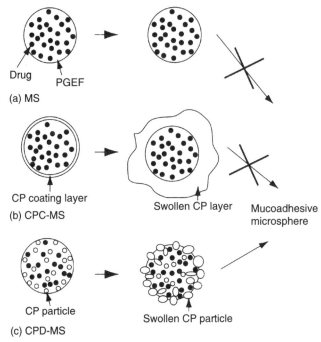

Fig. 7 Schematic diagram on the structure of PGEF microsphere and its mucoadhesive property [14].

Table 1 MRT of MS and CPD-MS in the alimentary canal [14].

	MRT $(h)^{0-\infty}$	
	PGEF-microsphere	**CPD-microsphere**
Stomach	0.57 ± 0.03	1.52 ± 0.05
Region I	0.80 ± 0.13	1.72 ± 0.06
Region II	1.16 ± 0.16	3.75 ± 0.04
Region III	2.61 ± 0.08	5.58 ± 0.07

Average $\pm$ SEM
Region I: Stomach — upper intestine
Region II: Stomach — middle intestine
Region III: Stomach — lower intestine

4.6 Intestine

Drugs administered to the intestine as the lower end of the alimentary system are expected to reach the entire body. Because drugs absorbed from the distal portion of the intestine can avoid the initial passage effect, it is necessary to attempt to control migration of the drug to the upper part of the intestine. For example, a double-layered suppository that has an ammonium salt of carboxyvinyl polymer at the front tip could be used.

The front tip swells in the intestine and increases in volume. At the same time, the mucoadhesive nature of the insert prevents the sustained release drug from migrating upward into the upper reaches of the intestine. Thus, in theory, the drug remains in its original location [15]. There is commercially available a suppository to cure hemorrhoids by increasing the healing effect near the inflamed area. This suppository is a mixture of an ordinary component with a mucoadhesive. According to the manufacturer's data, even 60 min after insertion it remained within 10 cm of the anus.

5 CONCLUSION

It is expected that the number of studies on mucoadhesive drugs will increase in the future. Currently, there are many studies on Carbopol. This material is used in a drug for oral mucosa inflammation. Other drugs that contain HPC as a mucoadhesive are also under investigation. Other materials such as rubbers from plants also are seen as having good potential for stability. It is hoped that the administration of physiologically active peptides through mucosal membranes will be more actively used as a result of the newly developed bioadhesion gels. It will undoubtedly improve the quality of life of patients.

REFERENCES

1 Peppas, N.A. and Buri, P.A. (1985). *J. Control. Release* **2**: 257.
2 Mortazvi, S.A. (1995). *Int. J. Pharm.* **124**: 173.
3 Chickering, D.E. and Mathiowitz, E. (1995) *J. Control. Release* **34**: 251.
4 Mortazavi, S.A., and Smart, J.D. (1995). *Int. J. Pharm.* **116**: 223.
5 Murrani, A.M., Farr, S.J., and Kellaway, I.W. (1995). *J. Pharm. Pharmacol.* **47**: 581.
6 Kuroishi, T., Asaga, H., and Okamoto, I. (1984). *Yakuri to Chiryo* **12**: 4055.
7 Yamamoto, M., Okabe, H., Kubo, Y., Naritomo, T., Ikura, H., and Suzuki, Y. (1984). *Kiso to Rinsho* **18**: 791.
8 Needleman, I.G. and Smales, F.C. (1995). *Biomater.* **16**: 617.

9   Machida, Y., Masuda, H., Fujiyama, N., Iwata, M., and Nagai, T. (1979). *Chem. Pharm. Bull.* **27**: 93.

10 Machida, Y., Masuda, H., Fujiyama, N., Iwata, M., and Nagai, T. (1980). *Chem. Pharm. Bull.* **28**: 1125.

11 Masuda, H., Sumiyoshi, Y., Shiojima, Y., Suda, T., Kikyo, T., Iwata, M., Fujiyama, N., Machida, Y., and Nagai, T. (1981). *Cancer* **48**: 1899.

12 Woolfson, A.D., McCafferty, D.F., McCarron, P.A., and Price, J.H. (1995). *J. Control. Release* **35**: 49.

13 Ito, R., Machida, Y., Sannan, T., and Nagai, T. (1990) *Int. J. Pharm.* **61**: 109.

14 Akiyama, Y., Nagahara, N., Kashihara, T., Hirai, S., and Toguchi, H. (1995). *Pharm. Res.* **12**: 397.

15 Tokumura, T., Machida, R. and Nagai. T. (1985) *Yakuzai-gaku* **47**: 42.

Section 6
Transdermal Patches

KENJI SUGIBAYASHI AND YASUNORI MORIMOTO

1 INTRODUCTION

Drugs placed on the skin can be classified into two types. There are those that focus on the skin (epidermal) itself and others that use the skin as a conveying agent. These vary in function, with the former systems (transdermal therapeutic systems-TTS) used to treat the epidermal system specifically. The latter, absorbed through the skin and distributed throughout the body, are called transdermal drug delivery systems (TDDS). The drugs used include various types of liquids, ointments, pap, plaster, and tapes. These can also be divided into coating and patch types. The patch type can maintain a constant, measured amount of drug delivered and application is simple. Thus, it is suitable as a drug delivery system for the entire body.

2 MANUFACTURING OF PATCHES

Patches are made into various structures and forms [1]. They are largely divided into reservoir, matrix, pressure sensitive (PSA) and tape types (see Fig. 1). In the reservoir type, the reservoir in the form of a liquid or a gel is covered by a supporting layer and a drug-containing membrane (many times, it is a sustained-release drug membrane). On the sustained release drug membrane, an adhesive layer is laminated. In the matrix type, drugs are included in a semisolid or solid matrix. This type needs no sustained-release drug layer because it shows less drug leakage compared to the reservoir type. Thus, the adhesive layer exists in direct contact with the matrix. In the pressure sensitive tape type, the adhesive layer itself contains the drugs. Although a pap drug cannot strictly be called the pressure sensitive adhesive type, from the viewpoint of using a mucoadhesive poly(acrylic acid) hydrogel as the drug layer, it might be classified as a pressure sensitive adhesive type. The decision on structure and shape of drugs depends on the properties of the drug involved. However, in many cases, determinations are made based on the technology at the time and the patents of the manufacturing companies involved.

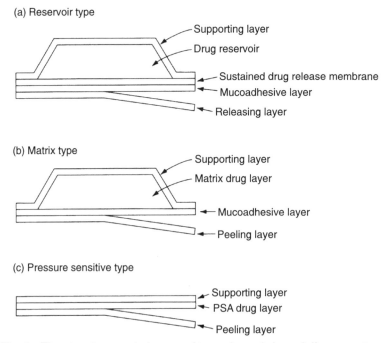

Fig. 1 The structure and shapes of transdermal drug delivery systems.

3 CHARACTERISTICS OF PATCHES

Most currently used medical drugs are either injectable or taken by mouth. Drugs applied on the skin can be used when injectable or oral drugs cannot be used or when skin-based administration is more effective. Therefore, when the advantages and disadvantages of patches are discussed, it is important to include the drugs used. One advantage of injectable drugs is their swift effect, even if the drug's effect is short lived. Even last for about several hours. If several days of administration is needed, cumbersome repetition of these methods is required. Physicians and other highly trained health personnel, are also required except in the case of insulin injections. Hence, one must visit a healthcare facility and quality of life will be affected.

In comparison, oral drugs can be self-administered and quality of life is then not affected. Some of these drugs can also be manufactured for sustained release in order to make the drug more efficacious. Unfortunately, oral drugs are inappropriate for those patients who either cannot eat or have had side effects that target their alimentary system. Most of the shortcomings of injections and oral drugs can be minimized or overcome by using transdermal patches.

On the other hand, the major shortcoming of transdermal patches is poor skin absorption of drugs, with the problem especially acute with water-soluble drugs. There is a similar concern regarding oral drugs. However, in general, the bioavailability of oral drugs is higher than with transdermal patches. It is also possible that transdermal absorptivity may change as a result of injury or swelling of the skin. Furthermore, increased irritation and sensitivity are problematic side effects. It is thus important to understand the characteristics of transdermal patches with respect to the nature of the specific skin on which the drug is to be applied.

4 CURRENT TREND OF TRANSDERMAL PATCHES

The history of transdermal patches for whole body drug delivery is brief. It has been only 15 years since scopolamine and nitroglycerin were first commercialized in the U.S. Since then, many drugs have been commercialized, including patches of isosorbido, cronidine, estradiol, nicotine, fentanyl, and testosterone. The development of turbuterol, and epersione has almost been completed. In particular, the reputation of the nitrogly-

Table 1 Developed and commercialized transdermal delivery systems and their characteristics.

General names Corresponding products	Developed or sales companies	Shape of products	Characteristics of products
Scopolamine (for motion sickness)	There is no need of absorption aid due to high transdermal property		
Transderm-Scop	Alza/Novartis	Reservoir type	A mineral oil is used as the reservoir. It uses a controlled release drug membrane. It is used once a day. It is patched behind the ear where absorption is rapid. There is a side effect if the patient forgets to remove the patch.
Kimete Patch	Myun Moon Pharm. (Korea)	Reservoir type	A mineral oil is used as a reservoir. It also contains squalene.
Nitroglycerin (to treat angina)	There is almost no need of absorption aid due to a high transdermal property. In order to avoid drug tolerance problem, new delivery methods are being investigated. Replacement drugs are also effective. Many similar products have been commercialized (there is a need in establishing standard on biological equivalence).		
Transderm-Nitro	Alza/Novartis	Reservoir type	A silicone oil is used as the reservoir. It uses a controlled release drug membrane. Nitroderm (France) and Nitroderm TTS (Novartisforma) are imported and are the same.
Nitro-Dur II	Key/Schering	Matrix type	Improved product of Nitro-Dur. Diafusor (Pierre Farbe) is imported and the same.
Nitrodisc	G. D. Searle	Matrix type	Controlled release property is achieved by microsealed system. It contains mineral oil, isopropyl palmitic acid, and poly(ethylene glycol).

Deponit	Lohman/Schwarz/Wyeth	PSA tape	Controlled release property is achieved by utilizing multilayers with different diffusivities.
Minitran	3M	PSA tape	It is the smallest tape drug among nitroglycerin TTS. Introduction to Japanese market is being evaluated by Zeria
NTS Patch	Hercom/Bolar	Matrix type	It contains isopropyl palmitic acid and diethylhexyl phthalic acid. The release gradually decreases. Transdermal NTS (Warner Chilcott Lab.) and Nitrogen (Kremer Urban) are the same products
Nitrol	Paco Pharm./Adria Lab.		It contains lauryl alcohol and isopropyl myristic acid as an absorption aid
Helzer	Nichiban/Taiho	PSA tape	The same as the one stated below.
Millisrole	Nippon Chemicals. Co.	PSA tape	The same as the one stated below.
Basorator	Sanwa Chemicals Co.	Ointment	It adheres to the skin after coating the backing
	Sanwa Chemicals Co./Rhone Poulanc Rola	PSA tape	
Minitro	Nisshin Seifun/Boehringer	PSA tape	
Isoolbido nitrate	Similar drug tolerance property as nitroglycerin		
Frandle tape S	Toa Eiyo/Yamanouchi	PSA tape	It is the improved product of frandol tape. In order to maintain the activity, it is in crystalline form.
Apatia tape	Teikoku/Teikoku	PSA tape	
Antap tape	Teisan/Teijin/Fujisawa	PSA tape	The irritation is reduced by using hollow fibers.
Isopit tape	Toko/Mitsui	PSA tape	It is the smallest among isosorbitol nitrate TTS.

(Continued)

Table 1 Continued

General names / Corresponding products	Developed or sales companies	Shape of products	Characteristics of products
Sawadol tape	Sawai	PSA tape	
Nitoras tape	Owaki/Takada	PSA tape	
Penety	Sakisui Chemicals Co./Nippon Shoji	PSA tape	
Liphatac	Medusa/Meiji Seika	PSA tape	
Skin sensitivity appears on serval tens % of the patients (effectiveness was emphasized for its development)			
Clonidine (for hypertonia) / Catapres TTS	Alza/Boehringer Ingelheim	Reservoir type	Reservoir contains a mineral oil. It also uses a controlled release drug membrane.
Estradiol (to treat low body level estrogen)			
Estraderm TTS	Novartis	Reservoir type	Reservoir contains ethanol. It also uses a controlled release drug membrane. The use frequency is twice per week. Etoraderm TTS is imported to Japan.
Fematrix	Ethical Pharm./Solvay	Matrix type	It mixes estradiol and norgestorel. It is a hydrophilic matrix.
Femtran	3M (New Zealand)		Climara system adopted.
Nicotine (for smoking habit)			
Nicoderm	Nicotine concentration is simulated Alza/Hoeschst Marion Roussell		
Prostep	Elan/AHP (US), Boehringer Ingelheim (Canada)	Nicodan (Holland), Nicodisc (Spain), Nicolan (Sweden), and Niconil (England, Ireland) are all the same product.	

Drug (product)	Manufacturer	Type	Remarks
Fentanyl (pain killer for late stage cancer patients) Duragesic	Alza/Janssen	Reservoir type	Reservoir contains ethanol. The drug and ethanol are dispersed into hydroxyethyl cellulose gel. The controlled release drug membrane allows 2 × per week use. After the use, the disposal is problematic.
Testosterone (male hormone) Androderm Testoderm	TTI/SmithKline Beecham ALZA	Reservoir type	
Terbuterol (a drug for asthma)	Nitto Denko/Hokuriku	PSA tape	Application area is small.
Ebelison (a drug for muscle ache)	Sekisui Chemicals Co./Mitsuo/Eizai	PSA tape	An innovation is seen on the stability of the drug.

cerin patch is excellent and many similar products have been commercialized. Table 1 lists the names of patches, the names of the companies under which the drugs were developed or sold, and the shape and characteristics of the drugs [2].

Patches for local use include steroidal antiinflammatory ones with drugs such as betamethasone valeric acid or prednisone, and nonsteroidal type antiinflammatory ones with indomethacin and ketoprofene (not only pap drug but also tape type were commercialized), and lidocaine, which is used to reduce pain before an injection. Antiitch tape is also available as an over-the-counter medicine.

A wide variety of drugs are available as transdermal patches, with many of them capable of relatively high transdermal permeability. In particular, patches developed the earliest, such as scopolamine and nitroglycerin, have high transdermal permeability. However, relatively recent commercial products, estradiol and fentanyl, do not show high transdermal absorptivity. Therefore, ethanol is used in the patch as an absorption aid. Even if ethanol is used, Estraderm does not show sufficient absorption in several tens percent of the patients. Therefore, patches that include estrogen and luteotropic hormone are currently under investigation. Such mixed drugs will also be important in the future.

To use more drugs transdermally, it is important to choose drugs that have high transdermal permeability without the use of absorption aids. It is also desirable to develop powerful transdermal absorption acceleration systems. There are many problems to be solved in the future, including screening of absorption aids, development of gels that optimized the function of absorption aids, and evaluation of the matrix to maintain stability and high drug efficiency.

The drugs themselves also have problems. Transdermal patches of nitroglycerin and isoolbido nitrate have the problem of drug tolerance. Namely, over time, drug resistance develops for nitrate and nitrite drugs and the desired effect is reduced. In a controlled clinical experiment of heart attack, the drug resistance is reduced if a rest period is used, hence, 0-th order absorption and constant concentration in the blood, characteristics of the currently sold nitrate drugs, is disadvantageous. For these nitrate drugs, there are other replacement drugs. These drugs also influence effectiveness and resistance. Thus, the efficiency of nitrate transdermal patches must be evaluated in its entirety. In the case of cronidine transdermal patches, skin sensitivity became a problem. It is necessary to interrupt the use, often in the middle of treatment, because several tens percent of patients will become locally sensitive.

Transdermal patches of testosterone also have problems A fraction of the testosterone is metabolized in the skin and converts to dihydrotestosterone. This metabolic side product can cause swelling of the prostate gland. Testosterone patches are placed on the body where the rate of absorption is highest, but this unfortunately can cause the metabolic product, dihydrotestosterone, to create side effects. Improved patches can be placed in areas where metabolized dihydrotestosterone is minimized. Reduction of transdermal absorption by the changed application area can be overcome by the addition of an absorptive aid.

Today's drugs have much to teach us about future transdermal delivery drugs.

5 NEW APPROACHES FOR TRANSDERMAL PATCHES

As already described here, absorption acceleration factors along with selection of drugs that have high transdermal absorption must be involved in any evaluation of the validity of transdermal delivery patches [3]. Useful methods for absorption acceleration include: chemical modifications (prodrug); use of chemical absorption accelerators; physical methods iontophoresis, phonophoresis, electroporation). Absorption aids such as ethanol and isopropyl myristine have already been commercially used. However, more effective, safer and wider applicability to drugs is currently under investigation. Iontophoresis, which accelerates drug absorption by using electrical stimulus, is especially hopeful for facilitating transdermal delivery of peptides. In this method, not only the development of drugs with a high transdermal delivery rate but also the possibility of injection replacements may be possible. It also has the benefits of on-off drug delivery. It still has problems with legal approval because it is a combination of a device and drugs. It is hoped that this issue can be resolved. See Figure 2 for more information.

Skin irritation is still an unsolved and important problem. Irritation is caused by both drugs and absorption accelerators. To reduce irritation, glycerin and oil-soluble vitamins have been reported to be effective. Irritation differs depending on the properties of the different mucoadhesives and the substrate. Research on base materials and additives with low skin irritation will be more active in the future. It is important when one

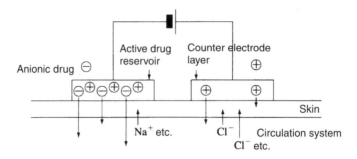

Fig. 2 Principle of iotophoresis.

investigates drugs, additives, and design delivery systems to take into account problems and to attempt to resolve them so better transdermal patches, will be available. Further development of transdermal delivery patches will further contribute to medical care.

REFERENCES

1   Sugibayashi, K. and Morimoto, Y. (1994). *J. Controlled Release* **29**: 177.
2   Morimoto, Y. and Sugibayashi, K. (1992). *Kizzu* **2**: 12.
3   Sugibayashi, K. and Morimoto, Y. (1996). *Recent Res. Devel. Chem. Pharm. Sci.* **1**: 63.

Section 7
Ointments for Antiinflammatory Drugs

TOSHIO INAKI

1 INTRODUCTION

The skin, the body's protection system, consists of a multilayered epidermis, the corium (commonly called the derma), and subcutaneous. If the body recognizes a drug to be a foreign material, it will not be absorbed. However, materials that have high solubility, such as iodide, ether and acetone, will be absorbed.

The absorption part from the skin depends on the nature of the compound to be absorbed. Nonetheless, many are absorbed through the outer epidermal keratinous layer. This is because the keratinous layer has 100–1000 times the surface area of specialized epithelial tissue at the body's surface, or some glands (e.g., sweat (sudoriferous glands). Mercury will be selectively absorbed through hair follicles. Drug properties, such as molecular weight, distribution coefficient, and dissociation constant affect their distribution throughout integumentary tissues as well as in bodily fluids following absorption by the body.

Transdermal absorption depends on tissue conditions and the physicochemical properties of the drugs. For example, the keratinous layer normally contains 5–15% water and maintains appropriate skin flexibility.

211

If the surface is confined (ODT) and the water content of the keratinous layer is increased, drug absorptivity increases. This tendency to increase is greater for water-soluble ester than for the hydrophobic ester (e.g., salicylate) [1]. If the keratinous layer is debonded, absorptivity increases depending on the degree of debonding. Thus, the keratinous layer is considered an absorption barrier [2].

Because absorptivity of drugs depends on physicochemical properties, such as diffusivity and distribution coefficient of the matrix, selection of the matrix is especially important for the development of highly efficient drug delivery systems. In general, requirements for these drugs are

i)   that the mixed drugs be stable,
ii)   less irritating to the skin,
iii) create no changes in appearance, and
iv)   no discomfort to the patient.

Gels are used as matrices for drugs. The characteristics of gels include:

i)   the main drugs must completely dissolve and the dissolved product must be transparent,
ii)   the gels must be flexible, not stick and be comfortable,
iii) they must be able to be used on hairy areas, and
iv)   drug absorptivity at the desired locations must be acceptable so that the drug is effective.

Hence, a gel matrix can be summarized as a matrix that is beautiful to look at, fresh, that provides for excellent drug absorption.

A gel matrix for an ointment is classified as a suspended matrix. Depending on the mixed liquid, it ca be divided into hydrogel and lyogel. However, currently available commercial products are mostly hydrogels, which use water-soluble polymers. Drugs for external uses are mostly organic compounds. There are many compounds, including corticosteroid, that are very difficult to dissolve in water. Dissolving these compounds is one of the problems that must be solved before drug efficiency can be improved. Because gel matrices can incorporate dissolution aids such as alcohol, they are attracting attention as a possible drug delivery system for difficult-to-dissolve drugs.

Accordingly, a drug delivery system design utilizing the characteristics of gels is becoming important to development of ointments. In the following, drug delivery system development for indomethacin (hereinafter abbreviated as IND), an anti-inflammatory drug with low solubility to water, will be described. With this example, the importance of gels as the matrix for transdermal drug delivery system will be explored.

2   DEVELOPMENT OF DRUG DELIVERY SYSTEMS

To develop a drug delivery system, it is important to evaluate the physicochemical properties of IND and an ointment matrix that can incorporate IND.

2.1   Solubility

Indomethacin does not dissolve easily in water. For optimal effect, therefore, any system developed to deliver IND must be able to dissolve it. When IND solubility was investigated as a function of the pH of water, it was $0.1 \mu g/ml$ at a pH of 3.0. This increases to $403.2 \mu g/ml$ at a pH of 7.0. Because pKa of IND is 4.3, the majority of it dissociated at $> pH 7.0$. Hence, if it is considered that IND is absorbed via passive transport, developing a water-based drug delivery system for IND will be difficult [3].

The solubility of IND in frequently used ointment matrices and organic solvents is low. However, it dissolves in solvents such as Macrogole 400, benzyl alcohol, and esters of fatty acid (see Tables 1 and 2). In particular, IND dissolves well into the diesters of fatty acids and the mixture has low irritation to the skin. Thus, these esters are suitable for drug delivery system matrices for IND. The solubility of IND to alcohol reduces when the number of carbo atoms increases. However, its solubility to ethanol is $18 mg/ml$, which is sufficient (see Fig. 1). Hence, it is probably best to use alcohol as the base material with an ester added for any IND a drug delivery system.

2.2   Stability

The stability of IND is in a close relationship with the pH of the matrix. Indomethacin is most stable at pH 4.5. However, if other pHs are used for drug efficiency and safety reasons, caution is necessary (see Fig. 2).

Table 1 The solubility of indomethacin in ordinary drug delivery system matrices.

Matrix	Solubility
Poly(ethylene glycol) 400	f.s.
Benzl alcohol	f.s.
Tetrahydrofuran	f.s.
Dimethylsulfoxide (DMSO)	f.s.
N,N-dimethylsulfoxide	sp.s.
Methyl salicylate	sp.s.
Salicyl glycol	s.s.
Propylene glycol	s.s.
Olive Oil	s.s.
Oleyl alcohol	s.s.
Oleic acid	s.s.
Lactic acid	s.s.
Glycerin	p.i.
Squalan	p.i.

f.s.@ freely soluble
sp.s.: sparingly soluble
s.s.: slightly soluble
p.i.: practically insoluble

Table 2 The solubility of indomethacin in esters.

Component name	Solubility (mg/ml)
Ethyl butyrate	18.0
Ethyl caproate	11.0
Ethyl caprate	6.1
Ethyl laurate	4.8
Ethyl myristate	3.4
Isopropyl myristate	1.9
Octyldodecyl myristate	1.1
Diethyl succinate	26.1
Diethyl adipate	34.5
Diisopropyl adipate	18.1
Diethyl sebacate	22.9
Triacetin	17.7
Tributylene	10.0

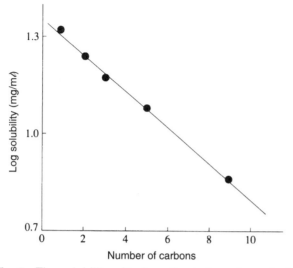

Fig. 1   The solubility of indomethacin in various alcohols.

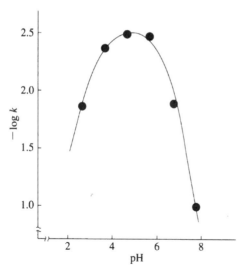

Fig. 2   The relationship between the decomposition rate constant of indomethacin and pH.

2.3 Absorptivity

An alcohol must be mixed in the matrix for solubility. If the solubility of IND is measured in the presence of an alcohol, as shown in Figs. 3 and 4, the concentrating action due to the evaporation of the alcohol increases absorptivity. When an ester is added, the degree of absorptivity increase varies depending on the type of ester used.

2.4 Viscosity Enhancer

As already described, it is necessary to add ethanol to any IND drug delivery system because IND dissolves very little in water. If the viscosity of polymers that swell in an ethanol aqueous solution is evaluated, the majority of the compounds as shown in Table 3 do not increase the viscosity of the solution. However, crosslinked poly(acrylic acid) copolymer, carboxyvinyl polymer (hereinafter abbreviated as CVP), showed high viscosity, suggesting that it might be an effective viscosity enhancer for an ointment. The CVP also exhibits high viscosity, thereby enhancing the effect of the poly(ethylene glycol) 400 to which IND solubility is high and

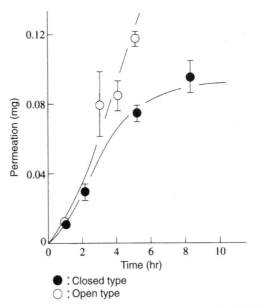

Fig. 3 Transdermal absorptivity of indomethacin in 50% (v/v) ethanol aqueous solution.

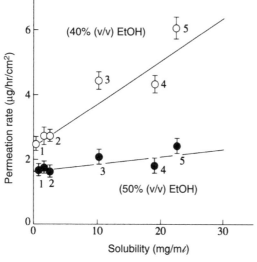

1: Octyldodecyl myristate
2: Isopropyl myristate
3: Tributylene
4: Disopropyl adipate
5: Diethyl celacate

Fig. 4 The relationship between esters with different solubility and transdermal absorptivity of indomethacin.

Table 3 Viscosity enhancement of various gelation agents in ethanol aqueous solution.

Gelation agent	Property
Arabic rubber	l.v.
Tragacanth gum	l.v.
Xanthan gum	n.s.
Calcium carrageenan	n.s.
Polycetin	l.v.
Sodium Alginate	n.s.
Propylene glycol alginate	b.s.
Carboxymethyl cellulose	l.v.
Hydroxypropyl cellulose	s.
Hydroxypropylmethyl cellulose	l.v.
Poly(vinyl pyrrolidone)	l.v.
Poly(vinyl alcohol)	l.v.
Bee gum	n.s.

n.s.: not swelling
l.v.: low viscosity
s.: stringy
b.s.: bad stability

an aqueous solution of propylene glycol, which is often used as a matrix for external-drug use.

Thus, CVP is hopeful as a useful matrix for an ethanol drug delivery system. Although CVP increases viscosity, it also contains carboxylic acid. Hence viscosity changes, depending on the pH of the matrix and the carboxylic acids bond with the cationic component in the matrix because the IND also contains carboxylic acid.

If cations such as organic amines and sodium hydroxide are added as neutralizing agents, the appearance becomes ointment-like. However, when these ointments were examined for their transdermal absorptivities on guinea pig, it was found that absorptivity depends on the type of cation (see Fig. 5). This indicates that the cation influences the effectiveness of the drug in addition to its pH-adjusting role. When the IND concentration in a platelet is measured by changing the amount of coating of an IND gel on the abdominal skin of a mouse, the IND in the blood showed a high value as the gel-coated IND was increased (see Fig. 6), indicating fast IND diffusion via the gel. Gels affect use, comfort and appearance, but they

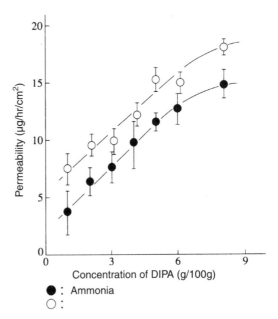

Fig. 5   Effect of counter ions on transdermal absorptivity.

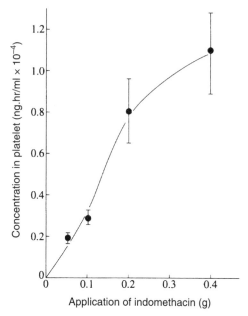

Fig. 6 Relationship between the amount of drug applied and the concentration in platelet.

also affect the effectiveness of drugs used. Accordingly, gels are widely used as the base matrix for external-use drugs.

3 FUTURE TRENDS

In this subsection, the usefulness of gels for transdermal drug delivery system was discussed. However, there are many more application examples that use other prescription drugs. For example, when isoprofen suspended gel matrix was administered to the intestine of a mouse, the blood concentration of the drug was found to be much higher than would have occurred with oral administration [4]. Other examples include: i) usefulness of an insulin gel matrix when administered vaginally; ii) application of high viscosity gel matrix drug delivery system on oral mucosa, and iii) a demonstration by Saettone and Cioannaccini [5] of greater strength and sustained effect of tropicamide if it is used with gel matrix rather than in the traditional eyedrops or eye ointments.

Gels have been used frequently as the base material for drug delivery systems. They have also been used by Tanaka *et al.* [6] to study the phenomena of phase transition of hydrogels and by Okano *et al.* [7] to study thermoresponsitivity. Hence, the application areas for gels are wide open. It is not an overstatement that gels are essential components in any future development of drug delivery systems. It is essential to study existing gels from various points of view. However, it is equally important to actively develop new gels in order to provide many gels for medical usages.

REFERENCES

1   Wurster, D.E. and Kramer, S.F. (1961). *J. Pharm. Sci.* **50**: 288.
2   Flynn, G.L., Durrheim, H., and Higuchi, W.I. (1980). *J. Pharm. Sci.* **70**: 52.
3   Inagi, T., Muramatsu, T., Nagai, H., and Terada, H. (1981). *Chem. Pharm. Bull.* **29**: 1708.
4   Hirano, E., Morimoto, E., Takeeda, T., Nakamoto, Y., and Morisaka, K. (1980). *Chem. Pharm. Bull.* **28**: 3521.
5   Saettone, M.F. and Cionnaccini, B. (1980). *J. Pharm. Pharamacol.* **32**: 519.
6   Tanaka, T. (1986). *J. Phys. Soc., Jpn.* **41**: 542.
7   Okano, T., Kikuchi, A., Sakurai, Y., Takei, Y., and Ogata, N. (1995). *J. Controll. Rel.* **36**: 125.

Section 8
Application of Chitosan Medical Care

SEIICHI AIBA

1 INTRODUCTION

Chitin is a polysaccharide that is an important reusable bioresource similar to cellulose [1]. Chitin widely exists in nature in the outer covering of crab, shrimp, and insects, as well as in the cellular walls of fungi. Its annual natural production is estimated to be 100 billion metric tons. Crab and shrimp shells consist of chitin, calcium carbonate, and protein, with chitin at 10–40%. The chemical structure of chitin, as shown in Fig. 1, is a homopolysaccharide with (1–4) bonded 2-acetoamido-2-deoxy-b-D-glucose (N-acetyl-b-D-glusosamine). In general, separated chitin is not completely acetylated. If chitin is deacetylated by concentrated alkali, it becomes chitosan. Similar to chitin, it is difficult to prepare chitosan with a 100% pure amine group. Thus, if the material dissolves into dilute acetic acid it is called chitosan and the concentration of the residual acetyl group is ≈5–40%. Manufacturing of chitin and chitosan is performed according to Fig. 2 [2].

In Japan, both are manufactured using waste crab and shrimp shells from a maritime product factories. Chitosan is used mostly as a flocking agent for water treatment, in addition to small amounts for wound

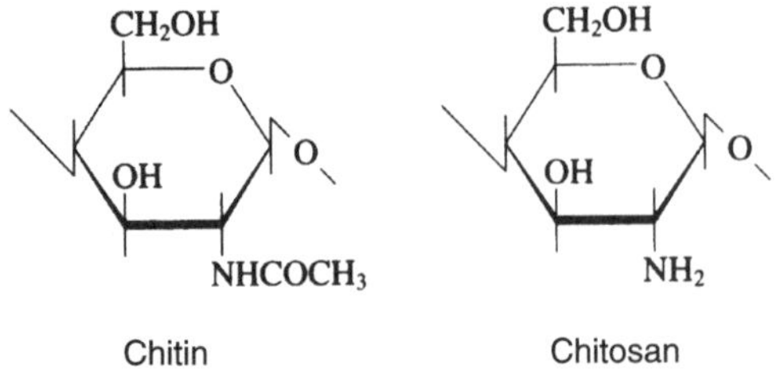

CH₂OH   CH₂OH

Chitin   Chitosan

Fig. 1 Chemical structures of chitin and chitosan.

dressings, endotoxin removal agent, cosmetics, food additives, antifungal deodorant fibers, chitosan beads for chromatography packing, and fertilizer [1]. However, as their various properties have been discovered, development efforts have increased. In this subsection, the general properties of chitin and chitosan, preparation of gels, and their application examples will be described.

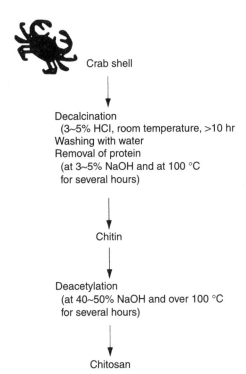

Crab shell

Decalcination
(3~5% HCl, room temperature, >10 hr
Washing with water
Removal of protein
(at 3~5% NaOH and at 100 °C
for several hours)

Chitin

Deacetylation
(at 40~50% NaOH and over 100 °C
for several hours)

Chitosan

Fig. 2 Manufacturing methods for chitin and chitosan.

2 GENERAL PROPERTIES OF CHITIN AND CHITOSAN

Solvents known to dissolve chitin are a mixture of formic acid and dichloroacetic acid, a mixture of trichloroacetic acid and dichloroethane, a mixture of dimethylacetamide and lithium chloride, methane sulfonic acid, calcium chloride dihydrate saturated methanol, and 10% sodium hydroxide solution [4]. Chitosan is much more soluble than chitin. It dissolves in aqueous solutions of many organic and inorganic acids.

Upon dissolution, chitin and chitosan can be molded into various shapes. It forms a film when coated onto a glass plate and then dried [5, 6] and a fiber if continuously spun in a coagulation bath [7, 8].

The obtained fibers can be used to make a nonwoven cloth. If the material is dropped into a coagulation bath, beads can be obtained [9]. By using the polycationic nature of chitosan, crosslinking is possible due to ionic bonding with polyanion or multivalent negative ions to form water insoluble gels. This method can be applied for beads and microcapsule preparation [10]. Various chitin and chitosan derivatives have been studied [11–13] to learn about changing the solubility and degree of crystallinity or addition of functions. Localized irritation has been studied for cosmetic application purposes. The results indicate that there is no toxicity and that both are highly safe materials [14]. From a small mammal (mouse) study of quasiacute toxicity, it was found to be as nontoxic as sugar and salt. For chitin, studies have indicated no toxicity when it is applied as absorptive threads and wound dressings [7, 15].

3 PREPARATION AND APPLICATION OF CHITIN GELS

Over the past years Japan has been among the research leaders worldwide regarding the level of studies on chitin and chitosan. Some products have already been commercialized and actually used [7, 16–18]. This is in part due to the functionality of chitin and chitosan as medical materials. Although chitin and chitosan are not inexpensive ingredients to use, when they are used for high value-added products, raw material expensive medical materials ratio makes the cost negligible. The characteristics of chitin and chitosan as medical materials are moldability [7, 16], bioabsorptivity [7], acceleration of wound healing [7], inhibition of bleeding [7], and immune response recovery [19]. In this subsection, the preparation of gels and their application as medical materials will be described.

Although there are only a limited number of solvents for chitin, if chitin is dissolved in these solvents and coagulated by a coagulating agent, various molded products, such as fibers, films and sponges, can be obtained. For example, if chitin powder (1 g) is dispersed in a mixed solvent of dimethylacetamide (50 ml) and N-methylpyrrolidone (50 ml), to which LiCl (5 g) is gradually added, it forms a viscous solution after stirring overnight [20, 21]. If this solution is cast on a glass plate and immersed into a coagulation bath of 2-propanol, the solution becomes a gel-like film. Upon washing with water and drying, a film is obtained. This chitinous film as a low degree of crystallinity in comparison to the chitin in a waster crab shell, and with high water uptake and the ability to form chitin gel when immersed in water. The characteristics of this film are summarized in Table 1. It has the potential to serve as a matrix for sustained drug release systems because its high absorptivity and high molecular weight materials can permeate relatively well. Gel beads are formed by adding the chitin solution dropwise into a coagulation bath. By controlling the conditions, beads 0.5–2 mm in diameter can be prepared.

A sponge-like chitin film is obtained by adding a blowing agent to a chitin solution and removing the solvent and blowing agent upon casting [16]. It absorbs water quickly and becomes soft because it is porous. This chitin sponge has already been commercialized as an artificial mucosa. It is used for healing wounds of the operation of the mucosal membrane in the nose.

It is also possible to prepare a chitin gel using a recently found calcium chloride dihydride saturated methanol [22]. If chitin powder is suspended in this solvent and stirred, it becomes a viscous solution. Upon casting this solution onto a glass plate and deforming, a gel-like chitin film is obtained by immersing the casting into a mixed solution of 20% sodium

Table 1 The characteristics of chitin film.

Heat treatment conditions (°C, time)	Water uptake (%)	Permeation coefficient (cm^2 s^{-1} at 37°C)	
		Urea ($\times 10^{-6}$)	Vitamin B12 ($\times 10^{-7}$)
Real treatment	220	4.3	8.2
120°C for 2 h	170	3.4	5.8
120°C for 5 h	130	3.3	4.6
145°C for 2 h (cuprofan)	110	2.8	2.8
	100	1.9	1.7

nitrate aqueous solution and ethylene glycol to remove the calcium. If this film is lyophilized, a porous chitin film is obtained. Effectiveness as wound dressings and drug delivery systems has been evaluated for these chitin gels.

A chitin gel can be prepared by methods other than those already described. While deacetylated chitin is chitosan, chitin can be obtained by acetylating chitosan. Chitosan is dissolved in an acetic acid solution and diluted to 50% concentration by methanol. If acetic anhydride is added to this solution at 3–10× the number of moles of the amine groups present, the solution solidifies in several minutes and forms a transparent gel [23–26]. Time necessary to solidification depends on the concentration of chitosan, methanol, and acetic acid anhydride [27, 28]. This gel can also be made into a thin film and the film permeates materials well [25, 26].

As a hybrid material with other polymers, a semiinterpenetrating network polymer (semi-IPN) made of chitin and poly(ethylene glycol) by photopolymerization of a poly(ethylene glycol) macromonomer [29]. A gel with water content at 60–81% and high strength as a crosslinked hydrogel (2.4 MPa) has been obtained.

Without using dissolved chitin, a sponge-like gel can be obtained using a crushed chitin made of b-chitin from a squid. This is used for wound dressing and good results have been reported [18].

4 PREPARATION AND APPLICATION OF CHITOSAN GELS

Chitosan exhibits different properties than chitin due to the presence of the amine group. First of all, chitosan has the properties of a polyelectrolyte. The degree of swelling of a gel made from chitosan is controlled by both pH and ionic strength. It is also suitable for fixation of drugs because the amine group is reactive. The amine group has high reactivity with the body and it activates platelets [17].

Because chitosan dissolves into an acidic aqueous solution, it is necessary to crosslink it to prepare a hydrogel. Generally, glutaraldehyde is used as a crosslinking agent [30, 31]. Ethylene glycol diglycidyl ether can also be used [32]. Additionally, by forming a polymer complex [10, 33, 34] or interpenetrating polymer networks [35–37] with an anionic polymer, a hydrogel can be prepared. As a special example, there is a gel formation that can be made by mixing a metallic oxide and chitosan solution [38].

Chitosan gels have often been studied for use as a drug delivery system matrix. After addition and fixation by glutaraldehyde to a mixed solution of an amino derivative of 5-fluorouracil (5FU) (see Fig. 3) and chitosan, crosslinking and microemulsification were performed. The 5FU fixed chitosan gel microparticles with diameters of 250–350 nm were obtained [30]. From these gel microparticles, only 5FU was released but no release was detected from the 5FU derivative. The release rate of 5FU is faster in ester type carbamoil type. The release rate can be controlled by changing the degree of acetylation of chitosan. In addition, the release rate was reduced to one-third that of the original value by fixing an acidic polysaccharide through polyelectrolyte complex formation.

The polyelectrolyte complex formed from chitosan and the anionic polymer is insoluble in water. Hence, it forms a hydrogel without chemical crosslinking. As anionic polymers, acidic polysaccharides, such as carboxymethyl cellulose, carrageenan, xanthane gum and alginic acid, are often used. The following is an example with κ-carrageenan. To a carrageenan solution that contains chitosan powder (pH 9), a potassium chloride solution with theophylline dispersion was added dropwise. A carrageenan capsule was made. This capsule was treated in a pH 2.1 solution and the chitosan was charged to form a complex with carrageenan [33]. The release of theophylline from this capsule is not influenced by ionic strength in the pH range from 1.2 to 8.0 and is released at a rate of 10%/h. Chitosan complex gels do not generally swell in neutral and acidic solutions.

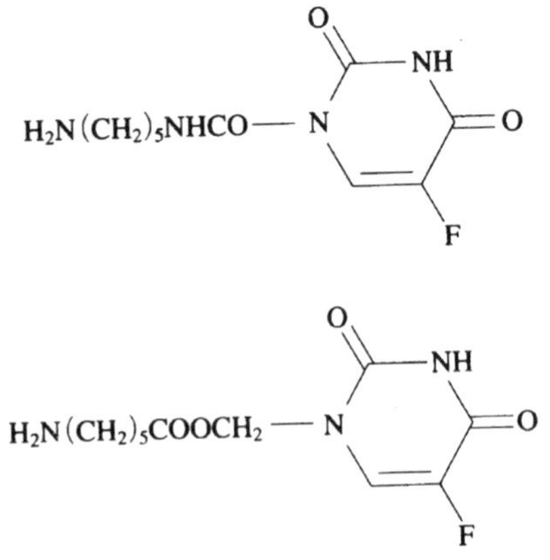

Fig. 3 The structure of 5FU.

However, the complex from chitosan and carrageenan swells greatly at pH 10–12 [39]. Similarly, the complex from chitosan and xanthane gum swells greatly at pH 10–11 [34].

Without using an anionic polymer, chitosan crosslinks ionically with polyphosphoric acid and forms a gel. The release behavior from a methotrexate-containing gelatin capsule, which is coated with this gel, has been investigated in a model stomach or intestinal fluids [40].

The chitosan gel started dissolving in the stomach fluid at pH 1.2, with little drug release at 2 h. In an alkaline intestinal fluid, the dissolution and swelling are controlled and the sustained release of chitosan gel is prominent.

The swelling and shrinking of the gel by change in a pH also takes place with the chitosan gel crosslinked with glutaraldehyde. In this case, it swells when the pH is acidic, and shrinks at neutral and alkaline pHs. In the case of a copolymer between chitosan and gelatin, swelling suddenly reduces at pH 6–7 [37]. Sustained-release chloramphenicol and other drugs have also been investigated.

The amine group of chitosan activates platelets and coagulates blood. If this property is used, chitosan can be used as a blood clotting agent. The chitin fiber, which was treated with a concentrated alkali and deacetylated, absorbs both blood and gels. Clotting takes place due to the activated platelets. Compared with collagen and oxidized cellulose, remarkable results have been reported using this material. Clinical trials have demonstrated sufficient effects [17].

Chitosan gel has been used as a dental material [38]. The gel contains hydroxyapatite and is used for dental surgery as a fixation material for the jawbone and teeth. If hydroxapatite and 2–6% calcium and zinc oxides are mixed with a chitosan solution, it gradually gels in $\approx 2 - 8$ min. Depending on the composition of the powder, gel time can be controlled. Characteristics include the ability to be injected into the missing part of the bone and the ability to gel. Interesting results might be obtained from experiments using animals.

5 CONCLUSION

Applications using the good biocompatability of chitin (which is an amino polysaccharide produced naturally) and the utilization of the amine group in chitosan have been discussed. Even without gel formation, chitin and chitosan are used extensively in the medical field and they offer great future potential [9,15,19,41,42]. It is hoped that the bioactivities of chitin and chitosan will be further elucidated and that they will be effectively used.

REFERENCES

1   Assoc. of Chitin and Chitosan Research (1995). *Chitin and Chitosan Handbook*, Giho-do Publ.
2   Takiguchi, H. (1995). *ibid.*, pp. 204–209.
3   Hashimoto, M. (1995). *ibid.*, pp. 204–209.
4   Nishi, N. and Tokura, S. (1995). *ibid.*, pp. 256–282.
5   Urakami, T. (1995). *ibid.*, pp. 402–437.
6   Nishiyama, A. (1995). *ibid.*, pp. 460–473.
7   Kibune, (1995). *ibid.*, pp. 324–354.
8   Nakagawa, Y. (1995). *ibid.*, pp. 474–480.
9   Seo, H., Itoh, Y., and Miyazawa, G. (1995). *ibid.*, pp. 506–537.
10 Shioya, T. (1990). in *Application of Chitin and Chitosan*, Assoc. of Chitin and Chitosan Research, ed., Giho-do Publ., pp. 155–174.
11 Sakai, K. (1995). *Chitin and Chitosan Handbook*, Assoc. of Chitin and Chitosan Research, ed., Giho-do Publ., pp. 209–218.
12 Suehi, T. (1995). in *ibid.*, pp. 219–226.
13 Kurita, K. and Ishii, S. (1995). in *ibid.*, pp. 228–254.
14 Mita, K. (1987). *Development and Application of Chitin and Chitosan*, Kogyo Gijsutsu-kai, pp. 248–269.
15 Minami, S., Yubiwa, T., Shigemasa, T., Okamoto, Y., Tanigawa, T., and Saiki, H. (1995). in *Chitin and Chitosan Handbook*, Giho-do Publ., pp. 178–202.
16 Tsurutani, R., Yoshimura, A., Tanimoto, N., and Kimura, (1995). *Maku* **20255.**
17 Tsurutani, R. (1996). *J. Fiber Sci. Soc., Jpn.* **5227.**
18 Shigemasa, Y. and Minami, S. (1994). *Kagaku Kogyo* **45463.**
19 Higashi, I. (1995). in *Chitin and Chitosan Handbook*, Assoc. of Chitin and Chitosan Research, Giho-do Publ., pp. 145–162.
20 Aiba, S., Izume, M., Minoura, N., and Fujiwara, Y. (1985). *Br. Polym. J.* **1738.**
21 Aiba, S., Izume, M., Minoura, N., and Fujiwara, Y. (1985). *Carbohydr. Polym.* **5285.**
22 Tokura, S. (1995). *Kobunshi* **44112.**
23 Hirano, S. (1988). *The Last Biomass: Chitin and Chitosan*, Assoc. of Chitin and Chitosan, ed., Giho-do Publ., pp. 21–50.
24 Hirano, S. and Yamaguchi, R. (1976). *Biopolymers* **151685.**
25 Hirano, S., Tobetto, K., Hasegawa, M., and Matsuda, N. (1980). *J. Biomed. Mater. Res.* **14477.**
26 Hirano, S., Tobetto, K., and Noishiki, Y. (1981). *J. Biomed. Mater. Res.* **15903.**
27 Aiba, S. (1988). *Report of the Institute for Product Sci.* **11233.**
28 Aiba, S. (1990). *J. Fiber Sci. Soc., Jpn.* **46558.**
29 Kim, S.S., Lee, Y.M., and Cho, C.S. (1995). *Polymer* **364497.**
30 Ouchi, T. and Ohya, Y. (1993). *Kagaku to Kogyo* **46798.**
31 Shantha, K.L., Bala, U., and Rao, K.P. (1995). *Eur. Polym. J.* **31377.**
32 Aiba, S. (1992). *Report of the Institute for Product Sci.* **1231.**
33 Tomida, H., Nakamura, C., and Kiryu, S. (1994). *Chem. Pharm. Bull.* **42979.**
34 Chu, C.H., Saiyama, T., and Yano, T. (1995). *Biosci. Biotechnol. Biochem.* **59717.**

35 Yao, K.D., Peng, T., Feng, H.B., and He, Y.Y. (1994). *J. Polym. Sci., Part A, Polym. Chem.* **321213.**
36 Yao, K.D., Peng, T., Xu, M.X., Yuan, C., Coosen, M.F.A., Shang, Q.Q., and Ren, L. (1994). *Polym. Int.* **34213.**
37 Yao, K.D., Yin, Y.J., Xu, M.X., and Wang, Y.F. (1995). *Polym. Int.* **3877.**
38 Itoh, M. (1995). *Seitai Zairyo* **1326.**
39 Sakiyama, T., Chu, C.H., Fujii, T., and Yano, T. (1993). *J. Appl. Polym. Sci.* **502021.**
40 Narayani, R. and Rao, K.P. (1995). *J. Appl. Polym. Sci.* **581761.**
41 Tokura, S. (1995). *Food Chem. Monthly, Jpn.* **1119.**
42 Yasutomi, Y., Nakakita, N., Shioya, N., and Kuroyanagi, N. (1993). *Nessho* **1918.**

Section 9
Sustained Drug Delivery by Gels

MASAKATSU YONESE

1 INTRODUCTION

Controlled-release drugs, for example, the many capsules and drugs that dissolve in the intestine, have been used for many years. In recent years, drug delivery systems (DDS) that use the minimum necessary amount and sustain the release of the drug for a prolonged period of time at the necessary location have been in development. Early DDS adopted the approach of controlling release by adjusting the dissolution rate, osmotic pressure, and interaction between the drug and matrix. Even today, these methods are the main ones. However, there are studies to further advance drug delivery systems that can deliver the necessary amount of the drug at the necessary time in a targeted manner to the needed area of the body. Gels and liposomes are used for these studies. These DDS have three mechanisms of drug storage, controlled release rate, and activation of release. Gels possess these three functions. Especially today when fundamental studies on the properties and functions of gels have been advanced, gels are expected to be very useful base materials for DDS [1–3].

230

2   POLYMERS AND GELS USED FOR DRUG DELIVERY SYSTEMS

The role of polymers used as an aid in drug delivery systems is far-reaching. They include use as: i) a bonding agent to adjust the strength and hardiness of drug delivery systems (if a polymer insoluble to water is used, collapsing can be inhibited); ii) a collapsing agent using a water-soluble polymer, iii) shape holding agent, and v) a masking agent for bitterness and drug odor as well as use as a coating to protect the drugs and ambient moisture. These drug delivery systems have made advances that use these properties of polymers. There are stringent restrictions on polymers that are used for drug delivery systems due to interactions with the body, with safety, biocompatibility, biodegradability, and antiblood clotting issues necessary to be considered. Polymers and gel raw materials that are often used for drug delivery systems are listed in Table 1.

3   GEL FUNCTIONS AND CONTROLLED DRUG DELIVERY

Agar and gelatin form crosslinking by hydrogen bonding and exhibit a sol-gel; transitions by temperature change. Sudden changes in the visco-

Table 1   Application examples of polymers and gels used in drug delivery systems.

	Natural polymer	Synthetic and part synthetic polymer
Shape holding agent	Starch	Crystalline cellulose
Bonding agent	Arabic gum, sodium alginate (NaAlg), agar, gelatin, starch, dextrin, tragacanth gum	Methyl cellulose (MC), ethyl cellulose (EC), sodium salt of carboxymethyl cellulose (NaCMC), hydroxypropyl cellulose (HPC), poly(vinyl alcohol), (PVA), poly(vinyl pyrrolidone) (PVP), CaCMC, HPC
Collapsing agent	Starch	EC, HPC,
Coating material	Gelatin, casein, shellac	hydroxypropylmethyl cellulose (HPMC), PVA
Base material for capsule	Gelatin	
Base material for suppository	Gelatin	Poly(ethylene glycol) (PEG)
Ointment		PEG, NaCMC
Replacement for plasma	Dextran	Hydroxymethyl starch, PVP

elasticity, solute holding function, and diffusion accompanying this transition are properties useful for controlled drug release systems for capsules, coated tablets, and suppositories.

In recent years, studies on the swelling-shrinking of hydrogels by solution compositions and temperature changes or on mechanochemical reactions with which electrical energy is converted to mechanical energy have advanced an understanding of the mechanisms of gel functions [4]. Using these functions, it is possible to develop new tablets or DDS with sensor functions (see Fig. 1) [5]. Gel properties that are useful for the DDS application are: i) the ability to form coatings and microparticles; ii) ability to act as a matrix; iii) sol-gel transition; v) blocking effect towards diffusion; v) swelling and water holding; vi) swelling-shrinking phase transition; and vii) mechanochemical reaction.

3.1 Blocking Effect in Gel Diffusion

A solute diffuses through free water in a gel. Due to the screening effect and the presence of bound water, the diffusion rate will slow. It is necessary to understand the diffusion coefficient D in the gel and the release rate of the solute when any DDS is to be designed [2]. If the volume fraction of the free water, namely the porosity fraction, is ε, D can be given by the Mackie and Meares equation [6].

$$f = D/D_0 = (\varepsilon/(2 - \varepsilon))^2 \tag{1}$$

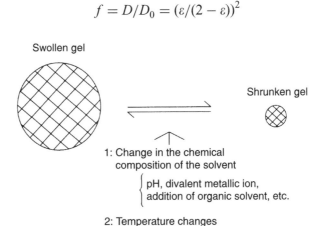

Swollen gel

Shrunken gel

1: Change in the chemical
   composition of the solvent
   $\left\{ \begin{array}{l} \text{pH, divalent metallic ion,} \\ \text{addition of organic solvent, etc.} \end{array} \right.$

2: Temperature changes
3: Electric field
4: Light

Fig. 1 Swelling-shrinking of gels by environment changes.

where D_0 is the diffusion coefficient in water. For example, when D is $10^{-9} m^2/s$, it takes $< 10 min$ until a solute completely diffuses out of a gel sphere that is $0.1 cm$ in diameter. If $f = 0.1$ or 0.01, the release time is $100 min$ or $1000 min$, respectively. According to Eq. (1), to obtain $f < 0.1$, it must be $\varepsilon < 0.5$, and for $f < 0.01$ the $\varepsilon < 0.1$, namely a gel with very small porosity, or very small water content must be used. Of course, if the diameter of the sphere reduces, the release time is further reduced. It is necessary to take advantage of a sudden change in the degree of swelling during the swelling-shrinking phase transition, formation of surface density layer, or an interaction between the drug and matrix, in order to control the release of the solute using the water content of the gel. When polymer drugs are released any drug smaller than the mesh size of the gel networks will pass while a larger drug cannot pass. It is necessary to consider the relationship between the mesh size and the size and shape of the drug [7].

3.2 Application of Swelling-Shrinking Phase Transition of Gels

If the phase transition properties of gels are used, it becomes possible to design DDS that allows drug release in response to chemical stimuli (pH, ion composition, glucose concentration, etc.) and physical stimuli (temperature, electric field, light, ultrasound, etc.). For the drug release control accompanying the phase transition or the structural changes of the matrix, the following behaviors are used as shown in Fig. 2. They include: (a) release control by diffusion; (b) following the squeezing of water by shrinking (squeezing effect); (c) the inhibition of squeezing by the formation of a dense skin during shrinking (skin effect); and (d) control of dehydration by use of a coating layer. In (b), inhibiting release requires that the control efficiency of the release be improved by applying a coating as in (d).

Development research on the DDS that uses phase transition of gels has been actively performed with thermoresponsive gels. Significant results have been reported for the development of base gel materials and the drug delivery system that use these gels. Although *in vitro* research is now underway, actual applications will become possible in the future. Polyacrylamide type polymers are thermoresponsive. As shown in Table 2, polymers that possess wide transition temperatures have been discovered. Under phase transition temperature the polymers will swell and above the transition they will shrink.

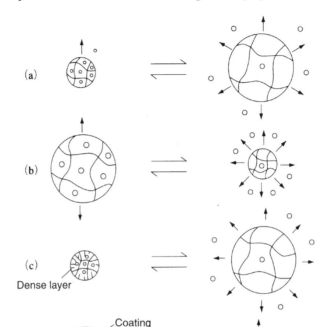

(a) Release control by difference in the diffusion coefficient
(b) Release control by squeezing of water (squeezing effect)
(c) Release control by formation of dense surface layer (skin effect)
(d) Release control by coating and squeezed water

Fig. 2 Swelling-shrinking of gels and drug delivery control.

Modification of thermoresponsive polymer gels regarding phase-transition temperature and gel strength is being attempted for potential use in DDS applications. The transition temperature of copolymer gels with isopropyl acrylamide and butyl methacrylate (IPAAm-BMA) reduces as the BMA content in the composition increases. At the 95/5 composition, the copolymer shrinks $> 20°C$. Bae *et al.* used this gel and reported that insulin permeation through the gel membrane generally stops at 20–30°C and that indomethacin is released pulsewise in response to pulsewise temperature changes between $20 \sim 30°C$ [8]. See Fig. 3 for more details.

Table 2 Thermoresponsive gels.

	Transition temperature ($^\circ$C)
Poly(vinylmethyl ether)	38.0
Poly(N-ethyl acrylamide)	72.0
Poly(N-ethyl methacrylamide)	50.0
Poly(N-isopropyl acrylamide)	30.9
Poly(N-methyl-N-isopropyl acrylamide)	22.3
Poly(N-acryloyl piperidine)	5.5
Acryloyl-L-**prolinemethyl ester**	24
Copolymers of N-isopropyl acrylamide and butyl methacrylate[a]	
Copolymers of acryloyl-L-prolinemethyl ester and hydroxyethyl methacrylate	
copolymers of acryloyl-L-prolinemethyl ester and hydroxypropyl methacrylate[a]	
IPNs of poly(ethylene oxide) and poly(acryloyl pyrrolidone)[a]	
IPNs of poly(tetramethylene ether glycol) and poly(N-isopropyl acrylamide)[a]	
Association complexes of poly(methylacrylic acid) and poly(ethylene oxide)[a]	

[a]: The phase transition temperature varies depending on the composition of copolymers, interpenetrating networks (IPN), and association complex

As shown in Fig. 2(c), drug release is inhibited by the formation of a dense surface layer when shrinking occurs above 20°C. When swelling occurs the dense layer swells and the drug is released efficiently. Acryloyl-L-prolinemethyl ester (A-ProMe) gel exhibits phase transition at $\approx$24°C. Even upon shrinking, the dense surface layer will not appear and only the mesh sizes vary. However, the copolymer gels of A-ProMe and hydroxypropyl methacrylate form a dense surface layer upon shrinking. Figure 4 shows the drug release result of a testosterone-containing gel made of this copolymer when it is embedded under the skin of a mouse. It has been demonstrated that constant release is possible for as long as +54 weeks [9].

The development of DDS to control release of insulin in response to glucose (blood sugar) concentrations is rapidly advancing. The copolymer gel of IPAAm and 3-arylaminephenyl boric acid swells and shrinks in response to glucose concentrations. This copolymer gel has attracted research attention because it can be an on-off type release control gel [10].

3.3 Application of Blend Gels to DDS

Improvement of gels for DDS is attempted in areas of modification of polymer residue, copolymers, blend gel, and multilayer gel (coated gel). A blend gel made of a cationic polyelectrolyte and anionic polyelectrolyte

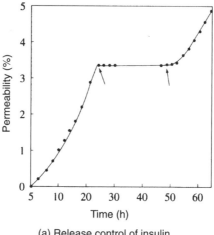

(a) Release control of insulin

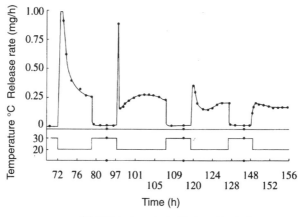

(b) Release control of indomethacin by
pulsewise temperature changes

Fig. 3 Drug release control using a copolymer of isopropyl acrylamide and butyl methacrylate.

respond to show swelling and shrinking behavior based on changes in pH and type or concentration of an added ion. A blend of gelatin and chondroitin sulfate (a blend of a protein and an acidic polysaccharide) shrinks in strongly acidic pH and permeability is drastically reduced [11]. In the body there are many examples, such as collagen and acidic mucopolysaccharide that demonstrate different functions. By imitating

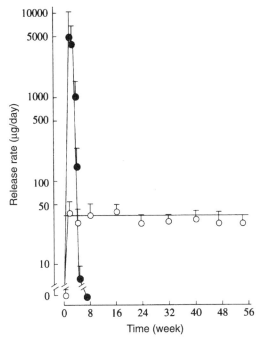

Fig. 4 Drug release behavior (*in vivo*) of testosterone-containing (○) a copolymer gel made of acryloyl-L-prolinemethyl ester and hydroxypropyl methacrylate and (●) acryoyl-L-prolinemethyl ester gel.

these examples, it is possible to prepare gels with unique characteristics. If the mixed solution of a protein and an acidic polysaccharide is made into an acidic pH, coacervation can be observed. Microspheres for which the diameter is controlled by adjusting pH and the degree of crosslinking are useful for drug delivery systems.

In a blend gel, there are interpenetrating network gels in which the networks of both polymers are intertwined without being chemically crosslinked to each other and a reptation type in which a linear polymer penetrated into the network of another polymer. If a transition between these two types of blends is considered, it is possible to develop a stimuli-responsive gel. For example, a reptation-type gel made of PVA gel to which alginic acid salt, Alg, is added changes into interpenetrating polymer networks upon the addition of Ca^{2+} ion by changing the Alg into a crosslinked network. If shrinkage during that transformation is used, drug release control is possible [12].

The number of application studies for gels to be used in DDS is increasing as the numbers from the Pharmaceutical Society, DDS, Society, and Association of Gel Research reveal. Table 3 summarizes the typical examples of gels for DDS.

Table 3   Application of gels for drug delivery systems.

Natural polymer	
Polysaccharide	
Chitin and chitosan	Sulfurdiazine-containing patch
	Chitosan microsphere made by emulsion solution dropping method (12 μm)
	Mucoadhesive chitosan microsphere (2.5, 25 and 50 μm) for alimentary canal
Alginic acid, Alg	Oral delivery drug by alginic acid beads
	Alg microgel beads (50 μm)
	Chitosan-coated Alg microbeads
	Chitosan-chondroitin sulfate modified alginic acid gel beads
Alg-calcium gluconate	Masking of bitter drugs
Alg-agar, Alg-carrageenan, and Alg-gellan gum complex	Delivery controlled polysaccharide minicapsule (mixed gel coated capsule)
Alg-chitosan complex	Release time controlled tablet
Alg-polylysine	Cell enclosed alginic acid-polylysine-alginic acid microcapsule
Agar	
Agar-pectin, Agar-Alg, Agar-Alg-ppy	Development of oral nongelatinous soft capsule
	Vitamin E-containing capsule
Pullulan	Theophylline-containing, pullulan-coated minicapsule
Dextrin	
Xyloglucan from tamarind	Transdermal drugs or drugs through mucosa
Dextran	pH-responsive carboxydextran microsphere
Protein	
Gelatin	Capsule
Serum albumin (bovine, human)	Bovine albumin microsphere as a drug substrate (membrane emulsion method)
Semisynthetic polymer	
Poly(lactic acid)	Calcitonin gene-related peptide sustained release drug
Poly(lactic acid-co-glycocholic acid)	Oral vaccine for type B hepatitis using microsphere
Synthetic polymer	
PVA	Matrix for suppository
Chitosan-PVA gel	Microsphere of chitosan gel-blended PVA gel

4   GELS IN DDS

Drug delivery methods include oral administration, transdermal absorption, injection (subcutaneous, intravenous, and intramuscular), sprays, insert (suppositories) implantation (subcutaneous, brain, bone), patches (oral or nasal, mucosa, and dentin) [13]. When gels are used for DDS, the shape and size considered will depend on the method of administration. For example, gel particles must be less than several hundred nm for injection. For this purpose, microcapsules, nanocapsules, microgels, coacervates, and microspheres are used.

Currently, gel application examples for DDS are mainly for sustained drug release systems. For oral drugs, examples include a microcapsule of theophylline, which is a drug for asthma treatment and a sustained release drug that is made of an ionic drug in a coated ion exchange resin. As base materials for transdermal drug delivery systems, gelatin and chitosan-gelatin blend gels are used. Mucoadhesive drug delivery systems offer a faster drug absorption rate than transdermal drug delivery systems and the control of the amount delivered is also easier. Thus, these are used for oral and dental diseases. For example, for periodontal diseases, such as alveolar pyrorrhea, a DDS using a tetracycline-containing copolymer of ethylene and vinyl acetate is adhered to the gum.

There are many attempts to insert or implant a gel that contains a drug. A DDS that consists of a sandwich of pilocarpine (for iris contraction) between ethylene-vinyl acetate copolymers, is an example of a DDS used for the eyes.

By adjusting gel membrane thickness, the release rate is controlled to 20–40μg/h, which allows the drug delivery system to be used continuously for one week without changing it. To control conception, there is a DDS that contains progesterone in ply(ethylene-co-vinyl acetate). This system is inserted vaginally and can be used for 1 yr because of the sustained release that is caused by slow diffusion. It is among the targeting DDS. Although this system currently must be removed after 1 yr, there is a study in which a biodegradable polymer, poly(lactic acid-co-glycol acid) is used.

Implantation is done subcutaneously, in the bone, or in the brain. For subcutaneous implantation, a silicone membrane is used. However, this must be taken out after completion of drug delivery. Currently, development is underway for a biodegradable DDS using poly(lactic acid-co-glycol). For implantation in the brain, a few methods have been tried. For administration of the Parkinson's disease drug, dopamine, poly(ethylene-co-vinyl acetate is used, and the anticancer drug calmostine as well as the Alzheimer disease drug betanecol use a biodegradable polyanhydride microsphere. For bone implantation, administration of gentamicin using poly(methyl methacrylate) or polyanhydride has been tried.

A DDS that utilizes gels has been attempted in various delivery systems. Currently, they take advantage mostly of sustained release function. However, it might be possible to develop DDS that delivers at the necessary time the necessary amount by using the sensor functions of gels.

REFERENCES

1 Sezaki, J. (1990). *Development of Medical Drugs*, vol. 13, *Drug Delivery Methods*, Hirokawa Shoten.
2 Yonese, M. (1990). *Pharm. Tech., Jpn.* **6391.**
3 Miyajima, K. (1996). *Proc. 13th Mtg. Soc. Phys. Chem. Mater. Properties* **1327.**
4 Yamauchi, A. and Hirokawa, N. (1990). *Functional Gels*, Kyoritsu Publ.
5 Okano, M. and Sakurai, H. (1990). *Organic Polymer Gels*, Chem. Soc., Jpn., ed., Gakkai Publ. Center, p. 66.
6 Mackie, J.S. and Meares, P. (1955). *Proc. Roy. Soc., Ser.* **A232498.**
7 Yonese, M., Kondo, M., Miyata, I., Kugimiya, S., Sato, S., and Inuka, M. (1996). *Bull. Chem. Soc., Jpn.* **69883.**
8 Bae, Y.H., Okano, T., Hsu, R., and Kim, S.W. (1987). *Macromol. Chem. Rapid Commun.* **8481.**
9 Yoshida, M., Asano, M., Kumakura, M., Katakai, R., Mashimo, T., Yuasa, H., and Yamanaka, H. (1991). *Drug Design Delivery* **7159.**
10 Fumiya, K., Miyazaki, H., Kataoka, K., Sakurai, Y. and Okano, M. (1997). *Proc. 8th Polym. Gel. Symp Jpn.* p. 15.
11 Yonese, M. and Nakagaki, M. (1975). *Yakushi* **95665.**
12 Yonese, M., Baba, K., and Kishimoto, H. (1992). *Polym. J.* **24395.**
13 Langer, R. (1993). *Acc. Chem. Res.* **26537.**

Section 10
Medical Sensors

ETSUO KAKUFUTA

1 INTRODUCTION

The recent progress in sensor technology has been remarkable. Of the many sensors now used in the medical field, they can be divided into diagnosis and cure types. As an example of the latter category, one sensor system monitors kidney dialysis. In this subsection, the concept of medical sensors and how gels are used for the construction of sensors will be explained. Furthermore, examples of recent research along with future potential possibilities will be considered.

2 SUMMARY OF MEDICAL SENSORS

Table 1 lists the main sensors used for diagnosis or cure. Moreover, classifications, measurement principles, and applications for diagnosis or cure also will be described. If commercial availability can be confirmed, it is so indicated. For enzyme sensors, the detection method (electrode) for enzyme reaction products is stated in the supplemental column.

In order to use sensors for medical purposes, many requirements must be met. These include:

Table 1 Major medical sensors.

Measured item	Classification of electrode	Principles	Applications	Supplemental
Potassium ion	Ion electrode	Glass film/Na$^+$	Diagnosis and cure (commercially available)	
Sodium ion	Ion electrode	Valinomycin (liquid film)/K$^+$	Diagnosis and cure (commercially available)	
Calcium ion	Ion electrode	Didecylphospholic acid (liquid film)/Ca$^+$	Diagnosis and cure (commercially available)	
Chlorine ion	Ion electrode	Ag/Cl$^-$	Diagnosis and cure (commercially available)	Oxygen electrode and hydrogen peroxide electrode
Glucose	Enzyme electrode	Fixed glucose oxidase	Diagnosis and cure (commercially available)	
Urea	Enzyme electrode	Fixed urease	Diagnosis and cure (commercially available)	Ampere meter
Gas dissolved in blood, breathing gas (O$_2$ and CO$_2$)	Composite electrode (oxygen electrode and carbon dioxide electrode)	Zirconia film/O$_2$ etc., CO$_2$ permeable membrane/pH electrode etc.	Diagnosis and cure (commercially available)	
Cholesterol	Enzyme electrode	Fixed cholesterol oxidase	Diagnosis (commercially available)	Oxygen electrode
Neutral fat	Enzyme electrode	Fixed lipoprotein lipase	Diagnosis	pH electrode
Creatine	Enzyme electrode	Fixed creatinitase	Diagnosis	Ammonia electrode
Pyruvic acid	Enzyme electrode	Fixed pyruvic acid oxidase	Diagnosis	Oxygen electrode
Ammonia	Ion electrode	Nonactin or monactin (liquid film)/NH$_4^+$	Diagnosis and cure (commercially available)	

i)   short response time (ideally, it should be < 1 min;
ii)   preparation of samples should be simple;
iii) quantitative analysis is possible;
iv)   sample quantity is less than several hundred μl;
v)   measurement sensitivity is μg-mg/ml; and
vi)   a long lifetime.

If it is used for cure monitoring purposes, safety is particularly emphasized. This is because the sensor is used in contact with the body of the patient or bodily fluid as part of a dialysis unit or an artificial organ.

3 APPLICATION OF GELS TO MEDICAL SENSORS

During the construction process of these forementioned sensors, both the type and method of gel application have been investigated. The research and development trend until 1980 is given in two detailed monographs [1, 2]. Work after this date used a science citation index as the database. The indexed items were *gel(s)* and *sensor(s)* (investigation I) and *gel(s)* and *biosensor(s)* (investigation II). There were 61 articles in investigation I and 13 articles in investigation II. Content was dominated by studies that used the sol-gel transition of SiO_2 type (or other metallic oxides). They occupied 82% of investigation I and 62% of investigation II. This trend increased especially in the 1990s. If only the words sensor(s) or biosensor(s) were used as the index, the former numbered 13,612 articles whereas the latter involved 837 articles. Thus, those articles using the combined indices are only 1% of the entire sensor research papers.

Oxides such as SiO_2 and TiO_2 are sols in alkali but gel readily if an acid is added to undergo sol-gel transition. This gelation mechanism is not simple but it can be briefly explained as follows. For example, in the case of sodium metasilicate (Na_2SiO_3), first it polymerizes to form colloidal particles with diameters from several nm to several tens of nm. Then the silanol groups on the surface of the colloidal particles dehydrate and crosslink three dimensionally to form a gel. Therefore, the sol-gel method can provide a thin, glassy film at or below room temperature. Also, via a coating method, miniaturization is possible. Many pores among crosslinked particles are formed and sensing materials can be incorporated in those pores. This method became very active recently partially as a result of thin films that have high strength and excellent chemical stability.

Table 2 Examples of recent research on medical sensors.

Type of sensor	Sensing material	Type of gel substrate	Literature
Glucose sensor	Glucose oxidase	Silica gel film	(3)
		Vanadium pentaoxide gel film	(4)
		Light transmitting silica gel (light sensor)	(5)
		Nafion® gel film (amperemetric sensor)	(6)
		Poly(vinylferrocene-co-acrylamide) redox gel film	(7)
Oxygen sensor	Ru(II)-diphenylfenanthrene ruthenium composite	Silica gel-coated fiber optic (light sensor)	(8)
		Zirconia film-coated silica gel (amperemetric sensor)	(9)
Carbon dioxide sensor	NASICON $(Na_{(1+x)}Zr_2Si_{(x)}P_{(3-x)}O_{12})$ (used as a solid-state electrode)		(10)
Sodium ion sensor	NASION (used as an ion selective, electric field-effect transistor)		(11)

Organic polymer gels used for sensors include alginic acid gel, alginic acid/chitosan complex gel, acrylamide gel, and N-isopropyl acrylamide (hereinafter abbreviated as NIPAAm) gel. Basically, a sensing material is dissolved in the sol, followed by gelation to form a thin sensor film. On the other hand, in one example, a Nafion gel film that had already been formed into a film was used.

The function of gels valued for sensors is the characteristic of moldability. Of those results of literature survey [3–11], sensors that showed potential for medical uses are summarized in Table 2.

4 FUNCTIONS OF POLYMER GELS AND THEIR APPLICATION TO SENSOR TECHNOLOGY

One of the excellent properties of polymer gels is that they can swell and shrink in response to changes in the environment (stimuli-responsive gels). This phenomenon was first reported experimentally by Tanaka [12].

Later, this field rapidly developed as *volume transition* of polymer gels. The authors also studied molecular-level understanding of the volume transition of polymer gels [13]. We concluded that the balance between the attractive and repulsive forces acts on the crosslinked polymer chains by changing the intermolecular interaction, such as coordination bonding, hydrogen bonding and hydrophobic interaction, a gel stretch (repulsive force), and contract (attractive force). Thus, control of volume transition was attempted using saccharide bond compatibility [14] and the enzyme reaction [15, 16] of lectin. As a result, a concept was proposed for constructing a biochemomechanical system [17–20]. This concept is a

$$(NH_2)_2C=O+3H_2O \longrightarrow 2NH_4OH+CO_2$$

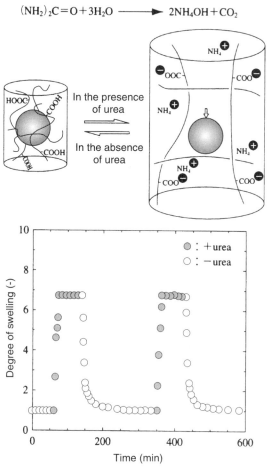

The pH of the system increases by the ammonium which is generated by the hydrolysis of urea. Then, the carboxyl groups dissociate, resulting in expansion of the gel by the coulombic forces

Fig. 1 Biochemical system made of urease inclusion gel.

functional, fixed biocatalyst that can convert biochemical energy into mechanical energy through the stretching and contracting behavior of gels.

Figure 1 shows a polymer gel made of a urease-containing copolymer made of NIPAAm and acrylic acid as an example of a gel that exhibits the biochemomechanical function. This gel reversibly stretches and contracts in the presence or absence of urea. To actuate behavior, a urea concentration of 1 mM (60 mg/ml) was sufficient. Therefore, this function makes it possible to use it as a urea sensor. In fact, Koopmann *et al.* [21] used this concept and developed a sensor for ionic saccharide detection. Unfortunately, there are no other examples on the study of this new type of sensors. Hence, it must await further research for the evaluation of its practicality.

5 CONCLUSION

The role of gels in medical sensors has been summarized and includes recent results. Particularly, the use of the volume transition function of a gel provides great potential for future sensor research. For example, there is an attempt to construct a drug delivery system using the sensor and matrix function of a gel. However, readers are referred to other chapters of this handbook for details.

REFERENCES

1   Suzuki, S. (1981). *Ion Electrodes and Enzyme Electrodes*, Kodan-sha Scientific.
2   Kiyoyama, T. (1986). *Chemical Sensor Practical Handbook*, Fuji Technosystem.
3   Tatsu, Y., Yamashita, K., Yamaguchi, M., Uamamura, S., Yamamoto, H., and Yoshikawa, S. (1992). *Chem. Lett.* 1615.
4   Gleser, V. and Lev, O. (1993). *J. Amer. Chem. Soc.* **1152**: 533.
5   Shtelzer, S. and Braun, S. (1994). *Biotechnol. Appl. Biochem.* **192**: 93.
6   Andrieux, C.P., Audebert, P., Bacchi, P., and Divisiablohorn, B. (1995). *J. Electro. Chem.* **394**: 141.
7   Bu, H.Z., Mikkerlsen, S.R., and English, A.M. (1995). *Anal. Chem.* **674**: 071.
8   Okeeffe, G., MacCraith, B.D., McEvoy, A.K., McDonagh, C.M., and McGilp, J.F. (1995). *Sensors & Actuators B.-Chem.* **29**: 226.
9   Peixoto, C.R.M., Kubota, L.T., and Gushiken, Y. (1995). *Anal. Proc.* **32**: 503.
10 Lee, D.D., Choi, S.D., and Lee, K.W. (1995). *Sensors & Actuators B.-Chem.* **25**: 607.
11 Caneiro, A., Fabry, P., Khireddine, H., and Siebert, E. (1991). *Anal. Chem.* **632**: 550.
12 Tanaka, T. (1978). *Phys. Rev. Lett.* **40**: 820.
13 Ilmain, F., Tanaka, T., and Kokufuta, E. (1991). *Nature* **34**: 9400.
14 Kokofuta, E., Zhang, Y.-Q., and Tanaka, T. (1991). *Nature* **35**: 1302.
15 Kokofuta, E. and Tanaka, T. (1991). *Macromolecules* **24**: 1605.
16 Kokofuta, E., Matsukawa, S., and Tanaka, T. (1995). *Macromolecules* **28**: 3474.

17 Kokofuta, E. (1992). *Prog. Polym. Sci.* **17**: 647.
18 Kokofuta, E. (1993). *Adv. Polym. Sci.* 157.
19 Kokofuta, E., Matsukawa, S., Ebihara, T., and Matsuda, K. (1994). *Amer. Chem. Soc. Symp. Ser.*, 548, pp. 507–516.
20 Kokofuta, E., Zhang, Y.-Q., and Tanaka, T. (1994). *J. Biomater. Sci., Polym. Edn.* **6**: 35.
21 Koopmann, J., Hocke, J., and Gabius, H.J. (1993). *Biol. Chem. Hopp. Seyl.* **37**: 41029.

Section 11
Encapsulation of Cells in Hydrogels

HIROO IWATA

1 INTRODUCTION

Artificial organs capable of metabolism and synthesis functions are very difficult to create. As a consequence, bioartificial organs capable of combining cells with the requisite functions and made of manmade materials has been attempted. The production of useful materials using cell-fixed column is drawing attention. Cellular fixation in a gel is going to be the key technology. When the science moves away from genetic compatibility, it is necessary to account for cross species fragility and demands made of animal cells. Incorporating animal (nonhuman) cells into hydrogels will be touched upon here and some examples will be provided.

2 ENCAPSULATION METHOD FOR LIVE CELLS IN HYDROGELS

In all encapsulation methods the polymer solution must be gelled by crosslinking the polymer chains after dispersing living cells in solution. Many water-soluble polymers possess high reactivity hydroxyl, carboxyl and amine groups on their side chains. By using a crosslinking agent that reacts with these groups, gelation can be readily achieved. Unfortunately, because crosslinking agents that react with side chains will also react with amine and carboxyl groups of proteins, cell toxicity is very high. Hence, different approaches are required. A summary of the methods thus far attempted is listed in Table 1. In the following, an actual example including the experimental method will be introduced.

2.1 Ion Complex

Alginic acid from seaweed forms a complex with a multivalent metallic ion and gels. Microencapsulation of islets of Langerhans cells, which produce insulin has already been accomplished and aggregated cells were produced [1, 2]. If islet cells suspended in an alginic acid solution are added dropwise to a culture solution that contains a metallic ion, the metallic ion diffuses into the alginic acid droplets. Furthermore, the ion forms a complex of gel beads in which the islet cells are enclosed.

If these beads of alginic acid and the metallic ion are embedded in the body, the metallic ion that forms the complex might exchange with monovalent sodium ions and dissolve. The multivalent ions also have cell toxicity. Hence, microcapsules must be created with a polyion complex of alginic acid [3, 4] and not a multivalent one.

Using the method already described here, gel beads made of alginic acid and calcium ions are manufactured. These beads are immersed in a solution of poly-L-lysine, a polycation, to form a polyion complex at the surface. The surfaces of the beads are again covered by alginic acid because a cationic surface has rather poor biocompatibility. Finally, the calcium ion, which forms the complex with alginic acid, is removed by treating the beads with citric acid and liquefying the alginic acid.

Table 1 Crosslinking methods that are used for encapsulation of cells in hydrogels.

Noncovalent bonding methods	Ion complex
	Polyion complex
	Hydrogen bonding
	Phase transition of polymer solution
	Use of a poor solvent in polymer solution
Covalent methods	Photodimerization reaction
	Oxidation of thiol group

Consequently, a three-layered structure of microcapsules made of alginic acid-poly-L-lysine-alginic acid is produced.

2.2 Formation of Hydrogen Bonding

A solution can be gelled by forming hydrogen bonds among polymer chains. Agarose, a component of agar used as a food source, is a polysaccharide consisting of {D-galactosyl-β-(1–4)-3,6-anhydro-L-galactosyl-α(1–3)}$_n$. The gelation temperature of an agarose aqueous solution is 25–40°C and the redissolution temperature of the gel is 60–70°C. There is significant hysteresis between gelation temperature and redissolution temperature. By utilizing this large hysteresis cells are incorporated into the agarose gel [5, 6].

Figure 1 shows the inclusion method for cells into agarose beads. Using a glass capped 50-ml glass tube, 0.15 g of low-temperature gelling agarose (Nakarai Chemicals) and 3 ml of nonserum culture are measured. After dispersing the cells into the agarose solution, 20 ml of paraffin fluid is added. The cap is then inserted into the glass tube and the tube is well shaken in order to disperse the agarose solution into the paraffin fluid with appropriate particle size the desired result. The glass tube is immersed in an ice bath and the agarose solution gels. After adding 20 ml of culture solution to the glass tube, the tubeis centrifuged at 2000/min for 15 min. The agarose beads will collect at the bottom. Figure 2 illustrates microencapsulated cells made using this method.

2.3 Use of the Phase Change of Polymer Solutions

The homopolymer and other copolymers of isopropylacrylamide exhibit low-temperature critical transitions near body temperature. If the solution temperature is increased, the polymer precipitates along the higher temperature side of the transition. Although it is not certain whether the precipitates qualify to be gauged a hydrogel, cellular encapsulation in polymer film has been reported [7].

2.4 Use of Low Cell Toxicity Solvents

Poly(2-hydroxyethyl methacrylate) (polyHEMA) is used for soft contact lenses and is a representative medical use hydrogel. Encapsulation of cells into polyHEMA or poly(2-hydroxyethyl methacrylate-co-methyl methacrylate) (polyHEMA-co-MMA)) will be introduced in the following.

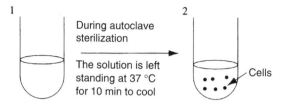

Cells previously separated are
added to the agarose solution
an then dispersed

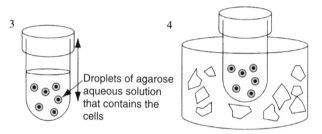

The glass tube is immersed in an
ice bath and the agarose solution
is gelled

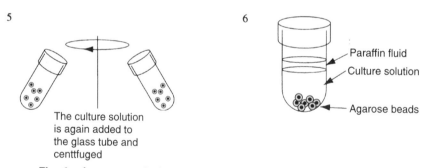

Fig. 1   An encapsulation method using cells in agarose beads.

Because these polymers are insoluble in water, it is necessary to use an organic solvent to encapsulate the cells. However, most organic solvents dissolve cell membrane fats and inflict irretrievable damage to a cell.

Crooks *et al.* used relatively small poly(ethylene glycol) (PEG200) as a relatively low toxicity solvent to dissolve poly(HEMA-co-MMA) [8].

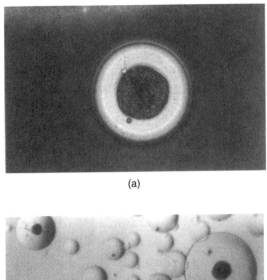

(a)

(b)

Fig. 2 Photomicrograph of pancreatic islets of Langerhans encapsulated in agarose beads.

Although the toxicity of PEG200 is rather low, it nonetheless harms the cell if it directly contacts it. Therefore, microencapsulation was performed using the device as shown in Fig. 3. From the nozzle in the middle the cell-suspended solution and from the outer nozzle poly(HEMA-co-MMA) dissolved in PEG 200 are injected into the culture solution. The culture solution extracts PEG200 and precipitates poly(HEMA-co-MMA) to form microcapsules. Again, although it is less toxic compared to other solvents, a significant number of cells die during the microencapsulation process.

Iopamidol, clinically used as a blood cell image enhancer, has been evaluated as an even less toxic solvent [9]. Cell toxicity is due mainly to osmotic pressure because of its use at high concentration. Because

Polymer solution

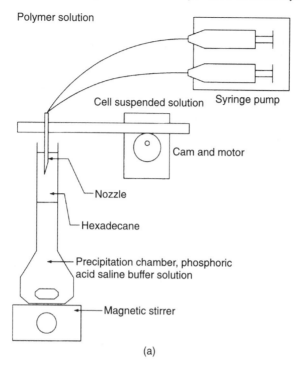

Cell suspended solution   Syringe pump

Cam and motor

Nozzle

Hexadecane

Precipitation chamber, phosphoric acid saline buffer solution

Magnetic stirrer

(a)

Cell suspended solution

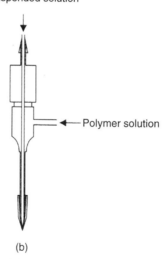

Polymer solution

(b)

Fig. 3 Encapsulation device used for encapsulating cells into poly-(HEMA-co-MMA) microcapsules.

iopamidol itself does not have the ability to dissolve cell membrane fat, toxicity is low. In fact, following encapsulation of insulin-producing cells, insulin production continued for more than one month, which indicates that living animal cells can be encapsulated. The shortcoming of this solvent is that it costs as much as several tens of thousand yen per 50ml.

2.5 Use of Photodimerization

Methods to introduce crosslinking between polymer molecules have been developed in the area of photosensitive resins. Although there are various photoreactions, photodimerization between side chain functional groups will be acceptable.

The poly(vinyl alcohol) developed by Ichimura and Watanabe has styrilpyridium as a side chain. This polymer undergoes gelation by dimerization of styrilpyridium groups upon irradiation of the visible light [10]. Thus, it satisfies two preceding conditions. There is an example in which islets of Langerhans were encapsulated using this reaction [11].

2.6 Formation of Hydrogels by the Oxidation of Thiol

Of the proteins (amino acids), cysteine has thiol (SH) on the side chain and thus it forms disulfide bonds intramolecularly or intermolecularly and contributes to the formation and stabilization of the higher-order structures of proteins. Because the formation of a disulfide bond by thiols is a

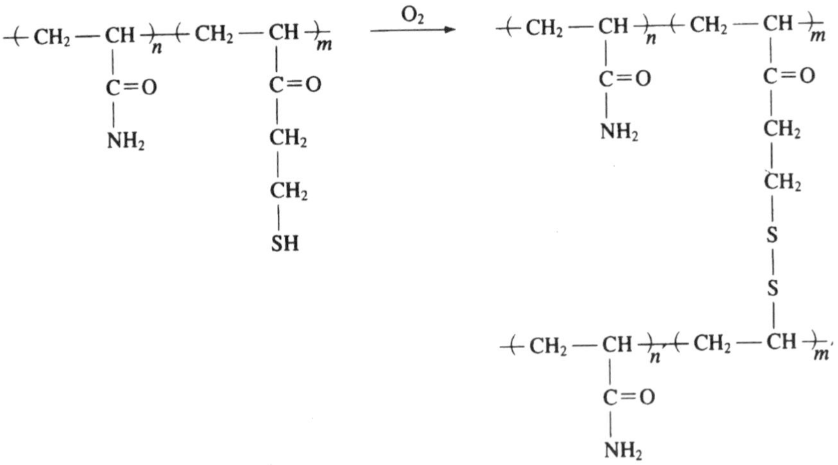

Fig. 4 Crosslinking reaction of acrylcysteamine).

normally occurring reaction in the body, it is considered nontoxic to cells. Using this reaction, cells were encapsulated in a hydrogel [12].

As shown in Fig. 4, a polyacrylamide hydrogel is formed by the progression of oxidation of thiols if poly(acrylamide-co-N-acrylcysteamine) dissolved in a culture solution to which cells are added is left alone in ambient air. Upon culturing the insulin producing cells encapsulated in this hydrogel, insulin was secreted for more than one month, indicating that the cells were alive in the hydrogel.

3 APPLICATION OF CELLS ENCAPSULATED IN HYDROGELS

3.1 Bioartificial Organs

If cells that produce hormone, are damaged and the cells can no longer produce a hormone, illness results. For example, the islets of Langerhans in the pancreas produce hormones including insulin, which is important in blood sugar regulation. If these pancreatic islets or are damaged or genetic problems create a malfunction, insulin is no longer produced and the result is an insulin-dependent diabetes mellitus. Thus, if the islets of Langerhans can be implanted, insulin-dependent diabetes mellitus could be cured. Likewise, many other illnesses related to glandular problems can be cured by implanting cells that produce the corresponding hormone.

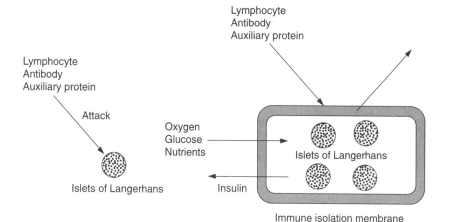

Fig. 5 The concept of immune isolation.

However, if the cells are implanted unaltered, the immune system of the recipient will attack the cells and reject them. Hence, implantation has to be done in a way that avoids rejection. One useful method involves immune isolation by hydrogels. Figure 5 depicts the concept of immune isolation. The cells are implanted after encapsulation in a hydrogel. The hydrogel must be able to pass necessary oxygen readily as well as nutrients to ensure the survival of the cells but be able to block the passage of immune cells that will attack the implanted cells.

The combination of donor and recipient creates two basic types of transplants—allogenic homografts and allogeneic homografts. Allogenic homografts indicate that donor and recipient belong to the same animal family, such as human to human, whereas allogeneic homograft means donor family and recipient family differ, for example, pig to human.

The requirements imposed on a hydrogel vary because the mechanisms of allogenic and allogeneic homografts differ. Because this handbook does not specialize in immune isolation, the readers are referred to a detailed monograph [13]. In this subsection, only the grafting of allogenic islets of Langerhans will be presented.

1500 agarose beads of encapsulated islets of Langerhans (shown in Fig. 2) were implanted in the peritoneal cavity of a mouse. The variations in blood sugar values before and after implantation are shown in Fig. 6.

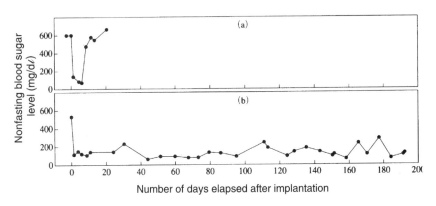

Number of days elapsed after implantation

(a) Implantation of 1500 exposed pancreatic islets in the peritoreal cavity of a diabetic mouse.
(b) Implantation of 1500 microcapsulated islets in the peritoreal cavity of a diabetic mouse.

Fig. 6 Diabetes cure in mouse with an allogenic homograft of agarose encapsulated islets of Langerhans.

The normal blood sugar level of a mouse is 100mg/dl. However, prior to implantation, the blood sugar was very high (500mg/dl) because the mouse is diabetic. Upon implantation of agarose-encapsulated islets, the blood sugar value returned to normal. During 200 days of observation, the blood sugar remained in the normal value range. In the same figure, the blood sugar values of a mouse with implanted islets that do not benefit from the protection of agarose gel coating are also shown as reference. In this case, the value temporarily returns to normal immediately after implantation. However, the blood sugar numbers returned to the high end only 10 days after implantation, suggesting that the implanted islets were rejected by the immune system of the recipient. These results indicate that agarose beads can protect the islets from a rejection reaction. Moreover, the islets continued to secrete insulin for a prolonged period of time to control blood sugar and the immune cells of the recipient were not able to reach and affect the allogenic homograft.

3.2   Prevention of Cell Aggregation

Many cells adhere to an extracellular matrix, such as collagen, or cell themselves adhere to each other. If cells that are removed from the body are cultured as they are, they aggregate. When the colony grows large, the supply of oxygen and nutrients to the center of the colony diminishes and the cells will die. Cells encapsulation is sometimes performed in order to prevent cells from touching each other and forming aggregations.

3.3   Protection of High Molecular Weight DNA

When a gene is analyzed, it is necessary to extract as high a molecular weight DNA as possible without harming it. The length of the chromosome gene is as long as several mm and can be readily sheared. Hydrogels are used to remove protein and lipids from cells and purify DNA. Cells are encapsulated in agarose beads. The protein and lipids are then extracted from the cells. The chromosomal DNA remains in the agarose beads because it is of extremely high molecular weight. Even if various operations are performed more rigorously, DNA in the gel will not experience shear force. At the end, DNA will be obtained by the digestion of agarose by the agarose digestion enzyme, agarase.

REFERENCES

1 Lanza, R.P., Ecker, D., Kuhtreiber, W.M., Staruk, J.E., Marsh, J., and Chick, W.L. (1995). *Transplantation* **27**: 1485.
2 Zekorn, T., Horcher, A., Siebers, U., Schnettler, R., Klock, G., Hering, B., Zimmermann, U., Bretzel, R.G., and Federlin, K. (1992). *Acta Diabetol* **29**: 99.
3 Lim, F. and Sun, A.M. (1980). *Science* **210**: 908.
4 Sun, A.M., O'Shea, G.M., and Goosen, A.F.A. (1983). *Progress Artificial Org.* 762.
5 Iwata, H., Amemiya, H., Matsuda, T., Takano, H., Hayashi, R., and Akutsu, T. (1989). *Diabetes* **28 (Suppl. 1)**: 224.
6 Iwata, H., Takagi, T., Amemiya, H., Shimizu, H., Yamashita, K., Kobayashi, K., and Akutsu, T. (1992). *J. Biomed. Mat. Res.* **26**: 967.
7 Iwata, H., Amemiya, H., and Akutsu, T. (1990). *Artificial Organs* **14 (Suppl. 3)**: 7.
8 Crooks, C.A., Douglas, J.A., Broughton, R.L., and Sefton, M.V. (1990). *J. Biomed. Mat. Res.* **24**: 1241.
9 Morikawa, N., Iwata, H., and Ikada, Y. (1994). *Proc. Ann. Mtg. Biomaterial Soc., Jpn.* **16**: 67.
10 Ichimura, K. and Watanabe, S. (1982). *J. Polym. Sci., Polym. Chem. Ed.* **20**: 1419.
11 Iwata, H., Amemiya, H., Hayashi, R., Fujii, S., and Akutsu, T. (1990). *Transpl. Proc.* **22**: 797.
12 Hisano, S., Morikawa, N., Iwata, H., and Ikada, Y. (1996). *Polym. Preprint, Jpn.* **45**: 372.
13 Iwata, H. (1995). *Kon-nichi no Ishoku (Transpl. Today, Jpn.)* **8**: 307.

CHAPTER 5

Farming and Agriculture

Chapter contents

Section 3   Application of Superabsorbent Polymers to Dry Land

Section 1

Characteristics of Superabsorbent Polymer (SAP)-Mixed Soil

SEIGO OUCHI

1 INTRODUCTION

The average annual precipitation in Japan is 1500 mm, a value that is rather fortuitous compared to other countries in the world. Despite this, there are occasions in the summer when agricultural products and trees may be affected by drought. According to a report by the Ministry of Agriculture, Soil Survey for Maintenance of Farm and Fertility [1], 16% (287,000 ha) of the ordinary agricultural sector and 19% (77,000 ha) of the orchards in Japan are experiencing inhibitory factors due to the water stress and thus improved water supplies are desired. Materials that are used to improve water holding are peat moss, perlite, and superabsorbent polymers (SAP). The water uptake of peat moss and perlite is several kilograms to several tens kilograms per 1 kg of dry product. The water is released easily upon pressurization as it is held by the capillary effect. On the other hand, the water uptake of SAP is several hundred kilo-grams/kilogram of dry product, and once the water is absorbed, it will not easily be released under pressure. Hence, SAP offers great potential for agriculture and greening, and extensive studies have been performed towards their application. Nevertheless, the current consumption in the areas of agriculture and greening is extremely low at several percent of total consumption [2]. The reasons for this are as follows:

i)   many factors, such as locality, soil conditions, climate, and growing methods, are interacting in a complex manner, which requires great time and effort;

ii) the amount of each use is low compared with that for industrial purposes; and

iii) the cost of SAP is high compared to the cost of the produced products.

Here, the word *greening* is used to indicate the planting of trees in the median and shoulders of roads as well as spraying seeds to prevent erosion.

2 CHARACTERISTICS OF SAP FOR AGRICULTURE AND GREENING

Table 1 lists the commercially available SAP for agriculture and greening [3]. The requirements on the use of SAP for disposable diapers and sanitary products are that they absorb large amounts of blood or urine quickly. However, for agriculture and greening, there are different requirements. Toyama [4] pointed out the following characteristics:

i)   there is no direct correlation between the water absorptivity of SAP and the growth of plants when SAP is used for water-saving purposes in agriculture;

ii) if the rate of water absorption is fast, soil aggregates are formed, which lead to difficult handling, however, SAP find powder is blown by wind during use;

iii) high thermal and radiation (visible and ultraviolet) resistance is required because of the outdoor use;

iv) to maintain good aeration and water permeation, the suitable particle size of the swollen SAP is approximately several millimetres, although a smaller particle size is preferred when it is used for the immersion method to better adhere to the roots; and

Table 1 Summary of superabsorbent polymers (SAP).

Main component	Name of product	Manufacturing company
Poly(acrylic acid) type	Aqua 100	Toagosei Chemicals
	Aquakeep	Sumitomo Fine Chemicals
	Aquamate AQ	Sekisui Plastics
	Aquareserve AP	Nippon Synthetic Chemicals
	Acrylihope GH-2	Nippon Shokubai
	Arasoab S	Arakawa Chemicals
	Viscomate PX-130	Showa Denko
	Diawet AL	Mitsubishi Chemicals
	IPC-01	Idemitsu Petrochemicals
	Sunwet IM-5000D	Sanyo Chemicals
	Poiz SA-20	Kao
Starch-graft copolymer type	Sunwet IM-1000	Sanyo Chemicals
	WAS	Nippon Starch Chemicals
Poly(vinyl alcohol) (PVA) type	Aquareserve GP	Nippon Synthetic Chemicals
PVA-poly(acrylic acid) type	Igeta Gel P	Sumitomo Chemicals
Polyacrylamide type	Geldia M800F	Nitto Chemicals
Isobutylene-maleic acid copolymer	KI Gel 201K	Kuraray
Cellulose type	Gel Fine H	Daicel Chemicals

v)   the ability to absorb water should last for a long time, preferably at least for several months and, if possible, for two to three years.

As a SAP for agriculture and greening purposes, the characteristics of a SAP made of vinyl alcohol and sodium acrylate as the main components (manufacturer: Sumitomo Chemicals, trade name: Igetagel P, hereinafter abbreviated as SIG) will be listed in Table 2 [5]. The monomer composition (molar ratio) of SIG is sodium acrylate : vinyl alcohol $= 6:4$. The highly absorbing sodium acrylate component is supported by the strong extension properties of the latter [6]. The characteristics of SIG are as follows:

i)   when water is absorbed, it is spherical;
ii) the strength of the highly swollen gel is very high and it will not fracture easily even with the application of external forces; in addition, it has high elasticity; and
iii) it has excellent thermal stability at high and low temperature as well as resistance to irradiation in the visible and ultraviolet regions.

The water-absorption capacity of SIG reduces as the concentration of the salt in the water increases (see Fig. 1). Because soil and irrigation water contain various salts, the water-absorption capacity of SAP will

Table 2 Summary of the characteristics of SIG.

General properties	Color	Air-dried product	Pale yellow
	Shape	Air-dried product	Fine particles
	Particle size, mm	Air-dried product	0.15~0.25
		At saturation water absorption	2~3
	Bulk density, g/cm^3	Air-dried product	0.85
	pH	0.5% Suspension solution	7~8
Ability to hold water (kg/kg dry product)	Air-dried product		<0.07
	Relative humidity	55%	0.10
	Relative humidity	75%	0.17
	Relative humidity	95%	0.24
	In deionized water	(10~30°C)[a]	500
	In deionized water	(pH 5~8)[a]	400~500
	In deionized water	(1-10-35.6 bar)[a]	500-430-380
	In tap water		340
	In 1% sodium chloride solution		72
	In 1% potassium chloride solution		76
	In 1% calcium chloride solution		25
Rate of water absorption (kg/kg dry product)	In deionized water	(5~10)[a]	420-500
Collapse tolerance (g/cm^2)	At saturation water absorption		500
	At 100 kg/kg water absorption		3000

[a]The water content upon holding 30 min over 74 μm shift after sufficiently absorbing water.

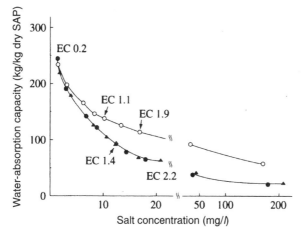

○: Sodium chloride; ●: Potassium chloride
△: Magnesium chloride; the concentration is,
respectively, from the left 0.01, 0.02, 0.04,
0.05, 0.06, 0.08, 0.10, 0.20 and 100 10^{-2} kg/kg

Water absorption capacity under water is
434 kg/kg dry SIG, EC: mS/cm

Fig. 1 Water-absorption capacity of SAP in salt solutions [13].

reduce to 10 to 20% of that in pure water. For example, the water-absorption capacity of SIG in pure water is 500 kg/kg. On the other hand, the water-absorption capacity of SIG at the Karuh River in Iran is expected to be around 100 kg/kg because the electrical conductivity (EC) of this river is 1.2 to 1.4 mS/cm and 75% of the contained cations are divalent [7]. Ouchi *et al.* [7] measured the water-absorption capacity of SIG in various irrigation waters and soils and reported that it is within the range of 100 to 200 kg/kg.

3 AVAILABILITY OF WATER HELD IN SAP TO PLANTS

How much of the water that is held in SAP is available to plants? Questions are often asked concerning whether or not the irrigated water or the water from the root may be absorbed by SAP and if the water once absorbed then is available to the plant. These questions will be addressed here.

Water in soil can be classified as gravitational water, capillary water, swelling water, and hygroscopic water (see Fig. 2). The water available to plants corresponds to the suction power of a water column 30–15,000 cm (equivalent to 0.003–1.5 MPa = pF 1.5–4.2), the capillary of 2–50 μm, and field water capacity, which is the water capacity (weight percent of water per dry soil) at the permanent wilting point. The water in the capillary smaller than these values is adsorbed water or bound water, and water larger than these values is gravitational water. In both cases, these are the types of water not available to the plants.

The water potential in SAP has been studied using relaxation time measurement by nuclear magnetic resonance spectroscopy (NMR), and for measurement on freezing and melting temperature, differential scanning calorimetry (DSC) was used [8, 9]. It has been found that the majority of water held by SAP (98–99%) is free water. Semibound water and bound water are known to be 1 to 1.5 kg/kg dry SAP.

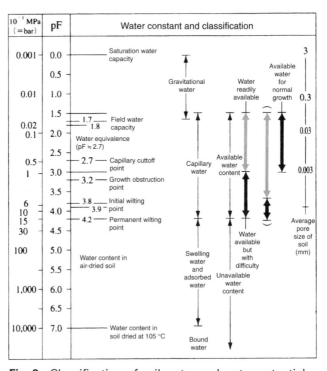

Fig. 2 Classification of soil water and water potential.

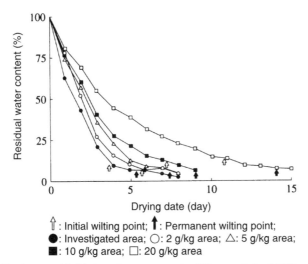

⇑ : Initial wilting point; ⬆ : Permanent wilting point;
● : Investigated area; ○: 2 g/kg area; △: 5 g/kg area;
■: 10 g/kg area; □: 20 g/kg area

Fig. 3   Drying and initial and permanent wilting point of SIG-mixed soil.

Ouchi *et al.* [5, 7] measured the water available to plants from the water held by SIG using the plant method and the saturation vapor pressure method and found that 98 to 99% of the water is available. Figure 3 shows the initial and permanent wilting point of a sunflower with SIG-mixed soil and nonmixed soil.

Figure 4 shows the measurement result of the maximum moisture–absorption coefficient (pF $4.46 = -2.89$ MPa). In any measured values, there is almost no difference with and without SIG, which indicates that most of the water held in SIG is available. As shown in Table 2, the water-holding capacity of SIG under 3.56 MPa centrifugal pressure is 380 kg/kg. This centrifugal pressure is greater than the permanent wilting point of plants (pF 5.5) and 75% of the water held by the SIG was regarded as being unavailable to the plant. The difference between these two may be caused by the steric structure of SIG, as the water in SIG is held in a fine lattice. There is little water release, even if sudden centrifugal force is applied due to the steric hindrance. On the other hand, the measurement methods that do not create steric hindrance (the plant and saturation vapor pressure methods) allow the majority of water to be released.

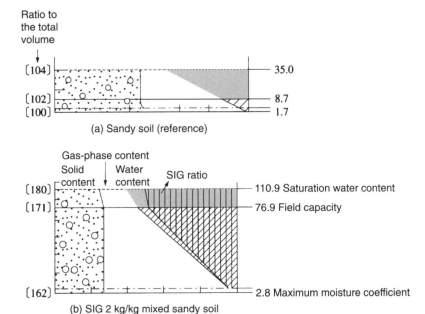

(a) Sandy soil (reference)

(b) SIG 2 kg/kg mixed sandy soil

Fig. 4 The maximum moisture coefficient, field capacity, and saturation water content of SIG-mixed soil.

4 CHARACTERISTICS OF A SAP MIXED SOIL

4.1 Changes in Total Volume by Wetting and Drying

When SIG absorbs and desorbs water, it swells or shrinks, the diameter changes by the factor of 10, and the volume changes by the factor of several hundreds. Figure 5 shows the change in total volume of SIG-mixed soil at wetting and drying processes. During the drying process, the rate of volume reduction is less than that in the wetting process, indicating that the volume that increased by the wetting and swelling processes experiences almost no decrease.

4.2 Change in the Distribution of Three Phases

The field capacity and distribution of three phases of a sandy soil and SIG-mixed soil are illustrated in Fig. 6. The physicochemical properties of the soil samples are listed in Table 3. The volume that is occupied by SIG, SIG ratio, is regarded as the water content and this value is added to the

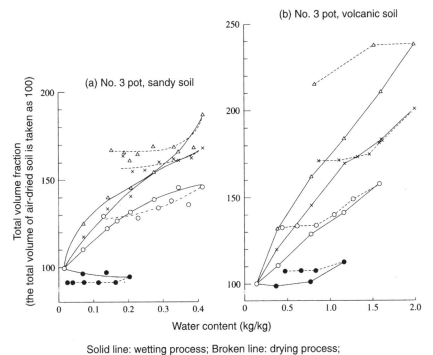

Solid line: wetting process; Broken line: drying process;
●: Reference (0) area; ○: SIG 2g/kg area;
×: SIG 5g/kg area; △: SIG 10 g/kg area

Fig. 5 Total volume changes of SIG-mixed soil by wetting and drying.

water content column. By mixing SIG, the ratio of the solid decreases. On the other hand, the ratio of the liquid phase dramatically increased and the available water increased by 10.6 times. In volcanic soil, the total volume increase and available water increase were not pronounced. Specifically,

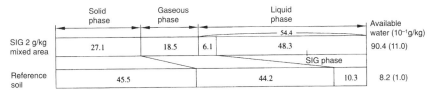

The total volume of the SIG-mixed area is 174% and the nonmixed (no additives) area is 103% when the volume of the air-dried reference area is taken as 100. The SIG phase is regarded as the liquid phase.

Fig. 6 Three-phase distribution and available water at the field capacity water content of SIG-mixed sandy soil.

Table 3   Physicochemical properties of sample soil.

Items	Sandy soil	Volcanic ash soil
Particle-size distribution (international method 10^{-2} kg/kg)		
Coarse sand	82.9	10.3
Find sand	11.2	34.9
Silt	3.8	40.5
Clay	2.1	14.3
Dissolution loss (10^{-2} kg/kg)	0.17	7.88
Soil property (international method)	S	L
Bulk density (rough packing, Mg/m^3)	1.30	0.52
pH (H$_2$O, 1:2.5)	5.6	5.9
EC (1:5, dS/m)	0.03	0.15
Saturation water content (10^{-2} kg/kg)	25.8	110.1
CEC (cmol (+) kg^{-1})	7.2	20.9
Phosphoric acid–absorption coefficient	501	2471

the effect of SIG addition on the available water is dramatic in sandy soil but significant results cannot be expected on the volcanic soil.

The soil-structure model of a SIG-mixed soil is depicted in Fig. 7 [10]. The total volume increased by the water absorption and swelling of SIG does not return to the original volume upon redrying. However, SIG

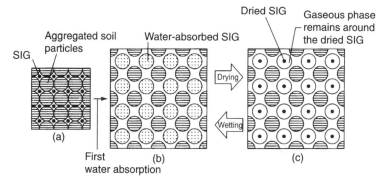

(a): SIG is mixed with soil and the SIG particles are dispersed among the soil particles.
(b): The total volume increased due to the water absorption. although the liquid phase increases, the gaseous phase does not change, resulting in a soft soil.
(c): SIG releases water and shrinks by drying. However, The total volume does not reduce noticeably. Thus, the majority of the SIG phase remains as the gaseous phase.

Fig. 7   The structure model of SIG-mixed soil.

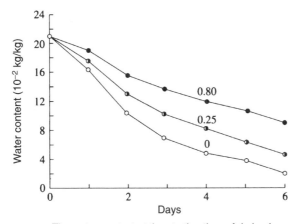

The water content at the starting time of drying is the same (21.1×10^{-2} kg/kg, 119g per pot).
The number in the figure is SIG-mixing ratio (g/kg)

Fig. 8   Evaporation of water from SIG-mixed sedimentary soil.

itself will dehydrate and shrink by drying. Hence, the gas phase increases by the amount of the excess reduction of the SIG volume over the total volume. Thus, SIG not only increases the water-holding capacity of soil but also aerates the soil by increasing the ratio of the gas phase.

4.3   Evaporation Characteristics from SAP-mixed Soil

Figure 8 shows the evaporation of water from the SIG-mixed soil with the same initial water contents [10]. As the SIG-mixed content increases, the amount of evaporated water decreases. Figure 9 shows the average water

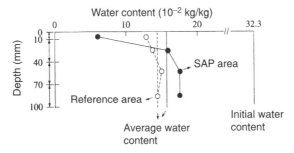

The mixing ratio is 2 g/kg and drying period is 3 days

Fig. 9   The degree of dryness at various depths of SIG-mixed soil.

content of a soil column a well as the water contents at various depths. The SIG-mixed soil dried more than the reference at the surface layer, whereas the deeper layers dried less [11]. This can be explained as follows:

i)   the cumulative evaporation of water from swollen SIG is less than the evaporation from the free-water surface, which is caused by the lack of water transport from the inner part of the polymer by the microphase separation of the polymer; and

ii)   the transport of the liquid-phase water decreased due to the reduction of particle-particle contact by the pores created by the drying and shrinking of SIG.

When water is sprayed onto a SIG-mixed soil, the penetration of water is slower than in the reference soil. This is caused by the water-holding capacity of SIG, which is larger than the soil itself.

4.4   Other Properties

4.4.1   Permeation Coefficient of Water at Saturation

Table 4 lists the permeation coefficient of water at saturation of a SIG-mixed soil [12]. In all cases, the permeation coefficient of water is high (greater than 10^{-3}) or regular (10^{-3}–10^{-5} m/s) and is slightly larger than the area without SIG.

4.4.2   Variation of the Soil Temperature

The variation of soil temperature of a SIG-mixed soil is smaller than the reference soil (see Fig. 10) [14]. This is because the water held increased by the mixing of SIG and, as a result, heat capacity increased.

4.4.3   Holding of Nutrients

Table 5 lists the elimination ratio of salts from the SIG-mixed soil [11]. The elimination ratio of the cations NH_4^+ and K^+ decreased by mixing

Table 4   The permeation coefficient of water of SIG-mixed soil.

Soil	SIG content (g/kg)		
	0	**0.2**	**0.5**
Sandy soil	3.7	10.4	6.3
Sedimentary soil	0.8	1.6	1.0
Volcanic soil	9.7	30.9	33.0

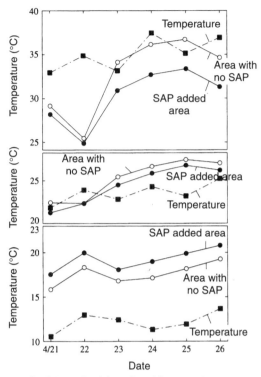

A mixture of poly(acrylic acid) type polymer
and pulp is mixed at a concentration of 25g/kg soil

Fig. 10   Variation of soil temperature in a superabsorbent polymer-mixed soil (Takeuchi [14])

Table 5   Elimination ratio of salts from SIG-mixed soil[a].

Soil	Treatment	NH$_4^+$-N	NO$_3^-$-N[b]	K	P
Sandy soil	SIG mixing	5.7	2.9	3.4	36.0
	Reference	7.4	2.7	4.5	17.7
	Difference[c]	−1.7	0.2	−1.1	18.3
Volcanic soil	SIG mixing	25.3	1.4	12.0	0
	Reference	32.4	2.4	16.2	0
	Difference[c]	−7.1	−1.0	−4.2	0

[a]Total elimination/initial amount (%). However, the elimination of elements from the soil that has no elements added has been corrected.
[b]NO$_3^-$ elimination is the ratio to the initial NH$_4^+$ value.
[c]SIG mixed – Reference.

Table 6 Absorption of nutrients by vegetables in SAP-mixed soil.

Vegetable	SAP content	Dry weight of the part above the ground (g/kg)	Amount absorbed[a]			Absorption ratio[b]	
			N	P_2O_5	K_2O	P_2O_5	K_2O
Tomato	0	4.2 (100)	71 (100)	29 (100)	55 (100)	41	77
	0.2	6.4 (152)	103 (145)	53 (183)	92 (167)	51	89
	0.5	9.8 (233)	132 (186)	78 (269)	139 (253)	59	105
Cucumber	0	4.7 (100)	171 (100)	45 (100)	139 (100)	26	81
	0.2	6.8 (145	168 (98)	62 (137)	183 (132)	37	109
	0.5	7.8 (166)	130 (76)	76 (169)	181 (130)	58	140

[a]Against the dry weight of the part above the ground.
[b]The ratio against the N absorbed (%).

SIG. Although the elimination ratio of the anion NO_4^+ did not show significant difference, the elimination of P was accelerated; P is difficult to move in soil and the degree of utilization is low. Thus, this increased elimination ratio may improve the usefulness of P. These phenomena are due mainly to the negative charge of SIG.

(4)   Absorption of Nutrients   The measurements of the absorbed amount of three nutrition elements (N, P, and K) for the tomato and the cucumber are listed in Table 6 [11]. The absorption of all three elements in the SIG-mixed area was higher than that in the reference area. Also, the absorption of P_2O_5 and K_2O in comparison to N was higher in the SIG-mixed area.

In general, a plant that undergoes water stress has higher N content than one that does not. The three essential plant elements are less than would be found in a healthier plant, where the ratio of P and K to N is lower than in a stressed plant. Accordingly, any plant grown in a SIG-mixed area has endured less water stress.

REFERENCES

1   Adachi, (1985). *Preservation of Soil, Water Quality and Agricultural Resources*, Hakuyu-sha, pp. 31–60.
2   (1989). Functional polymers. *Functional Materials*, February Issue, pp. 59–60.
3   (1992). Enhancement of water holding capacity of soil. Association of Superabsorbent Polymer Technology, pp. 12–15.
4   Tonyama, (1988). *Zairyo Gijutsu* **6**: 389.

5 Ouchi, S. *et al.* (1989). *J. Soc. Soil & Fertilizer Sci., Jpn.* **60**: 15.
6 Kitamura, S. *et al.* (1980). *Sumitomo Chemicals Technical Bulletin*, 1980-I. 1.
7 Ouchi, S. *et al.* (1991). *J. Soc. Soil & Fertilizer Sci., Jpn.* **62**: 487.
8 Masuda, F. (1987). *Superabsorbent Polymers*, Kyoritsu Publ., pp. 18–21.
9 Fushimi, T. (1990). Development of superabsorbent polymers. *Collection of Application Ideas*, Kogyo Chosa-kai, pp. 29–31.
10 Ouchi, S. (1992). *Pollution of Soil and Soil Modifier, Effective Use of Microbes in Soil*, Kogyo Gijutsu-kai, pp. 135–161.
11 Ouchi, S. *et al.* (1990). *J. Soc. Soil & Fertilizer Sci., Jpn.* **61**: 606.
12 Ouchi, S. Behavior and Influence on Plants of Superabsorbent Polymers made of Vinyl Alcohol and Sodium Acrylate (1994) pp. 72–80.
13 Yokoi, H. and Nakatani, N. (1982). *Physical Property Measurement of Soil*, Yokendo, pp. 270–276.
14 Takeuchi, Y. (1986). *Sakyu Kenkyu* **31**: 35.

Section 2

Application of Superabsorbent Polymers in Japanese Agriculture and Greening

SEIGO OUCHI

1 INTRODUCTION

Application of superabsorbent polymer (SAP) in agriculture and greening can be largely divided into the following two methods. The first method is to mix dried SAP into soil. Upon watering, SAP absorbs and holds water, showing the effect of increased water-holding capacity. Practical applications include:

i)   mixing SAP into the culture bed for rice, vegetables, and saplings;
ii) mixing SAP into sand and water dropwise to cultivate vegetables using the so-called *water-saving culture*; and
iii) mixing SAP into the culture bed for isolated growing of vegetables, growing of mushrooms, and other artificial beds.

The second method is to use the water-absorbed SAP as it is. Examples include:

276

i)   the fluid-seeding method in which seeds are planted directly in hydrogel;
ii)   planting seedlings of trees and vegetables with hydrogels adsorbed directly into the root;
iii) spraying the mixture of lawn seed, organic materials, and SAP onto the slope.

Superabsorbent polymers are also mixed into resins and processed into sheets for nurseries, artificial peat moss, vegetable wrapping sheets, and anti-frost film. For details, readers are referred to Chapter 3 of this handbook.

2   METHODS TO MIX SAP INTO SOIL

2.1   Mixing into Vegetable Culture Beds

If SAP are added to vegetable culture beds, the water-holding capacity and aeration improve and healthy seedlings are produced. The seedlings produced by this method will grow faster in the field and form vegetables earlier. It is especially useful for the culture methods which use a small amount of soil per plant, for example, soil block, paper pots, and market packs. The growth of cucumbers at various water frequencies is shown in Table 1. The greater the SIG content and the shorter the period between watering, the taller the plants.

The weights of both live and dried plants are also greater. If the growth of the plant at the top is evaluated by the ratio between the weight of the top plant and the height of the plant, then there is almost no difference with respect to the watering frequency. Specifically, the growth of the plant in the SIG-mixed area equals that in the reference area and the height and weight were in proportion to each other. Twelve other kinds of vegetables with proper watering showed the sample healthy growth effect [1]. The time necessary to the first flowering period for tomatoes was shortened and thus reproductive growth was also accelerated.

2.2   Water-saving Culture

In the water-saving culture, SAP polymer is added in sand, water is added dropwise, and vegetables are grown with a smaller amount of water than in conventional methods. It is especially useful in the dry areas or semidry areas where there is low precipitation. Figure 1 shows the results obtained

Table 1 Growth of cucumbers at various watering frequencies using SAP-mixed soil.

Watering frequency	SAP content (g/kg)	Plant height	Weight of fresh product		Weight of dried product		T/R ratio (dried product)	Drying ratio (%)		Amount of top dried product/height of plant (mg/cm)
			Top	Root	Top	Root		Top	Root	
Daily	0	(21.8)	(19.4)	(1.90)	(2.02)	(0.13)	15.5	10.4	6.8	93
	2	174[b]	156[a]	240[a]	154[b]	315[a]	7.6	10.3	9.0	82
	5	202[b]	190[a]	351[b]	188[b]	338[b]	8.6	10.3	6.6	86
Once every two days	0	85	83	118	76	115	10.2	9.5	6.7	83
	2	138[b]	128NS	173[a]	123NS	231NS	8.3	10.0	9.1	83
	5[a]	147	144[b]	267[b]	132[b]	246[b]	8.3	9.6	6.3	83
Once every three days	0	61	56	76	66	54	19.0	12.2	4.7	98
	2	95[b]	88NS	157NS	84NS	200NS	6.5	9.9	8.7	82
	5	125[b]	118[b]	263[a]	125NS	277NS	7.1	11.1	7.2	94
Once every four days	0	60	48	50	47	69	10.4	10.0	9.5	72
	2	97[a]	83NS	146NS	80[a]	262NS	4.7	10.0	12.2	76
	5	111[b]	92[a]	226[b]	90[b]	277NS	5.0	10.2	8.4	75

( ) is a measured value and the others are the ratio against this value (%).
[a]There is a difference at [b]: 1% and [a]: 5% confidence level in comparison to the reference sample, whereas NS indicates no difference within the confidence level.

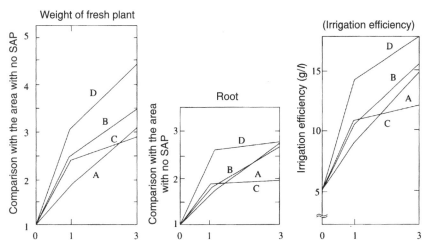

A-D indicates four different SAP
Irrigation efficiency = (weight of the top and root of live plant/total amount of water supplied).
Amount of water provided is the same in all areas and it was 1.03 l/m from a watering hose/daily.

Fig. 1 The influence of SAP mixing on the growth of turnips and irrigation efficiency.

by Takeuchi *et al.* [2] using turnips at Tottori Dunes. By mixing SAP into sand at the 1–3/kg level, the growth of the turnips was promoted. The plant at the top was by 1.7 to 3.1 times larger and the root 1.7 to 2.6 times heavier than the reference plants. The irrigation efficiency improved by 1.6 to 2.6 times. The results obtained in Mexico and Egypt are given later in this subsection.

2.3 Transplantation of Trees

Trees planted in medians, parks in cities, and in sand-blocking forests are generally left alone without irrigation except at the time of transplantation. The transplanted seedlings may die due to lack of water during the summer after transplantation because their root systems are not sufficiently established. In order to prevent this, SAP are beginning to be used at the time of transplantation.

Figure 2 shows the growth of new shoots and the number of leaves of laurel trees as a function of time when appropriate irrigation was provided [3]. The growth in the SIG 2 g/kg area was the most rigorous, whereas the

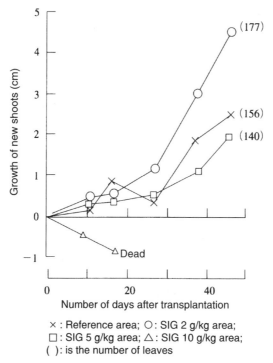

Fig. 2   Growth of new shoots and the number of leaves of laurel trees in SIG-mixed sandy soil.

5 g/kg area was less than the reference area. The trees in the 10 g/kg area died 17 days after transplantation. This is probably because of an excess amount of SIG that affected aeration of the soil.

The shoulder of a road was raised and SIG was mixed into soil to transplant Hirado Azaleas. The growth rate of the azaleas subjected only to natural precipitation is shown in Table 2 [3]. Six months after transplantation, the proportion of azaleas with less than 10% dead leaves is apparently higher than that of those in the reference area, indicating a higher survival rate. A similar result is obtained about fourteen months after transplantation when the azaleas with no damage are compared.

The proper concentration of SIG for trees is approximately 1–2 g/kg (g/l), which is much less than the 1–10 g/kg for vegetables. This is probably because trees have a large root ball and tree-root growth is slower than roots for flowers and vegetables, leading to oxygen deficiency.

Table 2   Growth of Hirado azaleas.

Date of investigation	Growth	SIG content (g/l)			
		0	**0.5**	**1**	**2**
October 28, 1987	○ Dead leaves <10%	68	92	99	89
(approximately 6 m later)	△ Dead leaves were 10–90%	23	7	1	9
	× Dead leaves were 90%	9	1	0	2
June 14, 1988	○ Healthy growth with no damage	49	68	88	82
(approximately 14 m later)	○ Slightly damaged	21	16	5	12
	△ Size of the bush is apparently smaller than at the time of transplantation	16	9	–	2
	× Dead: the dead tree remained	6	2	–	1
	Unknown: dead or stolen	8	5	7	3
	Number of samples investigated	190	100	96	98

2.4   Mixing SAP into Media of Bench Culture

In greenhouse agriculture, plant injury due to continuous cropping as a result of soil disease is a serious problem. In order to prevent this, a bench culture is created whereby the culture media are kept from contact with other soil. In this case, the amount of soil media is controlled, and hence the use of SAP is being considered. According to Watanabe [4], the number of days required for ripening strawberries shortened and the amount harvested increased as a result of mixing SAP into bench culture media. Okamoto *et al.* [5] also reported a crop increase when SAP was used for early or late harvesting of tomatoes.

When Koraishiba (low grass) is planted on the bench culture bed, it was considered necessary to use a medium thickness of 15 cm. However, Koshimizu *et al.* [6] demonstrated that approximately half of the traditional thickness was sufficient for healthy growth (see Fig. 3).

2.5   Fluid Seeding

Fluid seeding is a method in which seeds are mixed with hydrogel and after they germinate, they are planted in a field [7]. As the fluid, a SAP is used. The requirements for the fluid are:

i)   it possesses appropriate floating and suspending characteristics for the seed of interest;

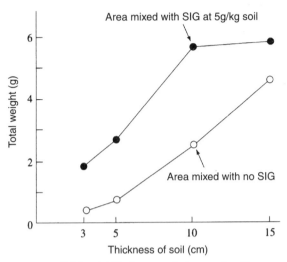

Total weight is the sum of stem, and roots.

Fig. 3   Growth of a lawn as a function of soil thickness.

ii)   it has appropriate fluidity; and
iii) it is harmless for the germination and growth.

Fluid seeding is used in Japan for fragile vegetables, such as spinach, garland chrysanthemum, and those vegetables with uneven germination as a result of seeding during a hot and dry summer, such as carrots. It is also used for flowers, such as the Turkish balloon flower.

Figure 4 shows the germination efficiency of a fluid-seeded carrot with varied irrigation frequency [8]. The fluid-seeded carrots with daily irrigation showed the highest germination efficiency and growth was also healthy. However, the areas irrigated only once every three days or not irrigated at all showed low germination efficiency and are not practical.

Thus, as is shown, appropriate irrigation is indispensable. On the other hand, it is known that if the thickness of the hydrogel in the soil exceeds 3 to 5 mm, the germination efficiency of spinach decreases [7]. This is because spinach requires approximately 60 to 100 *l* of water. Therefore, this method is not appropriate for crops that are planted over large areas.

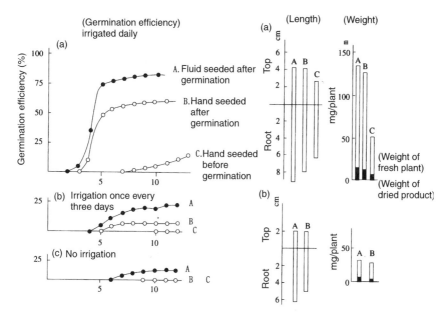

The amount of irrigation for (a) and (b) is 60% of saturation water capacity.
The germination of A and B was done by immersing the seeds in water
for one day and kept indoors for three days (the length of shoot is
approximately 5 mm).
To the fluid prepared by mixing 1 l of water and 2.5 g of SIG, 60 m l of the
seeds were mixed.

Fig. 4   Germination efficiency and growth of fluid-seeded carrots.

2.6   Seed Coating and Seed Immersion

In the seed-coating method a superabsorbent polymer is adsorbed around
seeds, dried and then seeded. This method is now gradually being used in
dry and semidry areas, and it utilizes even a small amount of precipitation
effectively.

According to Wu [23], the germination efficiency of eight kinds of
crops (wheat, corn, peanuts, cotton, nutmeg, tomato, watermelon and beet)
improved by 15% when the seeds sufficient for 1 μ (6.7 a) are coated by 25
to 125 g of SAP. In Hepei Province (China), the weight of dried corn 72
days after germination increased 45 to 118% over the reference area [9].

The survival rate of sweet potatoes after immersion into hydrogel
improved on an average of 23 locations by 23.7% [23]. It is also reported
that, if trees, such as willow, poplar, and oak, are to be transplanted, their

Table 3 Growth of grass by spray seeding.

Number of days after spraying	Number of grasses/25 cm^2						
	30	40	50	60	75	80	90
Area							
SIG area	12	11	0	6	6	6	0
Reference area	6	0	0	2	3	2	0

Samples outdoors   ↑ Irrigation   ↑ Stop irrigation
No irrigation afterwards
——→   ——→ Second experiment
First experiment

immersion into a hydrogel prevents drying during transportation and improves their survival rate [9].

In Japan, there is a report that growth of new leaves and roots is inhibited if trees are transplanted during an inappropriate season after the root ball has been immersed into a hydrogel, although it does help prevent wilting of the tree [10]. Thus, this method requires further evaluation to determine its effectiveness.

2.7 Spray Seeding

In this method, a mixture of lawn seeds and other materials, such as soil, organic compounds, fibrous materials, fertilizer and water, is sprayed onto exposed slopes along roads to prevent erosion caused by road construction. For this, SIG was added to the forementioned mixture at a concentration of 2 g/kg and sprayed onto a slope at 3 to 4 cm thickness. After one month, a layer was cut out and the grass was grown indoors with repeated drying and irrigation. The result is shown in Table 3 [11]. The result is that the SIG area has more grass and has a longer survival period compared with the reference area.

3 CONCLUSION

The most serious problem of SAP is its cost, which is 2200 yen per kilogram [12]. Assuming that it is used only near the plant (50% of the field), it requires 500 kg per 10 a, at a cost of 110,000 yen. The expected effect is the improvement of water-holding capacity, softness of the soil, and slight improvement of fertilizer holding. However, the effect lasts only about six months to one year. Hence, unless the end product is quite expensive, it cannot be justified. If the cost of compost is assumed to be 20,000 yen per one metric ton, the cost spent for SAP is equivalent to 11 t of compost per 10 a of field. This amount is several times the usual application amount of 2 to 3 t/10 a. In the case of compost, there is a slight improvement in water-holding capacity, improvement in microorganism flora, and a fair number of other elements. Judging from this, even though the price of SAP is lowered to one-fifth its current level, it is still impractical. Especially, the majority of countries with large areas of dry lands are developing countries where it will be more difficult to use such material due to its cost.

In Japan, SAP has been used in small quantities in the area of mostly high value-added and stable cost products, including culture of seedlings, artificial soil materials, fluid seeding, seed coating, and immersion. In the future, the use in these areas will continue.

In China, 52% of the total area, that is, 9.6 million km^2, is dry and semidry [13]. The land thus is used for seed coating and immersion methods. Because 1 kg per ha is used, the use in 1989 was 33 t ($= 0.33$ million ha). Even if only 1% of the land uses SAP, it will use 10,000 metric tons. This suggests great future potential.

REFERENCES

1 Ouchi, S. *et al*. (1990). *J. Soc. Soil & Fertilizer Sci. Jpn.* **61**: 606.
2 Takeuchi, Y. *et al*. (1984). *Sakyu Kenkyu* **31**: 100.
3 Ouchi, S. Behavior and Influence on Plants of Superabsorbent Polymers made of Vinyl alcohol and Sodium Acrylate (1994) p. 72.
4 Watanabe, K. (1982). *Report of Shiga Prefecture Agricultural Institute*, No. 117.
5 Okamoto, K. (1989). *Report of Shiga Prefecture Agricultural Institute*, No. 30 pp. 31–38.
6 Koizumi, H. (1986). *Proc. Assoc. of Lawn, Jpn.*, p. 67.
7 Sumitomo Chemicals. (1982). *Technical Bulletin on Fluidic Seeding Technique*, pp. 1–18.
8 Ouchi, S. *et al*. (1993). *J. Soc. Soil & Fertilizer Sci., Jpn.* **64**: 435.
9 Wu, D. (1990). *J. Produce, China*, Academia Sinica, Agriculture, Bejing, China, **119**: 22.
10 Maeda, H. *et al*. (1988). *Report of the Institute for Civil Engineering, Jpn.*, pp. 40–52.
11 Ouchi, S. (1992). *Soil Pollution and Soil Modifier: Effective Use of Microbes in Soil*, Kogyo Gijutsu-kai, pp. 135–161.
12 (1993). Construction Materials Cost Document, July.
13 Maki, T. (1993). *Frontier of Forestation of Desert*, Shin Nippon Publ., pp. 49–72.

Section 3
Application of Superabsorbent Polymers to Dry Land

RYOICHI TOKIUMI

1 INTRODUCTION

Japan is fortunate enough not to have major dry lands. However, there are many countries in the world with dry or semidry lands. Despite those hardships, agriculture and   are performed, deforestation continues, and there is an increase in desert area. There are reports in such areas where production of vegetables and fruits or greening by trees is performed using superabsorbent polymers (SAP). Here, some of those representative examples will be introduced.

2 MEXICO

Toyama *et al.* [1] reported attempts to grow vegetables in Baja Peninsula in Mexico using superabsorbent polymers. These attempts were based on their prior experience of vegetable growing in Tottori Dunes. The desert in Mexico is in Guerrero Negro, which is situated in the middle of Baja

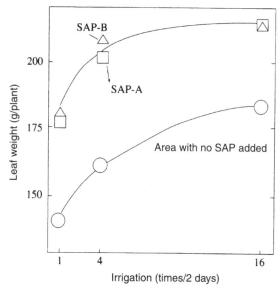

Fig. 1 Growth of Shanghai cabbage in SAP mixed sand as a function of the irrigation frequency.

Peninsula and is known as the world's largest salt field; the annual precipitation there is 78 m/m. Funded by a grant from the Ministry of Education, and after a preliminary investigation, a water-saving culture of vegetables using superabsorbent polymers was prepared. In the case of one vegetable, Qing Gang Cai (Shanghai cabbage), shown in Fig. 1 there was a significant increase in production. Thus, if the total amount of irrigation is the same, then the higher the amount of irrigation, the greater the production of the vegetable.

3   EGYPT

As part of the so-called *Green Earth Project* supported by Official Development Assistance (ORD) of the Ministry of International Trade and Industry, the "Joint Research and Development Project on Composite Water Holding Substances for Arid Regions," [2] was undertaken in Egypt from 1988 to 1994. This project was coordinated by the Japan Association for the Development of Deserts, following the creation of the Letter of Memorandum with the Ministry of Agricultural Land Development of Egypt. The project aimed at developing agriculture in dry lands

Fig. 2 Cultivation of melons and spinach on farms in the Bustan district (Egypt).

(water-saving agriculture), prevention of desertification, and technology transfer of advanced Japanese agricultural techniques through the development of a composite of a superabsorbent polymer and clay.

A joint research farm was established in the desert development area in the Bustan district located approximately 150 km northwest of Cairo, Egypt. Water-saving agriculture and deforestation experiments were carried out for three years, from 1990 to 1993, under the leadership of the Dry Land Investigation Center, Tottori University in Japan. The Society for Research and Development of Superabsorbent Polymers, established primarily by the makers of superabsorbent polymers, assumed responsibility for the development of the superabsorbent composite. The water-saving effect using superabsorbent composite was demonstrated, and a pilot plant for the composite with the capacity of 2 t/day was constructed and transferred to Egypt.

At the research farm (area approximately 8.4 ha, with five vinyl-covered greenhouses) in the Bustan district using the superabsorbent composite, melons, pak choi or spinach were grown in a greenhouse. In the field, kazalina pine and metasequoia were grown as wind-blocking trees; in addition, fruits such as oranges and grass as food for animals were grown.

Agriculture in Egypt was generally easier than at Tottori Dunes. One of the reasons may be the large difference between the daytime temperature and the nighttime temperature, which is low. When the superabsorbent composite was added by 0.1%, melons increased 30% by weight. Similarly, leaf vegetables increased by 10 to 30% (see Fig. 2).

4   CHINA

China is one of the major agricultural countries in the world, it has a population of 1.2 billion, and it is experiencing rapid economic development. Although China possesses vast farmlands, approximately 40% of the country is semidry land. In these areas, the dry season is from early spring to early summer, which is the time of seeding and transplantation of seedlings. Hence, the use of superabsorbent polymers might prove effective.

Lead by Academia Sinica, Agricultural Sciences, a national project entitled, "Investigation of Superabsorbent Polymers in Agriculture and Their Effectiveness," was undertaken from 1985 to 1990. Various agricultural organizations in the nation carried out application studies using several kinds of vegetables, fruits and trees. They made progress in achieving their goal of developing widespread applications.

Superabsorbent polymers are used mainly to coat seeds and seedlings during transplantation. The results of the studies have been organized in a monograph [3] by Wu who was the project leader. In 1987, a scientific and promotional film, *Superabsorbent Polymers*, received high acclaim and was translated into eight different languages. The Association of Superabsorbent Polymer Technology commissioned Academia Sinica and carried out a study, "Application of Superabsorbent Polymers in Agriculture," in the vast farmlands in Hepei Province, China. The crops used were the seedlings of cotton, corn, wheat, and grape.

Fig. 3   Samples of wheat seedlings (Hepei Province, China).

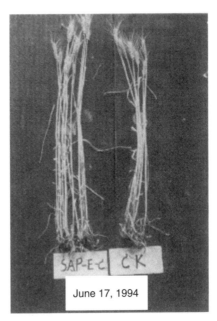

Fig. 4   Samples of harvested wheat (Hepei Province, China).

Fig. 5 Growth of summer-seeded cotton (Hepei Province, China).

Fig. 6 Harvesting of summer-seeded cotton (Hepei Province, China).

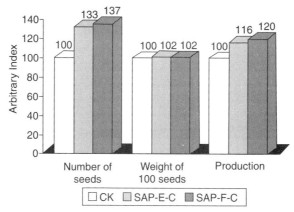

Fig. 7 Investigation of summer-seeded cotton production in a large testing area.

4.1 Seed Coating (Corn, Wheat, etc.)

At the ratio of seed : superabsorbent polymer : water $= 100 : 1 : 25$–50, the coating liquid is prepared. Avoid aggregation formation and dry to a homogeneous coating (see Figs. 3 and 4).

4.2 For Transplantation of Seedlings and Grafting (Saplings of Grape and Sweet Potatoes)

A colloidal solution at the ratio of superabsorbent polymer : water $= 1 : 100$–300 is prepared. Seedlings or grafts are immersed in the solution and briefly dried in the shade. They are immediately transplanted.

4.3 Fluid-seeding (Wheat, Spinach, etc.)

Seeds are germinated to the length of shoots less than 2 cm. These seeds are mixed into the colloidal solution of superabsorbent polymer and carefully sprayed on the field by a fluid-seeding machine.

The following is an example of an application study on cotton that was carried out in 1994 (see Fig. 5). A yellow-soiled farmland of Hepei Province was used to set up a large testing area of approximately $330\,\text{m}^2$. Treatment was applied to summer-seeded cottonseeds at the SAP concentration of 1.5% by weight of the seed. The seeding was performed May 10, 1994. Intermediate investigation of growth was conducted twice and final harvesting took place on November 28, 1994 (see Fig. 6). Figure 7 illustrates the investigation results of cotton production. As a result of using SAP, the production increased by 16 to 20%.

REFERENCES

1 Toyama, M. *et al.* (1986). *Sakyu Kenkyu* **33**: 24.
2 (1988–1993). Official Development Assistance (ORD) of the Ministry of Trade. The Joint Research and Development Project on Composite Water Holding Substances for Arid Region, Reports No. 1–No. 6, Association of Desert Development, Jpn.
3 Wu, D. (1991). *Superabsorbent Polymers and Agriculture*, Academia Sinica Press, Beijing.

CHAPTER 6

Civil Engineering and Construction

Chapter contents

Section 1
Water-Swelling Rubbers

TATSURO TOYODA

1 INTRODUCTION

Recent construction in urban areas to deepen existing tunnels and under rivers and sea to accommodate road systems in the face of serious land shortages has been aided by newer drilling techniques. As a result, water pressure demands on sealants have naturally increased and the sealants must be extremely strong. See Table 1 for these requirements.

Table 1 Requirements for sealants.

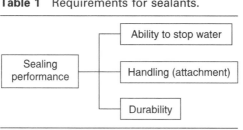

Table 2   Water-swelling rubbers in the construction market.

Application areas	Sealants
Tunnel	Sealants for sealed segments
	Water blocking plate
	Packing for bolt holes
Water purification plants and sewage	Sealants for box carbert
	Sealants for manholes
	Water blocking plate
	Sealants for fume pipes
Other secondary products of concrete	Sealants of water reservoirs for fire prevention
	Sealants for U-shaped groove
	Sealants for CC BOX

2   APPLICATION OF WATER-SWELLING RUBBERS AS SEALANTS

Table 2 lists those applications for water-swelling rubbers in the construction market. Most of the sealants used for tunnel construction are made of water-swelling rubbers. Construction and civil engineers and sealing materials manufacturers have been working together to develop effective sealing materials.

In areas other than for used as tunnel section sealants, water-swelling rubbers are use far less than nonswelling rubbers because they are very expensive. Thus, application is naturally limited to those situations in which sealant conditions are critical.

3   TYPES OF SEALANTS

Table 3 lists the types of sealants, which can be divided into nonswelling and water-swelling sealants. As can be seen by a comparison of water-swelling and non-swelling rubbers in Table 4, water-swelling rubbers are superior in water-sealant ability to nonswelling rubbers but expense is a drawback.

3.1   Nonswelling Rubber-Type Sealants

Of the nonswelling sealants, sponge rubber has been used for many years. Due to its flexibility, initial compression with even a low load allows the material to adjust to the contours of concrete structure surfaces. A sealant

Table 3   Types of sealants.

Non-swelling sealants		Unvulcanized butyl rubber sealants Solid rubber sealants Sponge rubber sealants Sealants
	Composite sealants	Solid rubber + sponge sealants Solid rubber + unvulcanized rubber sealants Solid rubber + sponge rubber + metal sealants Sponge rubber + unvulcanized rubber sealants
Water-swelling sealants	Superabsorbent type	Solid rubber sealants Sponge rubber sealants
	Urethane type	Urethane elastomers sealants Rubber blends sealants
	Composite sealant type	Water-swelling rubber + nonswelling solid rubber sealants Water-swelling rubber + metal sealants

Table 4   Comparison of water-swelling rubber and nonswelling rubber.

Type	Ability to stop water	Cost
Water-swelling rubber	○	×
Nonswelling rubber	×	○

rubber ring for a fume pipe is made by extruding and then molding solid rubber. For the concrete structure joints water-sealant plates are made of flexible elastomers and hard rubbers. In recent years, a composite sealant made of unvulcanized butyl rubber and sponge rubber has often been used because it takes advantage of these two rubbers.

3.2   Water-Swelling Sealants

3.2.1   Superabsorbent Polymer Type

This type of sealant is made by compounding a rubber with a super-absorbent polymer. The compounded materials are later vulcanized. Sodium acrylate polymer and copolymers of isobutylene and maleic anhydride are used as the superabsorbent polymer. In general, super-absorbent polymers are selected for their compatibility with the rubber that is used. If compatibility is poor, the superabsorbent polymer will dissolve into the water and the targeted swelling ratio will not be achieved.

3.2.2 Urethane Type

(1)   Urethane Elastomer-Type Sealants
These polymers are created by crosslinking a hydrophilic polyol using polyisocyanate and a water-swelling urethane elastomer results. An important attribute is that the swelling characteristics are affected very little by water.

(2)   Sealants from Blends of a Rubber and a Water-Swelling
Urethane Elastomer
These materials are rubber blends with water-swelling urethane elastomers. Because ordinary processing machines can be used, extrusion is possible.

3.2.3 Composite Sealants
These are sealants made of water-swelling and nonswelling rubbers. The properties of each type of rubber combine to improve water-stopping ability and lower costs.

4   SWELLING OF WATER-SWELLING RUBBERS

Figure 1 shows the swelling characteristics of a water-swelling rubber. A water-swelling rubber was immersed in tap water and the volume changes were followed as a function of time. After two days of immersion the

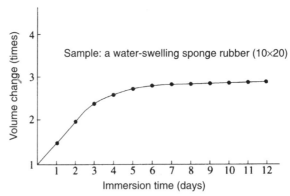

Fig. 1 Swelling characteristics of water-swelling sponge rubber.

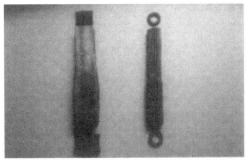

Water-smelling   Composite
rubber   water-smelling
rubber

Fig. 2   Water-swelling rubber.

original volume doubled and after seven days it reached saturation at three times its original value.

Figure 2 illustrates the swelling behavior of a water-swelling sponge rubber and a composite water-swelling rubber made of water-swelling sponge rubber and nonswelling solid rubber. They were partially immersed in tap water and swollen, thereby demonstrating the swelling behavior of the immersed portion.

5   WATER-STOPPING CAPABILITY OF WATER-SWELLING RUBBERS

5.1   Water-Stopping Properties of Water-Swelling Rubbers

Sealants basically seal by compression/repulsion. The sealing theory of a gasket is generally expressed by the following equation:

$$\sigma_g \geq m \cdot P_1$$

where σ_g is the effective contact area pressure of a gasket, m is the gasket coefficient, and P_1 is the water pressure.

Supplemental—Although the gasket coefficient (m) for a rubber sheet ($Hs = 90$) is 1.00 and for 3 mm-thick asbestos it is 2.50, it changes with surface conditions.

The compression repulsion force as shown in Fig. 3 is the repulsive force generated by the compression of a sealant. Figure 4 shows the relationship between stress relaxation and swelling pressure. Stress relaxa-

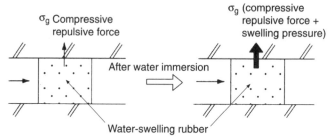

Fig. 3   Water-stopping principles.

tion of each nonswelling and water-swelling sponge rubber was determined by a stress relaxation testing machine.

The compression repulsive force of the nonswelling sponge rubber reduces over time. The compression repulsive force of the water-swelling rubber shows a temporary decrease at the beginning of measurement but an increase as time passes. This can be verified as shown in Fig. 1, in which compressive repulsive force is shown to increase as a result of the swelling pressure. Hence, water swelling is demonstrated to be useful for sealants.

5.2   Water-Stopping Experiment for Water-Swelling Rubbers

Figure 5 shows a testing device and Fig. 6 shows the test results on the water-stopping properties of a water-swelling rubber. For initial compression at 0% (compressive stress of 0 kgf/cm^2), water stopping pressure of 2 kgf/cm^2 was generated after five days.

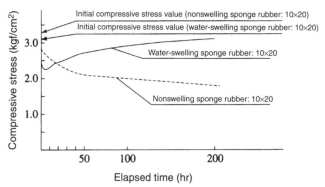

Fig. 4   Stress relaxation properties and swelling pressure.

Testing device

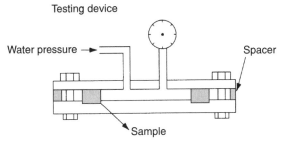

Water pressure →

Spacer

Sample

Test method
1. Water-swelling sponge rubber
2. Two compressions, 0 and 50%, were used
3. Immersion in water
4. Apply water pressure and examine the water leakage

Fig. 5 Testing device.

6 BASIC DESIGN OF SEALANTS

The properties of a sealant made of a water-swelling rubber were shown in the previous section. However, a higher performance sealant requires using the structure as described in section 6.1.

6.1 Addition of Sealant Groove

Figure 7 illustrates the structure of a sealant groove in which a water-swelling rubber is placed and swollen. The gasket's effective contact area

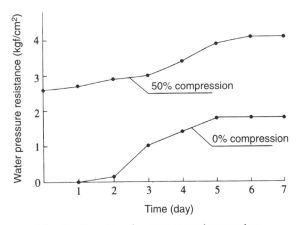

Fig. 6 Results of water stopping testing.

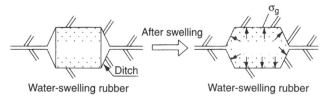

Fig. 7 An example of a sealant ditch.

pressure (compression repulsive force + swelling pressure) is confined to the groove and water-stopping function improves. On the segment of sealed construction technique, the groove is usually placed. However, due to mold design or construction difficulties, this may not be possible.

6.2 Use of Composites

If a ditch cannot be used on a concrete structure, composites made of water swelling and nonswelling rubbers will be effective. Figure 8 depicts an example of a composite sealant. The gasket's effective contact area pressure is confined within the nonswelling solid rubber and prevents diversion of the repulsive force, which improves the water-stopping function (there are many patents on this technology).

7 CONCLUSION

Application and usefulness of water-swelling rubbers as construction sealants have been described. Unfortunately, the market share of this material is low, even though most sealant used in the sealed technique adopt this material. This is due to expense. Because superabsorbent polymers and water-swelling urethane elastomers are costly compared to ordinary rubbers, water-swelling rubbers will be expensive. In the future, sealants manufacturers should lower their costs by increasing use of

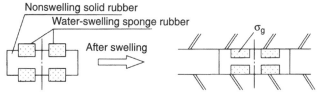

Fig. 8 An example of composite sealant.

superabsorbent polymers as well as blending them with nonswelling rubbers. It is hoped that engineers in the construction industry, and superabsorbent and sealant manufacturers cooperate and increase use of water-swelling rubbers.

Section 2

Prevention of Water Condensation

NOBUYUKI HARADA

1 INTRODUCTION

Japan has a very hot and humid climate. Prevention of water condensation is more important in Japan than in Western countries. Therefore, wood, a material with excellent water absorption and release characteristics, is generally used for housing. However, in recent years, there has been a greater emphasis on energy-efficient devices and an interest in doing something about moisture build-up in homes. This is particularly serious in apartments, where condensation on ceilings, walls, inside closets, and in window frames leads to fungal growth and mold and degradation of building materials. In this subsection, methods in which water condensation is prevented by using polymer gels, which have superior water absorption and release properties will be discussed.

304

2 MECHANISM OF WATER CONDENSATION AND METHODS FOR PREVENTING WATER CONDENSATION USING POLYMER GELS

Figure 1 shows the relationship between temperature and enthalpy as a function of moisture. When air at 20°C and with humidity at 70% encounters materials colder than 14°C, some of the water vapor condenses on the surface. Therefore, formation of water condensation depends on both the humidity and the air temperature. Basic approaches for prevention of water condensation involve:

i)   reducing the absolute humidity (water vapor pressure) of the air in the room;
ii) preventing the temperature of the surface from falling so that water condensation does not occur; and
iii) removing the condensed air.

Savings in both costs and labor are important. Therefore, effective methods to prevent water condensation involve effective water absorption followed by a release of these materials near the place where the water condensation has taken place.

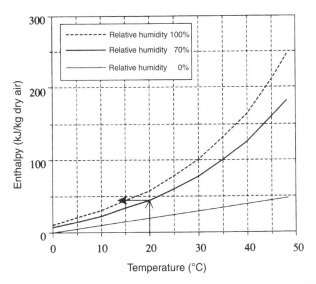

Fig. 1 Wet air isotherm. (Reference: totl pressure, 760 mmHg, 1 kg–dry air)

(a) Superabsorbent polymer sheet

(b) Superabsorbent polymer sheet
that has absorbed water and swelled

Fig. 2 Appearance of superabsorbent polymer sheets with and without water.

Polymer gels are materials that can absorb both moisture and actual water. Their excellent moisture and water absorption functions are widely used in sanitary products, agriculture, foods, and medicine. There have been many reports on those factors that influence moisture absorption behavior [1, 2]. In particular, many studies have been carried out on one of the best moisture absorption materials, superabsorbent polymers [3, 4].

An inexpensive and effective method uses a superabsorbent polymer sheet (see Fig. 2) [5] which contains polyacrylate type superabsorbent polymer made by Nippon Shokubai, to absorb moisture and water and release them when necessary. The mechanism of this method will be described.

3   CHARACTERISTICS REQUIRED FOR PREVENTING WATER CONDENSATION

The basic requirements for an excellent water condensation prevention material are:

i)   when the surface is below the dewpoint, it absorbs the condensed water on the surface;

ii) when the water-condensed surface climbs above the dewpoint, the moisture absorbed thus far is released and, at the same time, recovery of the function for repeated use takes place; and

iii) not only liquid water but also moisture in the room will be absorbed or released to minimize fluctuations in relative humidity.

Accordingly, the properties required for any material that can prevent water condensation include moisture absorption, humidity control, water absorption, and durability (antifungal prevention).

3.1   Moisture Absorptivity

Evaluation of moisture absorptivity can be made by measuring the moisture absorbed against the relative vapor pressure, namely, moisture absorption isotherms. The moisture absorption isotherms of a polyacrylate-type superabsorbent polymer are shown in Fig. 3.

The moisture absorption capability of a polyacrylate-type super-absorbent polymer is lower than that of inorganic moisture absorption agents, such as silica gel. In addition, moisture absorption significantly increased as the water vapor pressure increased. This is due to the excellent water absorption capability of the superabsorbent polymers. Hence, superabsorbent polymers are especially useful for moisture absorption at high humidity. Furthermore, by adding a moisture absorption agent, such as calcium chloride, a compound that has lower saturation

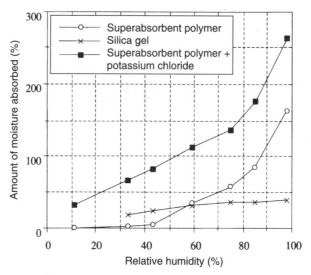

Fig. 3   Moisture absorption isotherms.

vapor pressure than the vapor pressure of the air, it is possible to obtain a moisture absorption agent suitable for both high and low humidity regions. However, in general, the water absorptivity of polyacrylate-type superabsorbent polymers decreases by ionic crosslinking. Hence, it is necessary to use a superabsorbent polymer that is resistant to metallic ions. For example, an anti-salt superabsorbent polymer [6] from Nippon Shokubai has excellent resistance to metallic ions. It absorbs a great deal of water even in the presence of multivalent metallic ions such as calcium. Thus it is possible to combine this polymer with a commercially available calcium chloride in order to reduce the absolute humidity in a room.

3.2   Moisture Adjustment Ability

Moisture adjustment ability can be evaluated by measuring the variation of relative humidity in a closed container in which the sample is placed at various temperatures. The humidity adjustment responsiveness of wood, which has excellent moisture absorption capability, and a superabsorbent polymer are compared. A 10-cm cube, which was preconditioned at a relative humidity of 60%, was placed in a closed container. The external temperature was varied from 25 to 35°C every hour and the relative

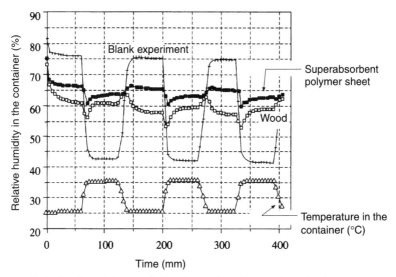

Fig. 4 Relative humidity variation in a closed container in the presence of moisture adjustment materials.

humidity and temperature in the container were recorded as shown in Fig. 4.

In an empty chamber it was demonstrated that the relative humidity of the chamber changed greatly as the temperature changed, varying from 75 to 42%. On the other hand, variations in relative humidity in the presence of wood or a superabsorbent polymer was approximately 3%. Hence, it is possible to use these materials to prevent water condensation in a room. The amount of the superabsorbent polymer used for the experiment per unit area was 1/30 of the wood used. Hence, by using the polymer, it is possible to develop a lightweight and thin interior decorative material. Figure 5 shows a printed superabsorbent polymer sheet on the wall of a closet. Water condensation in winter was significantly reduced in this apartment.

3.3 Water Absorptivity

The water absorptivity of a polyacrylate-type superabsorbent polymer can be adjusted at will by controlling the degree of crosslinking. Therefore, it is possible to prevent water condensation by tailoring the water absorptivity needed for the particular location of installation. Figures 6–8 show

Fig. 5 Example of an installation of a superabsorbent polymer sheet on a closet wall in an apartment.

the results obtained by the Japan Building Research Institute when superabsorbent polymers were used to prevent condensation. The water condensed at a constant temperature and humidity (heated side) was collected by the device shown in Fig. 6. The amount of condensed water was compared by changing the air temperature of the cooling side in cycles as shown in Figure 7. In the placebo experiment in which no superabsorbent polymer sheet was used, as much as $600\,g/m^2$ of

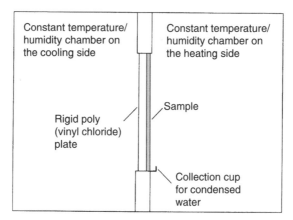

Fig. 6 A device to collect the condensed water in a constant temperature/humidity chamber on the heating side.

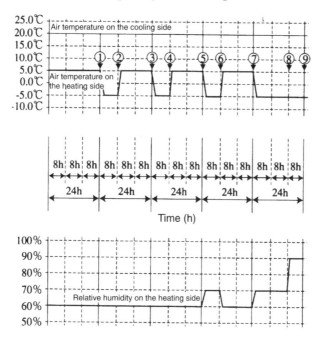

Supplemental • The ▼ symbol indicates the measurement
time of temperature, humidity, visual
observation of water condensation and the
weight of the condensed water.
• The time necessary to change the temperature
and humidity was approximately 0.5 –1h.
• The relative humidity of the cooling side was
not adjusted due to the low ambient
temperature.

Fig. 7 Temperature and moisture cycles used.

condensed water was collected. On the other hand, in the case of a superabsorbent polymer, no condensed water was observed, a consequence of the excellent water retention capability of the superabsorbent polymer. Figure 8 shows the amount of condensed water at the various measurement points as shown in Fig. 7.

3.4 Durability (Antifungal and Repeated Moisture Cycle Test)

When the antifungal property of a superabsorbent polymer sheet was evaluated (according to JIS-Z2911 standards) using a nonnutrient agar

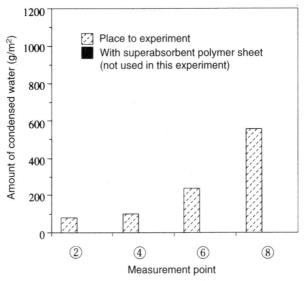

Fig. 8   Amount of condensed water at various measurement points.

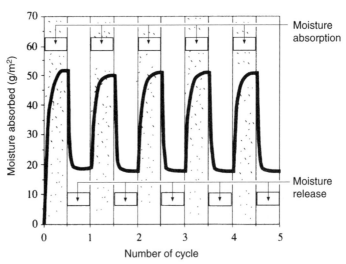

Sample: a superabsorbent polymer sheet (D-420)
Measurement condition:
  Moisture absorption at 25 °C, 90% RH for 4 h
  Moisture release at 25 °C, 60% RH for 4 h
Measurement times: 5 times

Fig. 9   Moisture cycle test.

culture, there was no fungal growth during the 28-day test period. However, under actual use conditions the adsorption of nutrients, which promote the growth of fungi, is possible. Therefore, it is useful to add an antifungal agent or apply a surface protective treatment. Figure 9 shows the results of a repeated moisture cycle test using a superabsorbent polymer. The moisture absorption and release characteristics were reported to be unchanged even during a long-term moisture cycle test that ran for one year.

4 APPLICATION EXAMPLE OF SUPERABSORBENT POLYMER SHEETS AS A PREVENTION MATERIAL FOR WATER CONDENSATION: APPLICATION TO SNOW DAM

According to the experimental results on "Underground Storage Using the Cooling Capability of Natural Snow" by The Association of Engineering Promotion, in which the humidity in the chamber was 90–95% for six months, the condensation of water was prevented by using superabsorbent polymer sheets on the ceiling [6]. When the superabsorbent polymer sheet was not used, stored vegetables rotted. However, the superabsorbent polymer sheet allowed storage of vegetables without any problems.

5 CONCLUSION

The application potential and an application example for a superabsorbent polymer used to prevent water condensation as a result of moisture

Fig. 10. Appearance of snow dam.

A large number of condensed water droplets at the portion where a superabsorbent polymer sheet is not used

Fig. 11 Underground storage ceiling.

Fig. 12 The ceiling of an underground storage area on which a super-absorbent polymer sheet is installed.

absorption and release characteristics have been discussed. Water condensation can be generated anywhere by differences in humidity or temperature. Using the properties of hydrogels to investigate how to achieve more comfortable living spaces is a goal of this author.

REFERENCES

1   Takizawa, A. (1972). *Polymers and Water*, Soc. Polym. Sci., Jpn., Ed., Sachi Shobo, Tokyo.
2   Hirai, Y. and Nakajima, T. (1988). *Hyomen* **26**: 123.

3 Kuwabara, N., Obata, N. and Tou T. (1987). *Kobunshi Ronbunshu* **44**: 15.
4 Ichikawa, A. and Nakajima, T. (1988). *Kobunshi Ronbunshu* **45**: 427.
5 Harada, N. and Shimomura, T. (1994). *Kogyo Zairyo* **42**: 30.
6 (1992). *Report on the Experiment on the Utilization of Cooling Energy of Snow*, Association of Engineering Promotion.

Section 3
Fireproof Materials

OSAMU TANAKA AND MITSUHARU OSAWA

1 INTRODUCTION

Earthquakes, lightning, fire, and one's father were historically cited as things the Japanese feared. In today's concentrated living environments, fires and earthquakes have the capability of destroying the social and societal fiber and creating a major catastrophe.

In Japan, the lessons learned from the complete destruction of several cities by fire during World War II, led to a building standard law that was established in 1950 in which fireproofing of cities was mandated. Since then, fireproof materials such as concrete are the major building materials. However, with a fourth modification of the building standard law in 1963, construction of skyscrapers became possible. Buildings made of steel rather than concrete then dominated.

Although steel does not burn, it does lose strength if it is exposed to the very high temperatures of fire. It is thus necessary to protect steel from these high temperatures in order to keep buildings safe. The material used to protect this steel from these high temperatures is a fireproof covering and any building made of any of these fireproof materials is considered to be fireproof.

316

Table 1 Steel temperature judgment standard in fireproof test.

	Steel temperature	
Heating time	**Average**	**Maximum**
Heating for 1 h	350 °C	450 °C
Heating for 2 h		
Heating for 3 h		

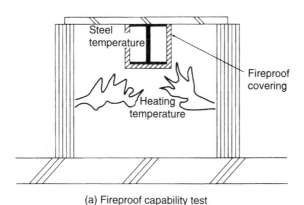

(a) Fireproof capability test

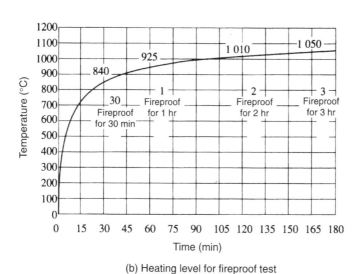

(b) Heating level for fireproof test

Fig. 1 Heating temperature for fireproof capability test.

Fireproof performances based on the purpose of a building, its size, area, and number of stories are specified in detail in Japan's building standards laws. The purpose of a fireproof covering is to prevent deformation and loss of lives when the building is exposed to fire for a certain period of time. In particular, if the material used is steel, it must be coated with a thermal insulation capable of absorbing enough heat that the temperature of the steel will not exceed 350°C. Both heating and steel temperatures have been specified by the fireproof tests used to formulate the building standard law. These are shown in both Table 1 and Fig. 1.

2   FIREPROOF COVERING MATERIALS

Fireproof covering materials in Japan before 1955 or so, were likely to be of sprayed asbestos form. However, as the buildings have become taller and larger, various fireproof covering materials began to be used, including sprayed, adhesive, tape and coating types.

Properties of fireproof covering materials include small thermal conductivity, abundant thermal absorption capability, and long-term durability. The materials that once dominated fireproof covering materials, sprayed asbestos, is a mixed material made of asbestos with Portland cement as the bonding agent. Asbestos is excellent for dispersion, fire resistance, and alkali resistance. It is a fibrous material with crystalline water that provides excellent thermal resistance and heat absorption. Unfortunately, because there is legitimate concern about the health problems caused by asbestos, other inorganic materials as shown in Table 2 are currently used.

Table 2   Materials used as fireproof coverings.

Portland cement	Carbon fiber	Bentonite
Alumina cement	Cepiolite	Aluminum hydroxide
Plaster	Wollastonite	Magnesium hydroxide
Rock wool	Mica	Calcium carbonate
Glass wool	Vermiculite	
Ceramic wool	Perlite	
Silica fiber	Talc	

3 APPLICATION OF SUPERABSORBENT POLYMERS

Burning is an aggressive oxidative phenomenon of flammable materials and gases that creates both heat and light. In order to suppress burning, it is necessary to:

i)   eliminate materials that can burn;
ii) suppress the temperature of the heat source below its flashpoint; and
iii) cut the oxygen supply.

Since ancient times, water has been used to extinguish fires. This is the most effective way to reduce the temperature of a heat source below its flashpoint and prevent the fire from spreading.

In general, fireproof covering materials have crystalline water in their structure or materials that exhibit endothermic reactions. Water-hardening mixtures of a cement with metallic oxides, such as aluminum hydroxide or magnesium hydroxide, are often used as fireproofing materials. The evaporation temperature of water is 100°C and as long as there is water, temperatures cannot exceed 100°C. All fireproofing has a time-based requirement in the standards, but those that have crystalline or free water available can actually prolong the time before the temperature of steel is increased. Generally, water-hardening cement spray contains 3–5% by weight free water as an equilibrium water content. This free water acts as an endothermic material. If this free water can be artificially held in large quantity, it can be used as an endothermic material. We developed an application of a superabsorbent polymer based on this concept. In the US and Europe, water is circulated in a steel tube during a fire to prevent the steel from increasing in temperature. A few buildings have adopted this method. However, the construction costs are significantly high, and thus, it has not generally been adopted.

If water is used as a fireproof covering material for the structural portion of a building, a major challenge is to develop materials that will not change that is, the amount of water will not be reduced for as long as 40–50 yr. In order to prevent the natural evaporation of water, it must be hermetically sealed by metallic material. Because water expands in volume and will explode if it is sealed over a long period of time, it is necessary to develop a thermally fusible covering material that leaks water at <100°C. For this purpose, a 3-layered covering material made of nylon, aluminum foil, and polyethylene has been developed. In the event the packaging material breaks, it will suddenly release water. A superabsor-

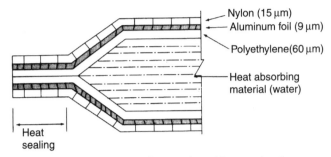

Fig. 2 Structure of heat absorbing packaging.

bent polymer chosen to be used as a heat absorbing packaging has solved this problem (see Fig. 2).

4 FIREPROOF COVERING MATERIAL, AQUACOVER

4.1 Heat Absorption Effect of Heat Absorbing Packaging

Heat absorption packaging adopts a continuous sheet shape that consists of an individual package 65 by 98 mm in size. In each package, 36.815 g of water and 0.185 g of superabsorbent polymer are included. Because there are 150 packages per each 1 m^2, the total amount of water per 1 m^2 is 5522 g/m^2 and the superabsorbent polymer is 27.75 g/m^2. The endothermic effect of these heat absorbing packages is obtained by calculating the amount of heat absorbed between an ambient temperature of 20°C and the temperature of 100°C which is

$$5522 \text{ g/m}^2 \times 1 \text{ cal/g} \times 80 \text{ °C} = 441.76 \text{ kcal/m}^2$$

and the amount of heat absorbed as water evaporates at 100°C is

$$5522 \text{ g/m}^2 \times 539 \text{ cal/g} = 2976.36 \text{ kcal/m}^2$$

and the heat generated by the superabsorbent polymer is

$$27.75 \text{ g/m}^2 \times 7 \text{ kcal/g} = 194.25 \text{ kcal/m}^2$$

which provides an endothermic effect of 3223.87 kcal/m^2.

4.2 Mechanism of Heat Absorption

A fireproof covering material that uses a superabsorbent polymer, Aquacover, is a 3-layered one composed of a heat absorption package, ceramic

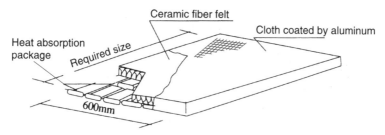

Fig. 3 Aquacover shape.

wool, and aluminum glass cloth (see Fig. 3). To avoid the increased temperature of steel, it is most effective to position the heat-absorbing package at the innermost position of the fireproof covering material. If the amount of water is the same but distribution is different, heat elevation characteristics differ. This finding was reported at the Symposium of the Society of Construction Science in October, 1990.

In a heat-absorbing package that is positioned in the innermost layer of a steel beam, the heat sealed joint will break gradually upon heating. Water then moves down and is absorbed by the ceramic wool at the bottom flange (see Fig. 4). During that time, heat is absorbed. Even after evaporation, the vapor remains for some time. Because of the vapor pressure in the inside of the fireproof covering material, it is believed that the heat is prevented from entering. The experimental result on a steel beam is shown in Fig. 5.

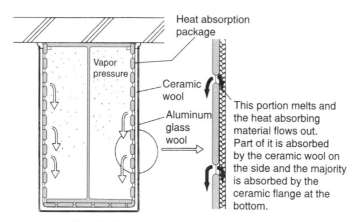

Fig. 4 The heat absorption mechanism.

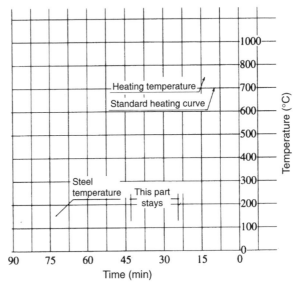

Fig. 5 Increased steel temperature during fireproof performance test of Aquacover.

5   FUTURE CHALLENGES

From the results of heating tests using steel beams, it has been found that the movement of water to the bottom is rapid where the current amount of superabsorbent polymer has been used. Thus, the steel temperature at the upper flange is high. If the water movement can be slowed down and distribute evenly throughout the heated sample, coating layer thickness can be reduced. However, as already described, superabsorbent polymers have high exotherms. Accordingly, adding water and developing a gel with a low exotherm are two issues to be addressed in the future.

Section 4
Sealed Construction Method

TAKESHI KAWACHI

1 INTRODUCTION

The sealed construction method is used to construct a horizontal tunnel. This method creates a great deal of soil, and depending upon the digging method used, this soil might be pasty and very heavy, which makes it difficult to move by truck. Hence it is necessary to modify the soil properties. One method employs superabsorbent polymers. In the following, plasticization of soil paste in a sealed construction method will be introduced.

2 TYPES OF TUNNEL DIGGING METHODS

Many infrastructures, such as railroads (including subways), roads, water purification systems, and sewage systems, are linear structures. To construct these structures underground, the construction methods listed in Table 1 are used. Explosives or digging machines for rocks are used for tunnels in mountainous areas. The exposed surfaces are then supported and digging continues. This method is limited to environments that are self-supporting, for example, rocky terrain. On the other hand, when this

Table 1   Classification of tunnels.

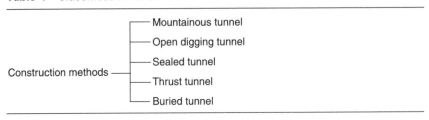

method is used to tunnel through nonrocky terrain, the sidewall must be supported during tunneling. After any tunnel is constructed, soil will be placed on top of it. This method is adopted when the structure is relatively shallow and low cost. However, because the effects of digging are felt on traffic in the area as well as the surrounding environment, urban areas especially feel the effects of construction. In the sealed construction method, a digging machine called a sealed machine is used to dig a tunnel horizontally in order to open a tube-like cavity. At the same time, segments of cement or steel are combined to create a support structure inside the tunnel cavity. This method can be used for various soil structures and reduces sinking, noise and vibration problems. Thus, it is now a dominant construction method in urban areas.

3   VARIOUS METHODS IN SEALED CONSTRUCTION METHOD

Table 2 lists various sealed construction methods. In the sealed construction method, a sealed machine digs through a tunnel and simultaneously keeps the ceiling from caving in. Various digging methods have been used of course, with the hand sealed method the first to be developed. This evolved into blind sealed, machine sealed, pressurized sealed, soil water sealed, soil paste sealed and soil pressure sealed methods. In sealed digging, the various methods are categorized based on how soil and water pressure issues are handled. [1, 2]

Figures 1 and 2 illustrate widely used soil water sealed and soil pressure sealed digging methods, respectively. In the soil water sealed method, the soil and water pressure at the digging front are counteracted by the applied soil water pressure, which stabilizes the digging front. In

Table 2 Classification of sealed construction methods.

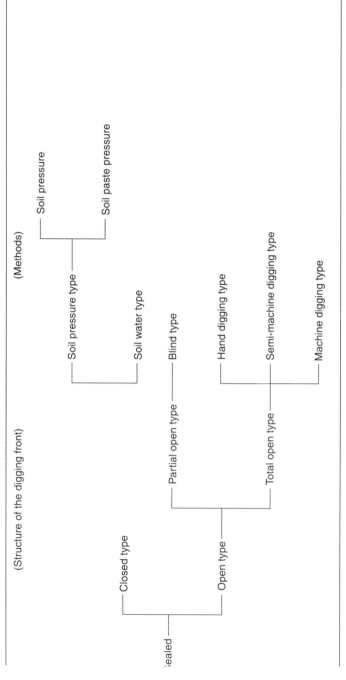

(Structure of the digging front)   (Methods)

ealed —

- Closed type
  - Soil pressure type
    - Soil pressure
    - Soil paste pressure
  - Soil water type
- Open type
  - Partial open type — Blind type
  - Total open type
    - Hand digging type
    - Semi-machine digging type
    - Machine digging type

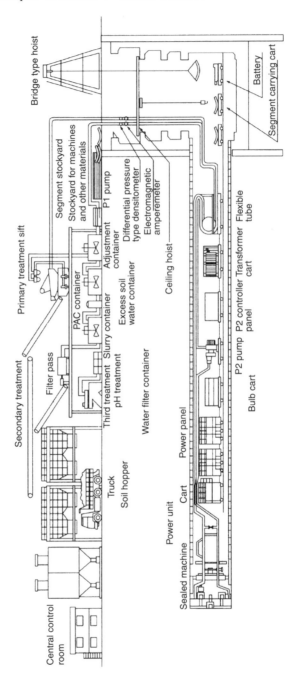

Fig. 1 Digging system for soil water sealed construction method.

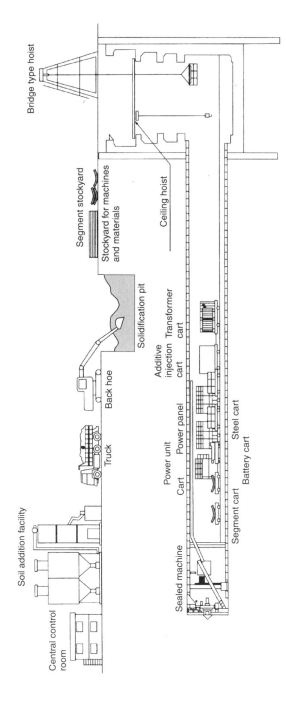

Fig. 2   Digging system for soil pressure sealed construction method.

this method, removed soil is mixed with water and, thus, it is necessary to separate the soil from the water. Soil separation methods include sifting and using a filter press for clay and silt. On the other hand, in the soil pressure sealed digging method, the plasticized soil near the cutting head is used in the cutting front chamber to stabilize the cutting front. The soil in the chamber will be carried above ground by a screw conveyer. Because of this, in the soil pressure sealed digging method, a large-scale soil separation system is not necessary. The soil paste sealed method, in which a high concentration of bentonite or superabsorbent polymer is injected in order to fluidize the soil and remove it smoothly, is a type of soil pressure sealed method. When the predominant soil is sandy, the sand sometimes plugs the chamber. Because sand is difficult to remove, this plugging phenomenon is dealt with by injecting air bubbles and then making the sand fluid. This air bubble sealed method is also a soil pressure sealed method.

The air pressure sealed method, a technique used to counteract the pressure exerted by the soil and water, was widely used in the past. However, because oxygen deficiency problems and a difficult work environment often occurred, it is seldom used today.

4   THE CHARACTERISTICS OF SOILS FROM SEALED DIGGING

The soils generated from the sealed digging methods currently used are listed in Table 3. The soil water sealed method disperses the soil in circulating water. The sand and gravel are separated by sifting and the fine particles, such as silt and clay, are separated by solid-liquid separation into dehydrated cakes. Hence, these soils can be carried by truck and need no further modifications.

On the other hand, soil generated by the soil pressure sealed method have been stirred, mixed and fluidized in the sealed chamber and then carried out by a screw conveyer. The degree of fluidization depends on soil quality because soil quality determines what additives can be used. In general, if a soil dominated by gravel is fluidized, bentonite or a low-quality pottery clay will be added. Thus, the mixture will have a consistency similar to concrete. For a sandy soil, creamy bubbles will be injected and the air bubbles fluidize the soil. Thus, a mixture of air bubbles and soil will be carried out. The air bubbles will disappear

Table 3   Water content of soils from sealed construction method.

Construction method	Sandy soil (solid content 25–40%)	Clay soil (solid content 40–80%)
Soil water method	Primary treated (separation of soil) soil	Soil aggregates solid contents +10% { Gravel 5–10% / Sand 20–35% }
	Secondary treated (filter pressed) soil	Cake 40–100%
Soil pressure method — Soil paste pressure	30–50% (+5–10%)	55–110% (+15–30%)
Air bubbles	27–45% (+2–5%)	45–100% (+5–20%)

Table 4 Improvement effect: characteristics, slump, and pH immediately following the addition of an additive and after transportation [3].

No.	Amount of the additive added			Immediately following the addition of an additive			After transportation	
	Polymer type (KS)	Calcium carbonate type (TF)	Cement type (SA)	Condition after mixing	Slump (cm)	pH	Conditions after transportation	Slump (cm)
0	–	–	–	Soil taken out of the tunnel (reference sample)	21.0	9.0	–	–
1	2	–	–	The soil is rather soft due to the low concentration of the additive	5.7	9.0	Probably because improvement is not sufficient, soil softened during transportation	6.5
2	3	–	–	Improvement is not sufficient; soil is aggregated and soft	5.0	9.2	There is no change caused by transportation	4.5
3	5	–	–	Improvement is almost appropriate; however, due to clay-like nature, the soil is not separated and is aggregated	2.0	9.0	Same as above	2.0
4	–	25	–	Improvement is insufficient; the soil adhered onto the mixing propeller and the screw became like a rod and transport was difficult	11.1	12.0	A slight improvement was noticed with transportation	9.5
5	–	–	25	The condition of the treated soil is relatively good; however, as with item 4, the soil adhered to the mixing propeller	3.1	11.0	There is no change with transportation	3.0
6	3	25	–	Improvement is extremely good; soil became well-separated particulate	0.5	11.2	Same as above	0
7	3	–	25	Same as above	0	10.6	Same as above	0.2

(a) Slump testing device

(b) Slump 3 cm

(c) Slump 25 cm

Fig. 3   Slump test of soil.

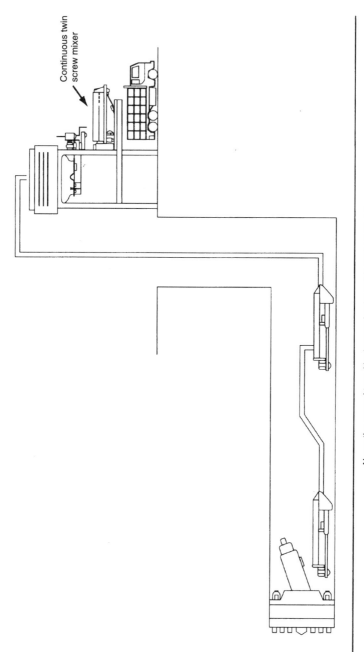

Continuous twin screw mixer

Above the ground type (the hopper for the soil is at the bottom)

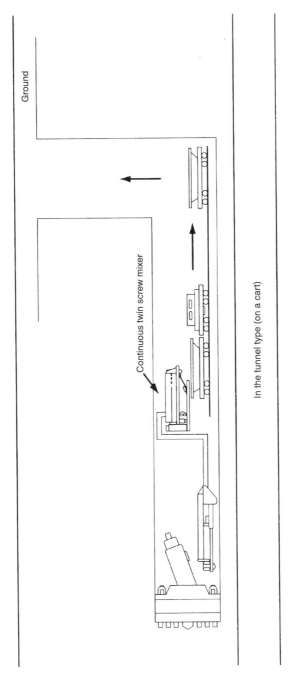

Fig. 4 An example of incorporating a soil modification mixer in a sealed digging system [4].

naturally or with the addition of an antiforming agent once the soil mixture is removed from the tunnel. The soil then can be carried easily by truck. The cutter on the cutting front rotates, plasticizes, and fluidizes clay soils. At that time, air bubbles may be injected to prevent the soil from adhering to the cutter, chamber and screw conveyer. Clay type soil is in fluid form and is thus very difficult to remove by truck. It is also necessary to modify the soil if it is to be used again. To improve soil that is weak or fluidized, calcium carbonate, water-soluble polymers, or superabsorbent polymers can be added.

5 IMPROVEMENT OF SEALED SOIL

There are three methods to improve sealed soil, in particular, soil that has been affected by the paste soil pressure sealed method. They are:

i)   physical treatment, such as natural sun drying or gravity dehydration;
ii) mechanical dehydration; or
iii) improvement by additives.

Of these, because method i) requires a great deal of space and many days for drying, it is seldom used. Soil paste is not fluid enough to use pressurized pump transport or filtering and the mechanical dehydration remains rare. The use of additives is easy to use, needs little space, and continuous treatment is possible. Thus, it is more often used. The additives used for this purpose are cement, calcium carbonate, and polymers. Polymers do not require incubation time and are effective nearly instantaneously. The soil can be treated continuously. Hence, this method can be a part of the soil pressure sealed method. In the case of polymers, changing the soil to basic pH can be avoided. Table 4 shows the experimental results of soil improvement following inclusion of polymeric materials. In this table slump or flow values are used to evaluate the fluidity of concrete or mortar.

Figure 3 illustrates testing conditions for soil using a slump-testing device for concrete according to standard JIS A 1101. The slump and flow values were both reduced by the addition of an additive. In the case of polymeric agents, strength is low in comparison to cement and calcium carbonate. As expected, the amount of additive varies depending on the soil to be treated but it is normally $3-5\,kg/1\,m^3$ of soil. For this type of treatment, both natural and synthetic polymer are commercially available.

If strength of the modified soil is desired, a polymer and calcium carbonate may be combined.

The addition of polymeric agents is effective in soil modification. For optimal benefit, of course, it is necessary to disperse the polymer uniformly. Currently used mixing methods are twin screw mixer, mixing shovel, and backhoe. Figure 4 illustrates an example of a sealed digging system to which a twin screw mixer is incorporated.

6   CONCLUSION

The modification method for soil removed by the sealed digging technique is now widely used. Future developments are anticipated. In order for this technique to be accepted more widely, it is necessary to develop auxiliary techniques to use soil removed from construction sites. Needed, in particular, are construction management and quality control, including the amount of additives added and mixing method used, as well as standardization in selection of additives and amount of additives. It is hoped that these problems can be solved based on actual use of materials and techniques.

REFERENCES

1   (1977). *Standard Methods for Tunnel Construction: Sealed Construction Methods*, Society of Civil Engineering.
2   Kuno, G., Yoshiwara, M., Ishizaki, H., and Omotake, Y. (1992). *Proc. 37th Symposium on Soil Engineering, Jpn*., pp. 1–6.
3   Okuno, O., Matsukiyo, T., and Miyazawa, T. (1990). *Gekkan Gesui-do* **13**: 39.
4   Ishizuka, H. (1992). *Gekkan Gesui-do* **15**: 76.

Section 5

Gelation of Waste Mud
(Gelation of Construction Waste Mud)

TAKESHI KAWACHI

1 INTRODUCTION

Waste mud is generated at construction sites. The annual quantity of waste mud generated was 1.5 million metric tons in 1993 according to the statistics reported by the Ministry of Construction and shown in Fig. 1. This figure is 20% of the total construction waste. Although construction waste mud does not contain toxic materials, it cannot be used as is because it is muddy. Thus, it is discarded as industrial waste. However, the shortage of disposal sites, long hauling distances and the increased difficulty of dumping waste in districts other than those where the waste was generated are becoming serious problems. Accordingly, the need to reduce weight and develop useful applications is extreme. Hence, even in the Construction By-Products General Projects (1991–1996) of the Ministry of Construction, high-level treatment and utilization are one of the main subjects. Both the government and industrial sectors are collaborating on new technological developments. Gelation of construction waste mud using water-soluble polymers or superabsorbent polymers

336

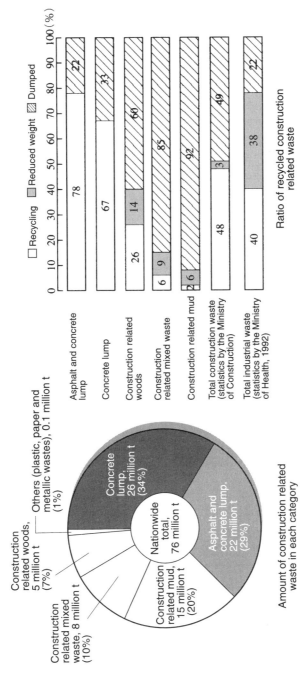

Fig. 1 Generation of construction waste (1993) [1].

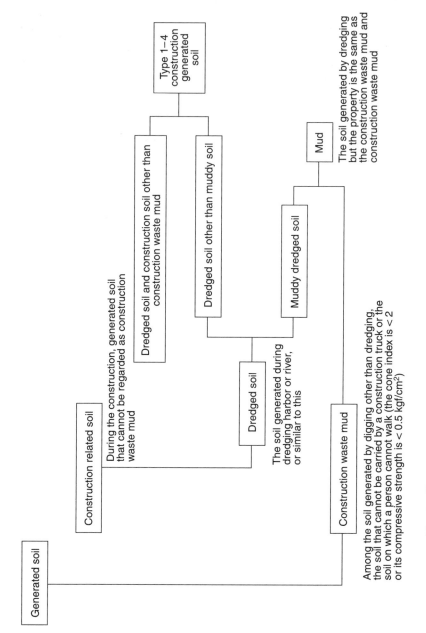

Fig. 2 The relationship between construction related waste mud and soil [2].

improves handling of mud by creating a more manageable soil-like appearance. Compared to the traditional mechanical dehydration, this is a much simpler process. However, under the current legal system, even if treatment is given, if the treated material is not value-added, it still must be discarded as industrial waste. Thus it remains similar to the dehydration approach. Therefore, it is hoped that waste recycling and proper treatment of waste will be more coordinated in the future.

2 TREATMENT AND DUMPING OF CONSTRUCTION WASTE MUD

2.1 Origin and Properties

Construction waste mud is a muddy material as defined in Figure 2. This soil, which cannot be carried by a construction truck has a cone index of $\lesssim 2$ or uniaxial compressive strength of $0.5\,kgf/cm^2$. Soil removal from harbors, etc. will not be included here. Construction that generates construction mud and its composition are listed in Table 1. Weak dug-out soil is the soil generated from slightly viscous soil. It becomes muddy by simply mixing and reducing its strength. Muddy water dug-out waste mud is soil generated immediately after use or excess waste mud that is produced by the underground continuous wall method or the earth drill method.

Figure 3 summarizes the continuous underground wall construction method. In this method, bentonite or a stabilizing solution that contains a water-soluble polymer at critical micellar concentration (CMC) is added when the work is being done underground. These materials will degrade

Table 1 Construction methods that generate mud and their characteristics [3].

		Composition	Solid content
Digging of weak ground		Soil (mainly clay and silt)	Over 50%
Digging of muddy water and mud	Continuous underground wall construction method	Bentonite, polymer, dispersion agent, soil	5–40%
	Sealed construction method	Bentonite, clay, forming agent, soil	Changes based on the method
	Various support method	Bentonite, polymer, dispersion agent, soil	5–40%
Improvement of ground (SMW, mixing, chemical injection)		Cement, bentonite, water glass, soil	20–50%, it can harden

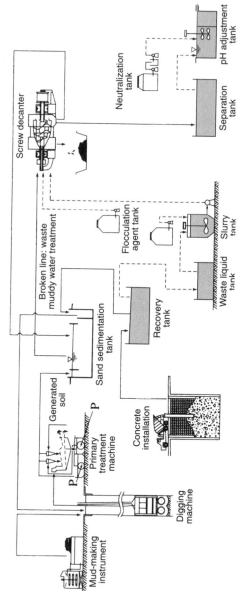

Fig. 3 An example of muddy water flow during continuous underground wall construction.

as a result of the contamination caused by cement during concrete installation or the mixing of sand. The degraded stabilizing solution is decanted and is called waste mud. Ground improvement mud is mud generated during the chemical injection method, the soil mixing wall (SMW) method, or the jumbo-jet special grout (JSG) method. The mud generated by these construction methods often contains a hardening agent and therefore, it hardens over time. However, depending on the amount of hardening agent or the intervening time, it might still be muddy. Table 1 shows the components of various waste mud, concentration range, and water contents. As shown in the table, ground improvement waste mud is alkaline. Cement can contaminate the waste mud generated by the muddy water digging process and if often shows basic property.

2.2 Treatment and Disposal of Construction Waste Mud

Generated waste mud is often treated at either the construction site or an intermediate treatment plant as shown in Fig. 4, followed by disposal at the final treatment plant. However, it is sometimes used for landfill. In this case, the distinction between waste mud and treatable soil is based on the forementioned criteria, that is, cone index of 2 or uniaxial compressive strength of $0.5\,\mathrm{kgf/cm^2}$.

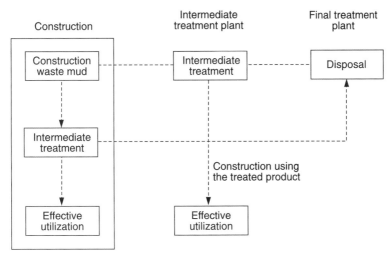

Fig. 4 A flow chart of the treatment and disposal of construction waste mud.

As shown in Table 2, the main intermediate treatment methods are machine treatment and chemical treatment. The machine treatment often utilizes compressive force, vacuum, and centrifugal force. For the treatment, machines such as a filter press, belt press, or screw decanter are used. In all machine treatments, the flow of the waste mud requires pretreatment in order to aggregate the waste mud. The water generated by the dehydration process is neutralized and disposed as sewage. The characteristics of the waste mud upon machine treatment depend on the dehydration machine and dehydration conditions. However, the water content is in the range as shown in Table 2 and moving it by truck is possible.

Chemical treatment involves adding a cement type solidification agent or a polymer. Although it is necessary to use a mixing machine for the chemical, it is a simpler method than the machine treatment one. It also requires a smaller treatment one. It also requires a smaller treatment yard. For intermediate treatment, a dehydration method, sun drying, is used. A large scale, highly stable operation does, however, require a large treatment yard. As rainy days also have to be considered, it is done only on an emergency basis. See the treatment flow chart in Fig. 5 for full details of the operation.

3 IMPROVEMENT TREATMENT OF WASTE MUD BY ADDITION OF A CHEMICAL AGENT

Treatment of waste mud by chemicals has been used for many years. The chemicals used are inorganic ones, such as cement and calcium carbonate, and organic ones, such as water-soluble polymers and superabsorbent polymers. These are widely used methods as they are quite simple. In the following, the current situation regarding use of these compounds will be introduced.

3.1 Inorganic Chemicals Methods

Waste mud has a high water content and is muddy. If the water contained is reacted with a hardener such as cement and a hydrated mineral is formed, the mud plasticizes or hardens. If cement is added to waste mud and cured, the strength created will depend on the amount of cement as shown in Fig. 6. Thus a hardening treatment utilizing cement to treat waste mud as well as sludge has been done for many years. However, in order

Table 2 Characteristics of dehydration machines [4].

Method	Belt press	Roll press	Filter press	High-pressure filter press	Vacuum dehydration	Centrifugal method (screw decanter)
Dehydration mechanism	Mechanical pressing force 30–100 kPa	Mechanical pressing force 30–100 kPa	Mud supply pressure 300–500 kPa	Mud supply pressure 300–500 kPa and filtration by pressure (1.0–1.5 Mpa)	Reduced pressure filtration (−30–60 kPa)	Centrifugal force (1500–3000 G)
Mud supply method	Continuous	Continuous	Intermittent	Intermittent	Continuous	Continuous
Applied specific resistance	10^9–10^{11} m/kgf	10^9–10^{11} m/kgf	10^{10-12} m/kgf	10^{11-12} m/kgf	10^{9-11} m/kgf	–
Aggregation agent	Inorganic + organic	Inorganic + organic	Inorganic	Inorganic	Inorganic	Inorganic + organic
Water content of dehydrated cake	100–700%	120–300%	50–100%	40–90%	150–300%	150–300%
Property of separated water	Clear	Clear	Clear	Clear	Almost clear	Depends on pre-treatment
Treatment capacity	350–500 kg/m·h	350–500 kg/m·h	7–25 kg/m²·hr	5–20 kg/m²·hr	15–20 kg/m²·hr	4–10 t/hr

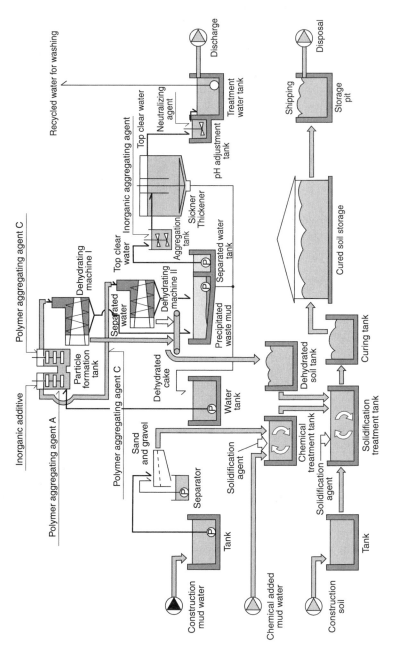

Fig. 5 Treatment flow chart of construction waste mud.

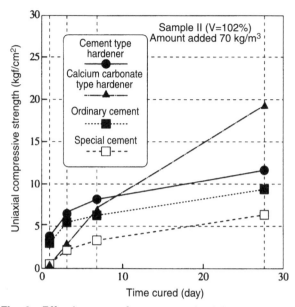

Fig. 6 Effectiveness of cement and calcium carbonate [5].

for this to be economical, it is necessary to use the least amount possible and mix it to homogeneity. Hence, in addition to improving the cement used, mixing methods have also been studied. Figure 7 depicts a widely used twin screw mixer used because it is efficient.

In addition to the use of ordinary Portland cement, special cements for waste mud also have been developed. The water in the waste mud is fixed as hydrate into etoringite, a special hydrated mineral that can fix a large amount of water. Another special cement that is not affected by the organic components in waste mud is also commercially available. Obviously, fairly high strength can be achieved depending on the amount of cement added. Other than cement, calcium oxide can also be used. Calcium oxide not only fixes water by reacting with it to become calcium hydroxide but also evaporates the water by exothermic reaction. This approach allows for hardening adequate for moving the material by truck. Various other inorganic hardeners are also commercially available. Although their compositions are unknown, the mechanisms of the effect are mostly due to hydration, etoringite and the formation and hydration of calcium oxide.

Treatment of waste mud by inorganic agents is simple and effective. However, a problem common with treatment is that the treated mud

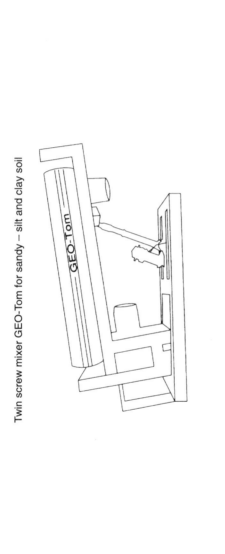

Twin screw mixer GEO-Tom for sandy – silt and clay soil

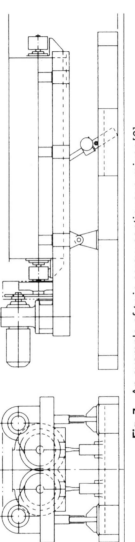

Fig. 7 An example of twin screw continuous mixer [6].

becomes alkaline. Depending on the treatment location or application area, alkaline water can be carried by rain, which then affects water conditions and flora in the surrounding environment. This trend is more obvious at the beginning of treatment but gradually decreases over time. Thus, it is necessary to pay attention to the treatment location and method at the beginning of the disposal of the final product. In some cases, wastewater observation and testing may be necessary.

3.2 Methods by Organic Agents

Although organic agents do not add as much strength as inorganic agents can they do not increase soil pH and thus they affect the environment to a lesser extent. Various agents are commercially available and these can be divided into water-soluble polymers and superabsorbent polymers. Figure 8 shows the effect of these additives. Flow value decreases with a smaller amount of additive than is the case with inorganic agents. The flow value here was determined using a mortar flow value instrument (JIS R 5201) and the spread was measured. The smaller the flow value is the better the improvement.

Although it depends on the water content of the waste mud, addition of a polymer at $1-3\,kg/m^3$ will make it feasible to move it by truck.

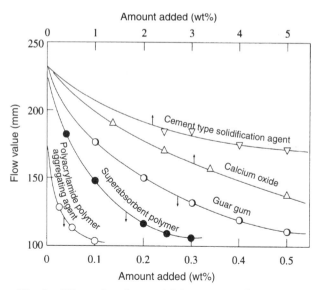

Fig. 8 Effect of various additives on the flow value [7].

Optimal effectiveness requires that the organic agent be evenly distributed. Thus, machine-mixing may sometimes be done. Although addition of calcium carbonate following addition of an organic agent was introduced recently, water fixation depends on the organic agent.

Organic agent-added soils can be used for tree planting. However, it is possible that the organic agent can be a source of bad odors or can increase the chemical oxygen demand (COD). It is therefore important to evaluate the selection of materials and the disposal method prior to application.

4 CONCLUSION

The current status of the treatment, disposal, and application of super-absorbent polymers as an intermediate treatment of construction waste mud have been described. The evaluation of treated soil and handling seem to be in the development stage. The shortage of disposal yards will surely continue. Therefore, depending on the development of technology as well as regulations, more innovative and focused technological progress may follow.

REFERENCES

1 (1995). *Countermeasure for General Construction By-Products*, Association of Advancement for Construction Waste Recycling.
2 (1994). *Application of Construction Generated Soil*, Civil Engineering Research Center.
3 Kawachi, T., Ogawa, Y., and Iso, A. (1994). *Proc. 5th Symp. of the Soc. of Waste Materials, Jpn.*, pp. 200–202.
4 (1994). *Introduction to Environmental Soil Engineering*, Society of Soil Engineering, Jpn., pp. 124–130.
5 Seki, S., Ogawa, Y., Akimoto, K., and Yoshinari, K. (1995). *Proc. 30th Symp. Soil Eng., Jpn.*, pp. 2221–2222.
6 Ishizuka, H. (1994). *Gekkan Gesuido* **15**: 76.
7 Miura, S. and Fujinaga, N. (1995). *Proc. 6th Symp. of the Soc. of Waste Materials, Jpn.*, pp. 231–233.

CHAPTER 7

Chemical Industries

Chapter contents

Section 6   Application for Electrophoresis

Section 1
Application of Gels for Separation Matrices

HIROAKI TAKAYANAGI

1 INTRODUCTION

A gel is a structure in which a liquid (in a broad sense, a gas is included) coexists with a solid that acts as a 3D matrix. The solute dissolved in the liquid can interact with the solid at the liquid/solid matrix interface. This interaction is determined by the type and properties of the solute, material and structure of the gel matrix, and the composition of the liquid. Hence, multiple solute components can be separated using differences in interaction. The separation can also be accelerated using external physical stimuli, such as the electric field, magnetic field, and pressure.

The properties and applications of a representative gel for separation will be discussed here. These applications include absorbate functions, and ion exchange resin and molecular sieves. It is also used for a liquid chromatographic (LC) substrate in microscopic applications that require a high level of accuracy. For separation, sheet and tube-type products may be used. However, they will not be discussed in this section.

There are various practical advantages gained by using gel particles for separation. Among these,

i)   gels can be repeatedly used with good reproducibility,
ii)   handling is easy and safe because they are solid, and
iii) various gels have been developed and commercialized and, thus, they are easily acquired.

Separation gels are sometimes used in a batch manner as a suspension in liquid. Ordinarily, however, they are used as packing materials in a column for a flow-through method. This type of separation is known as either the LC method or auxiliary techniques. Therefore, knowledge in the LC field is often used to improve the separation performance of these gels. For actual applications, readers are referred to the corresponding literature.

2 CLASSIFICATION OF GELS BASED ON THE SHAPE AND CHEMICAL STRUCTURE OF PARTICULAR GELS

2.1 Shape of Gels

Gels for separation usually have spherical or irregular particles and their sizes range from several μm to several mm. The particle diameter is chosen based on the separation purpose although there are instances when it is restricted by the gel material. When the purpose is the accurate separation and characterization of materials, it is practical to have particles from several μm to at most several tens of mm due to the requirements of LC substrates. On the other hand, if it is intended for purification of a specific material (fractionation or industrial purification), slightly larger particles, ranging from several hundred μm to 1 mm, are often chosen.

A gel of matrix solid for has a structure that allows it to permeate a liquid (solution) in which the target materials for separation are contained. This structure is a so-called network or porous structure. This structure helps diffusion of solute into the particles. Further, the size of the network functions as a molecular sieve and contributes to the separation of the materials.

2.2 Classification Based on Chemical Structure

The solid matrices for gels can be classified into several groups based on their chemical structures. This classification is not absolute but rather historical in nature. They are classified here largely into three types and their representative properties, that is, physical properties, such as particle shape, pore structure and network structure, and chemical properties, such as pH stability, will be described briefly.

2.2.1   Inorganic Compound Gels

Gels that belong to the inorganic compound category [1, 2] are represented by silica and alumina as well as compounds that include metallic ions such as apatite. They are all polar materials and are often used as adsorbents. We will choose silica as a representative example. The matrix consists of silicon dioxide. Silanol, a weak acid, covers the surface. It is rather ionic in nature and can form hydrogen bonds. Thus, silanol can adsorb polar groups that are dissolved in a liquid. As with the shape of silica, it is porous and spherical if the particle size is $< 30 \, \mu m$. However, particles larger than this tend to be irregular (crushed) particles.

In general, inorganic compound gels are porous and the pore size ranges from several nm to several tens nm and their specific surface areas are on the order of several hundred m^2/g. It is possible to add various functionalities because the surface of metallic oxides like silica can be chemically modified (see Fig. 1). For example, silica modified with long chain alkyl groups, such as octadecyl and octyl groups, are used as

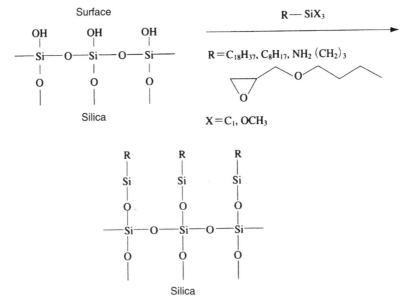

Fig. 1   Chemical structure of silica and surface treatment compounds.

hydrophobic particles for reverse phase chromatography. Other function-alities include amine groups (for ion exchange) and glycidyl groups. Inorganic compound gels are generally hard and have high mechanical strength but they tend to be only weakly acidic and alkaline. Some of them gradually dissolve.

2.2.2 Synthetic Organic Polymer Gels

These gels are made of synthetic organic polymers [3, 4]. Polystyrene crosslinked with divinyl benzene, poly(acrylic acid) and poly(acrylic acid) derivatives are well known materials. Polystyrene type gels are used for both porous adsorbents and ion exchange resins. Figures 2 and 3 illustrate the chemical structures of polystyrene and poly(acrylic acid)-type ion exchange resins, respectively. Ion exchange resins can be divided into

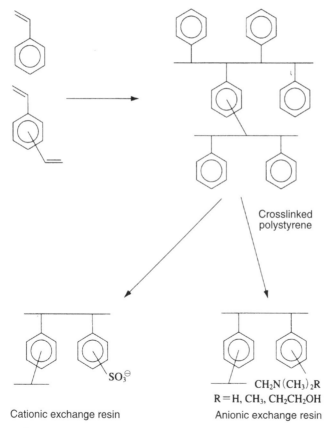

Crosslinked polystyrene

$SO_3^{\ominus}$

Cationic exchange resin

$CH_2N(CH_3)_2R$
$R = H, CH_3, CH_2CH_2OH$

Anionic exchange resin

Fig. 2 Chemical structure of polystyrene-type ion exchange resins.

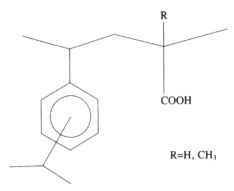

Fig. 3   The chemical structure of poly(acrylic acid)-type ion exchange resins.

strongly acidic cationic exchange resins (sulfonic acid type), weakly acidic cationic exchange resins (acrylic acid type and methacrylic acid type), strongly alkaline anionic exchange resins (trimethylammonium type), weakly alkaline anionic exchange resins (dimethylamine type), and others based on their functional groups. Other ion exchange resins include phosphoric acid type, phenolic type, and mercaptan type, all of which are cationic exchange resins and polyamine types, which serve as anionic exchange resins.

Chelated resin that can form a complex (see Fig. 4) and amphoteric ion exchange resins that have both positive and negative changes (see Fig. 5) are both similar to ion exchange resins. Other polymers possess reactivity characteristics as affinity substrates.

Synthetic organic polymer gels are synthesized by suspension polymerization or emulsion polymerization and their shapes are spherical.

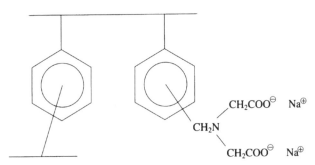

Fig. 4   An iminodiacetic acid type chelate resin.

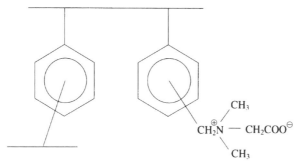

Fig. 5 A dimethylglycine type amphoteric ion exchange resin.

The particle sizes can be controlled from several μm to several mm and the size used depends on the application. Regarding the structure of gels, there is a network type, which is polymer chains to which liquid molecules permeate (it is sometimes called a gel type) and another type is a porous one in which a porous structure exists regardless of whether there is any liquid. The network size of the network type gels depends on the amount of crosslinking agent used. Molecular weights of 10^2–10^4 are common for the excluded volumes. On the other hand, the pore sizes of porous type gels can be controlled widely—from several nm to several μm.

The mechanical strength of organic polymer gels varies depending on the amount of the crosslinking agent, although they are more flexible when compared with inorganic type gels. They are also chemically stable in the presence of an acidic or alkaline environment but degradation can proceed in the presence of an oxidizing agent. Thermal resistance is usually inferior to that of inorganic gels because these gels are organic.

2.2.3 *Natural Polymer Gels*

Natural polymer gels (mainly polysaccharides) are made of dextrin, agarose, and cellulose [5]. They are often used to separate easily denatured proteins and nucleic acid. They form structures in which diffusion is possible only when they are swollen with water. The available network sizes are in the excluded volume limited molecular weights of 10^2–10^7. The particle sizes are mainly in the range of several tens μm–several hundred μm and they are either spherical or irregular. Unmodified gels (molecular sieve) and ion exchange resins are known. A gel with a group that reacts easily, such as an epoxy group, is known as an affinity substrate. This polymer is used by chemically reacting affinity ligand.

Natural polymer gels are generally soft. They easily deform or degrade under pressure changes. Gel structures can be degraded by either drying or freezing. There are other gels that decompose with the action of an acid, an alkaline, an oxidizing agent, or a microbe. Many of them require care in handling because of stability problems.

3 SEPARATION MECHANISMS AND SEPARATION AGENTS

Separation mechanisms can be classified into several modes. Although actual separation is achieved by the combination of multiple modes, characteristics of representative examples for each mode are introduced [6].

3.1 Molecular Sieves

This form of separation, sieving, uses the 3D structure of gels, that is their network size. Separation based on this principle depends on both network and solute sizes. This method has poor separation efficiency in comparison with other methods.

Sieves are usually tested as part of LC (gel filtration or gel permeation chromatography) for relatively small-scale separation tasks. The main applications include fractionation or elimination of salt for polymers such as proteins and peptides, separation of saccharides or oligosaccharides by ion exchange resin, and molecular weight fractionation of synthetic polymers by porous organic polymer gels [7]. The molecular sieve effect is fundamental to gels and is thus used for separation. Even when other modes are used, selection of network size or pores, which correspond to the size of the solute (or the impurity to be removed), can help to improve separation efficiency [3].

3.2 Adsorption

If the gel backbone is polar or can bond with hydrogen and the liquid in question is nonpolar, the polar group solute will adsorb onto the gel surface through hydrogen bonding or dipole interaction. Conversely, when the gel backbone is nonpolar and the liquid is polar, the solute with low polarity is adsorbed onto the gel surface by van der Waals forces. The former is called a normal phase and the latter is called a reverse phase.

Both methods are often used with gels. The adsorbed material can be separated and recovered because it can be redissolved into the liquid, depending upon the composition of the liquid, namely polarity.

Separation based on adsorption deals mainly with organic compounds having molecular weights of $< 10^4$. For separation using adsorption, porous organic polymer gels are mainly used. They are used for large-scale purification of fermentation-produced antibiotics and useful products obtained from plant extracts [3, 8]. Reverse phase-type gels are often used for high-performance liquid chromatography (HPLC). In particular, the silica modified with octadecyl group (C18, abbreviated as ODS) is widely used as the general-purpose solid phase.

3.3 Ion Exchange Materials

Ion exchange materials [8] represented by ion exchange resins are widely used in industry. Applications are broad and include desalination, high-level purification of water (super-pure water), and separation and purification of organic compounds and proteins. Synthetic polymer ion exchange resin gels used for industrial production can handle the purification of fermentation-produced amino acids (lysine) with a strongly acidic cationic exchange resin, the separation of nucleic acids for food products using a strongly alkaline anionic exchange resin, and the separation of a protein (basic bacteriolytic ones like egg white) with a weakly alkaline cationic exchange resin. A cellulose-type ion exchange resin can be used for the purification of lactic acid-type proteins like milk. In all cases, the crude solution that contains the target compound is loaded, ion exchanged and adsorbed and then the target material is desorbed by changing the condition of the external solution.

In a small-scale separation, especially for biopolymer separation such as for proteins, polysaccharide gel-type ion exchange materials can be used for many possibilities. At the HPLC level separation, the use of an organic polymer gel is most effective.

As a special ion exchange material, amphoteric ion exchange resin has unique separation functions. A snake-cage resin consists of a strongly alkaline anionic exchange resin with a linear poly(acrylic acid) (see Fig. 6).

This resin can be used to reduce the amount of sodium in soy sauce because it selectively holds salts, particularly sodium chloride (NaCl). Dimethylglycine-type resin is useful for removing sodium sulfate and sodium chlorate from an electrolysis salt solution [9]. Mercaptan-type ion

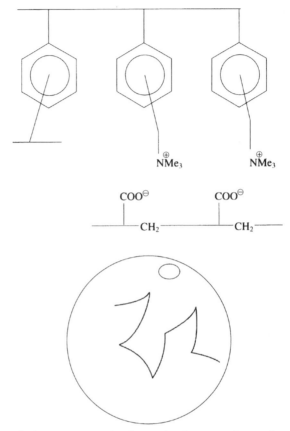

Fig. 6 Chemical structure and schematic diagram of a snake-cage resin.

exchange resin is used for capturing heavy metals such as mercury. It is often disposed of instead of reused.

3.4 Special Chromatographic Separation

A chelate exchange LC uses a transition metal ion for fixation. Separation is achieved by utilizing the difference in the ability to form a complex of each chelating agent. An example of large-scale application is the separation of glucose and fructose using a Ca^{2+}-type strongly acidic ion exchange resin. For small-scale application, gels for optical separation of amino acids and hydroxy acids using Cu^{2+} are commercially available.

Affinity chromatography is a special LC method that uses specific biointeractions. There is no single gel that is widely used. However, a substrate gel to fix affinity ligand is commercially available. Representative examples include polysaccharide or organic polymer gels that contain primary amine, carboxylic acid and epoxy groups [10].

3.5 Others

A substrate for a synthetic reaction is an example of a special application of a spherical gel. If a chemical reaction is repeated on the surface of a substrate gel, a side chain with a specific structure can be formed. If this side chain is removed from the substrate, a compound with a desired structure can be obtained. Using this approach, a polypeptide can be synthesized by reacting the amino acids sequentially. A chemically modified and crosslinked polystyrene is one of the materials used, although it is not used for actual separation. However, it can at the end of a process separate highly pure polypeptides and, thus, it is regarded as a separation substrate in the broadest sense.

REFERENCES

1 Brown, P.R. and Hartwick, R.A. (1989). *High Performance Liquid Chromatography*, Wiley, New York, pp. 145–188.
2 Unger, K.K. (1979). *Porous Silica, J. Chromatography Libr.*, *Vol. 16*, Elsevier, Amsterdam.
3 Verall, M. S. (1996). *Processing of Natural Products*, Wiley, Chichester, pp. 159–178.
4 (1996). *Diaion®, Ion exchange resin, Synthetic Adsorbate Manual I: The Fundamentals*, Mitsubishi Chemicals.
5 Mikes, O. (1988). *High-peformance Liquid Chromatography of Biopolymers and Biooligomers, Part A*, Elsevier, pp. 142–146.
6 Mikes, O. (1988). *High-performance Liquid Chromatography of Biopolymers and Biooligomers, Part B*, Elsevier, pp. 142–146.
7 (1996). *MCI GEL Technical Bulletin*, Mitsubishi Chemicals.
8 (1996). *Diaion®, Ion exchange resin, Synethetic Adsorbate Manual II: The Applications*, Mitsubishi Chemicals.
9 (1997). *Advanced Functional Microparticles and Their Auxiliary Technique*, Association of Functional Microparticles and Their Auxiliary Technique, Ed., Soc. Chemical Engineering, Jpn. pp. 180–183.
10 Giddings, J.C., Grushka, E. and Brown, P. R. (1989). *Adv. Chromatography*, **28**: 1.

Section 2
Application for Tissue Culture of Plants

YUICHI MORI

In recent years, production of flower seedlings, such as orchids and carnations, leafy plants, and vegetables and fruits, such as strawberry, potato and sweet potato, has been accomplished by tissue culture. This approach is convenient for mass production in a virus-free environment. First, plant cuttings unaffected by viruses are made in a sterile environment and then cultured in a liquid medium that contains a growth hormone, or on a solid culture such as agar. An adventitious (root-forming) embryonic plant is nurtured until it becomes a seedling. These seedlings are then transplanted onto an agar or gellan gum culture that contains nutrients, such as glucose, and cultured. When photosynthesis is desirable the seedlings are transplanted to a nonsterilized field culture. At this point the gel (such as agar), which contains sugars, must be hand removed. If the agar gel adheres to the seedlings, viral growth, which causes root rot and a plummeting survival rate will be increased. The process of removing agar from delicate seedlings by hand is associated with high labor costs and seedling quality problems caused by mechanical

362

damage. In particular, if gels such as agar are used as the culture substrate, the changes needed for growth of seedlings, the addition of any nutrients lacking in the culture, and removal of waste byproducts is very difficult. The main shortcoming of the agar culture is that the sol-gel transition of agar is extremely high. Thus, it is impossible to melt the gel without inflicting thermal damage to the seedlings.

As a solution to the traditional agar or gellan gum tissue culture hydrogels, we tried thermoresponsive hydrogel (TRHG). For example, a hydrogel made of a chemically crosslinked thermoresponsive polymer, poly(N-isopropyl acrylamide), shows a transition temperature of near 30°C. Below this temperature, TRHG absorbs water (culture liquid) and swells, whereas above this transition temperature, the internal water is expelled and the gel shrinks. Although the volume change of TRHG depends on crosslink density and culture composition, it can reach 10–50 times its original size. The process is thermally reversible. We evaluated the tissue culture of an orchid using microparticles of TRHG using this phenomenon.

When 1 *l* of culture liquid (sugar content at 4 wt%) is absorbed into 20 g of TRHG microparticles at the correct culture temperature of 22°C, which is below the transition temperature, particle fluidity is lost and the particles form a gel. Two hundred ml of this gel is placed in a glass container as shown in Fig. 1(a) and sterilized in an autoclave. Twenty-five orchid seedlings (cymbidium) 2.5–3.0 cm long were transplanted onto a gel and cultured at 22°C for approximately 3 mo (see Fig. 1 (a)). The growth of seedlings on the TRHG gel was at least comparable to that on the agar culture. Upon culturing for ≈ 3 mo the length reached 10–13 cm and then the glass was immersed in warm water at 36°C, which is above the transition temperature. The TRHG particles shrunk as shown in Fig. 1 (b). The TRHG gel changed into a sol to become a liquid, and the seedlings could be removed from the glass easily. At the same time, the TRHG gel could also be removed. The volume of the TRHG gel at this process at 36°C is 1/15–1/20 that at 22°C. Hence, the sugar included in the TRHG gel particles is mostly expelled from the particles and is then easily removed. In fact, upon completion of the culture, the sugar concentration was approximately 2 wt%. However, following elevated temperature treatment, the sugar content fell to 0.2 wt%. As already described, for agar cultures, it was necessary to remove the gel by hand. However, the use of thermoresponsive gels makes possible the significant

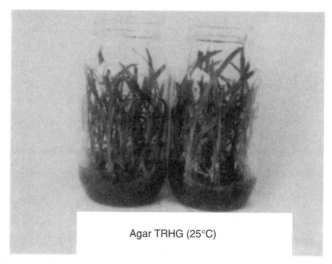

Fig. 1 (a) Seedlings of an orchid cultured for 3 mo at 22 °C using agar gel (left) and TRHG gel (right).

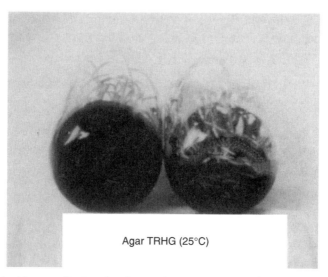

Fig. 1 (b) Upon culturing for 3 mo, the appearance of the agar gel (left: it does not melt) and TRHG gel (right: it melts) when the temperature was raised to 36 °C.

Agar TRHG (25°C)

Fig. 1 (c)   The seedlings were taken out of the culture container by melting the TRHG gel at 36 °C (right).

reduction in labor costs. Furthermore, sugar, which is included in TRHG gel, is excluded during this elevated temperature treatment process. Hence, when the seedlings were transplanted to a non-sterilized field, the viral growth was not a problem and the survival rate of the seedlings was excellent.

An unexpected effect was discovered when a thermoresponsive hydrogel rather than an agar gel was used as the culture substrate. The main problem for seedlings is viral contamination. In the case of agar gel, the gel shrinks during syneresis. Hence, any virus growing in the culture liquid is expelled from the gel. Because the roots penetrate the agar gel, the culture liquid coming out of the gel stays at the gel/root interface, which provides an excellent environment for viral growth. On the other hand, because the thermoresponsive gel substrate was not at saturation, the culture liquid was not present in gel particles. Because the root grows between the gel particles rather than in the particles, there is significant suppression of viral growth.

The orchid root cultured in agar gel did not exhibit hairy root growth whereas on the root of the seedling grown in the thermoresponsive hydrogel, a significant number of hairy roots were recognized. The growth of the hairy root is closely related to a plant's water stress. Water stress resistance of seedlings upon transplantation from a water-rich culture medium to soil is believed to influence strongly the survival rate of the seedlings. The fact that there were no hairy roots in the agar culture indicates that the water in this culture is free water and the seedlings have easy access to this water. On the contrary, in the case of the thermoresponsive gel, because was not at saturation, the degree of freedom of water in the gel is much lower than in the agar gel. This leads to the growth of hairy roots that seek out water. The low degree of freedom of the water in the thermoresponsive gel is also related to the forementioned suppressed growth of viruses. Because viruses have much higher metabolism than plants, they require water with high degree of freedom. It is more difficult for viruses to use the water in a thermoresponsive gel.

Of particular interest is the accelerated growth of the seedlings when they were transplanted to soil without removing the thermoresponsive hydrogels in which they were first grown. In general, plants required less water and other nutrients. High available water supply and low temperatures promote root rot and higher temperatures, with concomitantly an increased need for water and nutrients, also produce a different set of problems for nurseries.

Quite often, plants have excess water in winter or at night and they are water deficient in summer or during the daytime. If thermoresponsive gel particles adhere to the root or exist in the vicinity of the root, water is absorbed below the transition temperature and stored in the gel; this water is released to the plant at above the transition temperature. Hence, thermoresponsive gels regulate the water content of the soil optimally and reduce the major stress on seedlings when they are transplanted from a gel culture to soil.

In order to confirm the effect of the thermoresponsive gel described here, seedlings were cultured in the usual mixed compost and in a thermoresponsive, hydrogel-added, mixed compost. The concentration of the gel added was a 1 g/1 l and the temperature of the greenhouse varied from ≈ 13 to 30°C. As shown in Fig. 2, growth of the orchid seedlings was accelerated in the thermoresponsive, hydrogel-added compost.

Left: a mixed compost is used as soil
Right: a mixed compost + TRHG gel (Ig/U) is used as soil

Fig. 2   Growth of orchid in a nursary.

Accordingly, by changing the culture media from agar to a thermo-responsive gel, the same substrate can be used from culture to nursery, thereby allowing a continuous process and automation. It demonstrated significant reduction in costs and improvement in the quality of the seedlings.

Traditionally, farmers purchased seeds from horticultural distributors. However, seed germination is unstable and requires both effort and time. Hence, they began to purchase seedlings. In the future, the production of seedlings is expected to increase further and tissue culture technology will be more important. The traditional approach has been that of using materials originally produced for other industrial purposes in farming. However, there will be many materials specifically designed for horticultural purposes. The results described here are one of the early examples of the application of functional polymers. It is hoped that new applications areas will be created through interaction with other research areas.

The details of this research, "Cultural Method of Orchids using a Thermoresponsive Hydrogel," were reported on at the Nagoya International Conference, March 1996.

Section 3
Oleogels and Their Applications

TOMONORI GOM

1 INTRODUCTION

Today, we are surrounded by environmental pollution. In particular, the oil pollution of rivers, lakes, the sea and the land caused by tankers, factories, homes, and restaurants is a serious problem. At wastewater treatment plants a method to reduce oil during an early stage of water purification is being sought because it reduces the amount of bioactive mud.

To solve these problems oil treatment compounds such as adsorbents, gelation agents, and emulsifiers have been developed. A self-swelling type oleogel, which absorbs oil into the polymer networks, was proposed and received attention. Oleogels have been around for many years, with some of them having been proposed in the 1960s. These gels include a copolymer of alkylstyrene and divinyl benzene (Dow Chemical) [1] crosslinked methacrylate-type polymers, such as poly(t-butyl methacrylate) and poly(methyl methacrylate) (Mitsui Petrochemicals) [2–4], polynorbornene (Norsolor), and radiation-crosslinked polystyrene (Murakami) [5]. However, these compounds exhibit excellent absorptivity to a specific oil, but high viscosity oils, such as fat, cannot be absorbed and therefore the application is limited. Therefore, except for polynorbornene, the others were not commercialized. Later, oleogels made of crosslinked long-chain polyacrylates, which can be applied to a wide range of oils, were developed (Oleosorb, Nippon Shokubai). Due to the unique functions of these oleogels, applications to areas other than for simple oil treatment are being developed. In the following, the fundamental properties of oleogels and their application examples will be introduced.

2 DEVELOMENT OF OLEOGELS

2.1 Comparison to Hydrogels

Hydrogels, widely used today as a material for disposable diapers, are crosslinked polymers with, for example, many hydrophilic groups, such as sodium carboxylate. Water molecules are hydrogen bonded with carboxylic acid groups or incorporated into the networks spread by the static repulsive forces of the carboxylic ion themselves. On the other hand, oleogels are crosslinked polymers that contain oleophilic groups, such as long-chain alkyl or alkylaryl groups. The oil is incorporated in the networks by the interaction with the oleophilic groups. Unfortunately, oleogels cannot use the spreading effect of static repulsive forces as hydrogels can and so absorption of liquids is only several tens times that of its own weight, which is one to two orders of magnitude smaller than hydrogels. Table 1 lists the differences between oleogels and hydrogels.

2.2 Difference from Oil Gelation Agents

Oleogels and oil gelation agents are often regarded as being the same. However, due to their functional mechanisms, they should be classified differently. Oil gelation agents are classified as hot melt-type, solution addition-type, and synthetic polymer-type. They gel oil by orienting themselves in the oil and forming networks. Hot melt-type is represented by 12-hydroxystearic acid, which is now widely used as an oil treatment agent for home cooking oil. Oils can be gelled by the addition of a small amount. However, it creates a fire hazard because the oil treatment requires reheating of waste oil, which is flammable [6]. The solution addition-type, such as amino acids, also creates environmental problems because a solvent is involved. A synthetic polymer-type, such as a copolymer of styrene and butadiene, has limited applicability, with both

Table 1 Differences between hydrogels and oleogels.

	Hydrogel	Oleogels
Component	Crosslinked hydrophilic polymers	Crosslinked oleophilic polymers
Structure		
Functional mechanism	i) Hydrogen bonding between hydrophilic groups and water molecules ii) Spreading of the molecule by static repulsive force among ionic groups	Molecular interaction between oleophilic groups and oil molecules
Absorption ratio	Several hundred times	Several tens times

a large amount and the necessity of mixing it required. The gels formed by these gelation agents tend to be weak at high temperatures or when an external force is applied. Many lack long-term stability.

By contrast, oleogels can absorb oils without the need for heat or mixing of ingredients, or the addition of a solvent. The formed gel holds oil, creates a desirable shape, and is transparent. Table 2 compares the differences between oil gelation agents and oleogels.

2.3 Design of Oleogels

As already stated, the oil absorption capability of oleogels depends on the interaction between the oleophilic groups and the oil molecules. Therefore, it is extremely important to select the monomer to which oleophilic groups can be introduced in order to synthesize oleogels. Those monomers that can incorporate oleophilic groups include alkyl(meta)acrylate, alkyl(meta)acrylamide, α-olephane, alkylvinyl ester, alkylvinyl ether, and alkylstyrene. In this case, reducing the glass transition temperature (T_g) of

Table 2 Differences between oil gelation agents and oleogels.

Classification		Types	Mechanism of gel formation	Characteristics
Oil gelation agent	Hot-melt type	12-Hydroxystearic acid	Network formation by the molecular orientation in the oil	Inexpensive
	Solution addition type	Metallic soap etc. Benzyliden seorbitol Amino acids etc.		Insufficient holding of gel shape It is necessary to heat or mix
	Synthetic polymer type	Styrene-butadiene		
Oleogel		Polynorbornene	Spreading of preformed polymer networks	Excellent holding of gel shape
		Crosslinked alkylacrylate		Unnecessary for heating or mixing Slow absorption of highly viscous oil Expensive

the oleogel and enhancing molecular movement are important to effective absorption of highly viscous oils.

Another factor in the control of oil absorption capability is the selection of crosslinking conditions that are able to prevent oleogels from dissolving into the oil. The lower the degree of crosslink density, the higher the ability to absorb oil. In practice, however, it is necessary to design sufficient gel strength in a similar manner as in hydrogels.

3   FORM OF OLEOGELS

In the following, commercially available forms of oleogels will be introduced (see Table 3).

3.1   Water Dispersion Type

Water dispersion type oleogels from Oleosorb SL series (Nippon Shokubai) gels are commercially available. Oleosorb SL-130, with average diameter of 30 μm and resin content of 30 wt% can form a film easily by drying. They are used as a coating or paint on a film or paper. Oleosorb SL-160 with average diameter of 300 μm and resin content of 59% is used for column packing to remove oil in water.

3.2   Powder Type

Powder type oleogels called Norsolex (Nippon Zeon) and Oleosorb PW series (Nippon Shokubai) are commercialized. Norsolex is a white powder granule with particle sizes of 0.3–2 mm. It absorbs aromatic and

Table 3 Commercially available oleogels.

Type	Brand name	Company	Particle diameter	Characteristics
Water dispersing type	Oleosorb SL-130	Nippon Shokubai	30 μm	Resin content 30 wt%, possesses film formation capability
	Oleosorb SL-160	Nippon Shokubai	300 μm	Resin content 59 wt%
Powder	Norsolex	Nippon Zeon	0.3 $\sim$ 2 mm	Forms blocky cells
	Oleosorb PW-170	Nippon Shokubai	1 $\sim$ 2 mm	High rate of oil absorption
	Oleosorb PW-190	Nippon Shokubai	300 μm	Excellent fluidity, monodisperse spheres

halogenated solvents at a high absorption ratio. Thus, it is a useful oil treatment in mechanical plants and repair shops. Oleosorb PW-170 is a white granule with particle sizes of 1–2 mm that consists of primary particles with average diameter of 30 μm. It is designed to prevent the aggregation that can accompany the absorption of oil. It absorbs oil quite quickly.

On the other hand, Oleosorb PW-190, a white monodisperse particle with average diameter of 300 μm, shows excellent fluidity and process-ability. These powder type oleogels are sold in polypropylene bags. Other products use nonwoven cloth as a substrate to make an oil absorbing package or sheet.

4 FUNDAMENTAL PROPERTIES OF OLEOGELS

The fundamental properties of oleogels will be explained in the following.

4.1 Oil Absorption Ratio

The absorption ratio of oleogels is influenced by the difference between the solubility parameter of the oleogel and oil, and the molecular weight, steric structure and degree of crystallinity of the oil. In particular, oleogels exhibit high oil absorption ratio to those oils with solubility parameters of 6–10 $(cal/cm^3)^{1/2}$. It is difficult to absorb oils at a high absorption ratio with molecular weight of over 1000 or high degree of crystallinity. Table 4 shows the oil absorption ratio of oleogels to various oils.

Table 4 Oil absorption ratio of commercial oleogels.

Brand name	Norsolex	Oleosorb PW-190
(Oil absorption ratio) (g/g)		
Toluene	20	15
Kerosene	13	12
Trichloroethylene	34	25
Octanol	1	7
Dibutyl ether	4	10
Butyl acetate	2	9
Soy oil	2	8
Oleic acid	2	10

4.2 Oil Absorption Ratio for Mixed Oils

The oil absorption ratio of oleogels to mixed oils cannot be predicted simply from the oil absorption ratio to the individual oils. When a mixed oil approaches a polarity that makes it easily absorbable by an oleogel, the mixed oil may sometimes show higher oil absorption ratio than the individual sum. It seems that the mixed oil may be recognized as single oil. The selective absorption of specific oil was not observed. Figure 1 shows the relationship between the mixing ratio of decane and acetone with oil absorption ratio.

4.3 Rate of Oil Absorption

The time required for saturation swelling of oleogels is influenced greatly by the viscosity of a particular oil and the particle size (surface area) of the oleogel that is used. Hence, in order to reduce saturation swelling time,

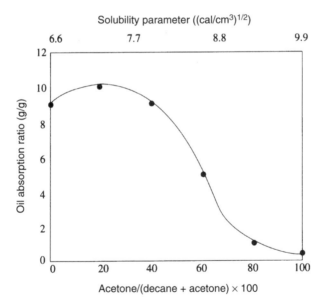

<Conditions> Oleogel: dry resin of Oleosorb SL-160
Temperature: 20°C
Absorption time: 24 h

Fig. 1 Oil absorption ratio of decane/acetone mixed solvent.

reduction of oil viscosity by heating and reduction of the particle size of the oleogel are both necessary.

4.4   Oil Retention Capability

If pressure is exerted for a prolonged period on an oleogel that absorbed oil, the amount of oil proportional to the pressure will be released. Figure 2 shows how much oil is released at a pressure equivalent to the swelling pressure of an oleogel that has absorbed toluene. In a comparison with hydrogels, it is demonstrated that oleogels have lower swelling pressures.

4.5   Selective Absorption of Oil from Oil/Water Mixture

Oleogels do not absorb water. Therefore, they can selectively absorb and separate oil from an oil/water mixture. However, the absorption of oil dissolved in water is controlled by the distributive equilibrium between water and oil rather than the adsorptive equilibrium of activated charcoal. Hence, it is not suitable for high-level purification of oil when oil concentration is low.

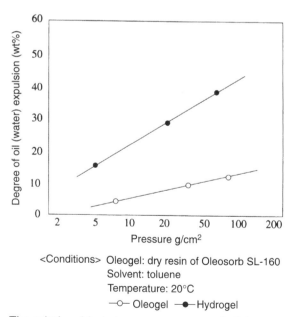

<Conditions> Oleogel: dry resin of Oleosorb SL-160
Solvent: toluene
Temperature: 20°C
—○— Oleogel —●—Hydrogel

Fig. 2   The relationship between pressure and oil (water) expulsion.

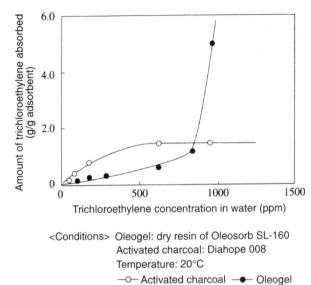

<Conditions> Oleogel: dry resin of Oleosorb SL-160
Activated charcoal: Diahope 008
Temperature: 20°C
—○— Activated charcoal —●— Oleogel

Fig. 3 Adsorption isotherms of trichloroethylene.

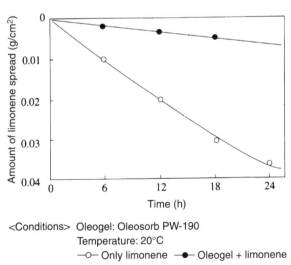

<Conditions> Oleogel: Oleosorb PW-190
Temperature: 20°C
—○— Only limonene —●— Oleogel + limonene

Fig. 4 Inhibition effect from spreading of limonene.

However, it has a high removal rate for oil suspended or floating at high concentrations. Therefore, by combining an oleogel with activated charcoal, it is possible to design a high-performance wastewater treatment system [7]. Figure 3 shows the adsorption isotherm of trichlorethylene.

4.6   Inhibition Effect of Spreading

Oleogels release oil slowly once it has been absorbed. Figure 4 shows the inhibition effect that occurs following spreading on limonene. The spreading per unit amount of oil absorbed by the oleogel is significantly smaller than would be the case if the oil had not been absorbed by an oleogel.

5   APPLICATION OF OLEOGELS

Applications of oleogels, including those currently being developed, will be introduced.

5.1   Oil Treatment Agents

With their ability to absorb oil selectively from an oil/water mixture, oleogels are used in various industries to remove solvents and impurities [8]. They are also highly valued for the process in which halogenated solvents or plant/animal oils that enter into wastewater during washing of metals and precision parts and for food processing must be removed [9, 10]. An application as a treatment filter for the oil mist generated by paint factories is also being considered [11]. In the future, application of oleogels to various separation technologies will be developed.

5.2   Sealant

A new sealant that combined an ordinary sealant and an oleogel is being designed. The oleogel will absorb the oil that is in contact with the sealing surface, which increases the sealing pressure and improves the sealing effect. Applications to packing and gaskets for automotive use are being studied [12].

5.3   Sustained Release Matrix

Using the property of oleogels in which oil is gradually released once it has been absorbed into the air or water provides for research on sustained release matrix applications. For example, it has been confirmed that multiple-component perfumes absorbed by an oleogel release a constant balance of fragrances. In addition, the gel maintains shape well. Therefore, it is used as a matrix for perfumes used in automobiles and homes [13–16]. Applications in horticulture for insecticides, insect repellents, antifungal agents, fertilizers, antifoaming agents, antisoil agents and solid fuel uses [17] are also being evaluated. When our understanding of distributive equilibrium of oil/oleogel/water phases is further advanced, the application areas for oleogels will increase.

5.4   Additives for Resins and Rubbers

It is possible to increase significantly the concentration of plasticizers, antiflammable agents, and oleophilic additives by adding oleogels to rubber or plastic. Figure 5 shows the increased amount of processed oil

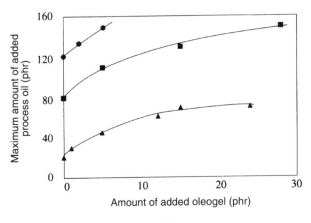

<Conditions> Oleogel: Oleosorb PW-170
Oil: Paraffin oil Sunpar 110 (Nippon Sun Petroleum)
Heated roll: 15 cm diameter, roll temperature 50°C,
   rotation speed 10.3 rpm/9.1 rpm
Initial mixing time: 1 min
Mixing time: 5 min
—●— Butadiene rubber (BR01)
—▲— Isoprene rubber (IR2200)
—■— Styrene-butadiene rubber (SBR 1502)
All are from Nippon Rubber Co.
Evaluation: The maximum amount of the process
   oil added to the rubber that can be debonded
   from the roll.

Fig. 5   Increased process oil gained by the addition of an oleogel.

possible when an oleogel is added to various rubbers. Prevention of bleeding of oleophilic additives [18], reduction of reflection [19], and improved processability, such as reduction in both mixing time and adhesion to mixing devices, are some of the benefits gained by adding oleogels. It is possible to design softer, antiearthquake, stretching, insulating, antiflammable, and antifreezing materials.

5.5 Other Applications

The volume and refractive index changes following absorption of oil by oleogels are used to detect oil leaks. These technologies are now commercially available [20, 21]. Oleogels are also used as liquid membranes for ion electrode membranes [22]. Oleogels have the potential to be useful in analysis and medical devices.

6 CONCLUSION

The application examples introduced here are only a few of the oleogel uses. Many properties of oleogel are not yet well understood and there are many technological problems to solve. In order to improve oleogels enough to create industrially high value-added materials, the following points need to be considered:

i)   increased oil absorption ratio;
ii) wider applicability of these oils, including the development of oleogels that can absorb polar solvents in the solubility parameter range of 10–15 $(cal/cm^3)^{1/2}$ and silicone oil;
iii) wider availability of shapes, including increased availability of various particle sizes, oleogel films, and composites with other materials; and
iv) reduction in costs.

By achieving these goals, the application studies on oleogels will further advance in the future.

REFERENCES

1   (1970). Tokkyo Kaiho 27081.
2   (1975). Tokkyo Kaiho 15882.
3   (1975). Tokkyo Kaiho 59486.
4   (1975). Tokkyo Kaiho 94092.
5   (1989). Tokkyo Kaiho 80441.

6   (1993). Yushi, **46**: 36.
7   (1993). Tokkyo Kaiho 15871.
8   (1995). Tokkyo Kaiho 251174.
9   (1992). Tokkyo Kaiho 15286.
10 (1992). Tokkyo Kaiho 145904.
11 (1992). Tokkyo Kaiho 90814.
12 (1992). Tokkyo Kaiho 341662.
13 (1991). Tokkyo Kaiho 272766.
14 (1991). Tokkyo Kaiho 287505.
15 (1992). Tokkyo Kaiho 261474.
16 (1991). Tokkyo Kaiho 297340.
17 (1992). Tokkyo Kaiho 159394.
18 (1993). Tokkyo Kaiho 214114.
19 (1994). Tokkyo Kaiho 57007.
20 (1993). Tokkyo Kaiho 203533.
21 (1992). Tokkyo Kaiho 149042.
22 (1993). Tokkyo Kaiho 322834.

Section 4

Application of Superabsorbent Polymers Gels to Oil-Water Separation (Use of Superabsorbent Polymer Containing Sheet)

KATSUMI KOBASHIMA

1 INTRODUCTION

Oil-water separation is an important technology in many areas as shown in Table 1 [1]. In particular, along with the need for environmental protection, other problems with oil-containing water treatment have also become important. The latest problem is how to deal with Freon and the problems associated with the replacement materials. In the parts-manufacturing industries, the degreasing process used for washing and rinsing solutions is becoming an important issue. Separation of small oil particles and emulsified oil is especially difficult and effective methods have not been developed. Here, application examples of hydrogels for oil-water separation using oil-water separation paper sheets developed by the authors will be given [2].

Table 1   Fields that require oil-water separation.

Condition	Countermeasure	Corresponding field	Supplemental
Water-oil	Dehydration, refining	Oil refining Refining of food oils and fats Purification of solvents Manufacturing of paints Related to lubricant, insulator, motor, and handling oils	Process Control Products/quality control
Oil-water	1) Countermeasures for environmental maintenance, protection of maritime pollution by oil leakage and oil-containing waste water, protection of river and soil pollution	Industries related to petrochemical industry Countermeasures for wastewater from ships Industries that use fats Processing of marine products Processing of livestock meat Related to food oils and fats Wastewater from facilities that provide meals, hotels and restaurants Related to machines, metals, and plastic products Countermeasure for construction wastewater Related to electronic parts manufacturing Surface treatment of metals (electroplating and painting) Cleaning and machine maintenance service Related to machine maintenance (car wash) Related to transportation (cargo equipment, deliveries) Related to disposal of waste products from oil and fat industries	Related laws Ocean pollution prevention law Water pollution prevention law
	2) Process management, countermeasures for cleaning the washing liquid, reduction of cost by prolonging the use time of the washing liquid.	Management of degreasing (electroplating, painting) Performance holding of lubricant (metalwork) Recycling of deactivated and degraded oils	Production management Product management Cost management

2 SEPARATION AND GELATION OF FINE OIL PARTICLES AND EMULSIFIED OILS

2.1 Coalescence Separation of Fine Oil Particles

When a mixture of materials with different properties, such as oil and water, passes through fine pores, the mixed state becomes unstable and particle size increases by association. This phenomenon is known as the capillary coalescence effect. If the coalesced oil adheres, liquid flow is halted.

The authors developed the capillary coalescer sheet shown in Fig. 1, which prevents adhesion of the coalesced oil droplets to the mater membrane of a hydrogel [3].

2.2 Function of Hydrogels

Oil that is deposited on a water-insoluble hydrogel, such as agar and connyaku, can be readily separated. This is because the water film on the hydrogel surface inhibits adhesion of oil onto the gel networks. Based on this mechanism shown as model 2) in Fig. 1, the authors attempted separation of collected oil by making the substrate of coalesced oil particles unstable.

Figure 2 illustrates the separation of adhered oil as water is absorbed in the superabsorbent incorporated sheet onto which oil is deposited prior to water immersion.

3 PRACTICAL ASPECTS OF OIL-WATER SEPARATION SHEET

3.1 Construction of Oil-Water Separation Sheet

Although various methods have been reported by the authors, an example shown in Table 2 will be introduced here. Oleophilic polypropylene fibers (pulp-like) were used as the material to entrap oil. A crosslinked poly-(ethylene oxide) hydrogel was used so that the material could desorb the entrapped oil. This crosslinked gel is insoluble in water, although it disperses into water-containing methanol colloidally. This suspension was deposited onto a sheet.

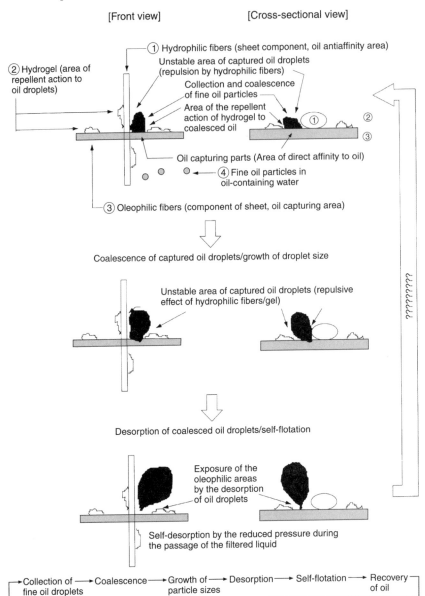

Fig. 1 Operation mechanism of oil-water separation sheet using a super-absorbent polymer gel.

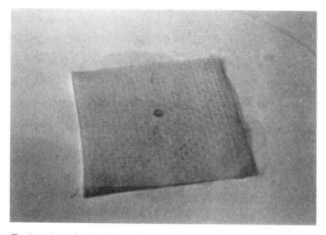

Explanation: On the sheet described in Table 3, oil was placed
and then swelling took place. Prior to the swelling, the adhered
oil penetrated the sheet. However, when the sheet was immersed
in water, the adhered oil was able to be separated.

Fig. 2   Oil separation on a hydrogel-containing oil-water separation sheet.

3.2   Evaluation of Oil-Water Separation Sheets

The general properties of oil-separation sheets are shown in Table 3. The
A test described in the table is the test method developed by the authors
(see Fig. 3). This method evaluates the oil-water separation sheets by how
much an oil droplet placed on top of the wet sheet permeates the sheet. If
there is no hydrogel included, the oil spreads throughout the surface. Thus,
it is possible to estimate qualitatively the oil-water separation by observing
the remaining state of the oil droplet.

3.3   Experimental Examples of Oil-Water Separation Effects

Light spindle oil was forced-mixed into a water-soluble surfactant
continuously. Using the forementioned sheet and the method as described
in Fig. 4, this oil-containing solution is treated. Figure 5 shows the
variation of the oil content at the exit (point A in the figure). The
condition of the oil droplets (sampled at points B and C in the figure)
before and after the treatment is illustrated in Fig. 6.

As already mentioned, this sheet functions not by simple filtration
separation but by coalescence due to the capillary effect. Therefore, it is
necessary to install a tank to separate the coalesced droplets by floating

Table 2 Composition of practical oil-water separation sheets.

Raw materials for sheet construction	Hydrophilic fibers (cotton): oleophilic fibers (PP: 2 d, 5 mm) = 1:1	
	Film former (PVA), 3% of the fiber weight	
Sheet formation	Two layers of round net + one layer of short net 7:3, average weight per unit sheet area = 120 g/m^2	
	A dispersant for paper (poly(ethylene oxide) type) 1–3 ppm	
Addition of water resistance	Pressure treatment by a hot roll at 110–120°C (fusing treatment of synthetic fibers)	
Enhancement of water,	Immersion solution	1) N-methylolacrylamide (crosslinking agent)	5.0 %
pressure, and oil resistance		2) Ammonium peroxodisulfate (polymerization catalyst)	0.4
(crosslinking in the sheet)		3) Ammonium chloride (condensation catalyst)	0.4
		4) Fluorine type oil resistant agent	0.1
	The sheet is immersed in the above solution (solution content 100%), dried and cured at 100°C for 30 s	
Processing	Formation of hydrogel	Crosslinked poly(ethylene oxide) (soluble in hydrophilic solvents)	2.0 %
		Cationic polymer (emulsified in methanol: water = 1:1 imixed solvent)	0.1
		The sheet is immersed in the above solution (water content 200%) and dried at 110°C for 30 s	
Construction of accelerated liquid spreading	A wavy sheet is sandwiched like cardboard (fusing temperature 140–150°C)	

Supplemental 1) Formation of hydrogel is done by commercially available polymer and it can disperse colloidally in a hydrophilic solvent
2) The enhanced water and oil resistance is achieved by condensation reaction within the sheet (see, authors, *J. Functional Paper Res.*, Jpn., No.6 (1979))
3) Solution immersion is carried out using industrial production scale machine
4) Addition of wavy sheet is to make commercial products. Product appearance: cardboard honeycomb structure

Table 3 General properties of oil-water separation sheet (composition of Table 2).

	Weight per unit area of sheet (g/m²)	Thickness (mm)	Density (g/cm³)	Wet strength (kgf/cm³)	water absorption (mm/10 min)	Oil resistance (time)	A test (s)
Original paper	140	0.53	0.26	0.8	100	0	0
Processed sheet	152	0.53	0.29	2.5	35	24<	180

Supplemental 1) Original paper: heat treatment only processed sheet: product with final treatment
2) Oil resistance test: when spindle oil droplet is placed on the sample, the time is measured until the oil droplet disappears
3) A test: Explained in Fig. 3 (see, the author, *J. Functional Paper Res.*, No.17 (1989))
4) Other testing methods: JIS-P

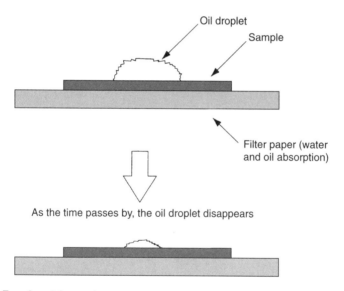

As the time passes by, the oil droplet disappears

Experimental procedure:
1: The sample is swollen by water
2: Deposit 0.5 cc of light spindle oil on the sample
3: Place the sample on a dry filter paper
4: Measure the time until the oil droplet on the sample disappears

Fig. 3 A simple method to evaluate oil-water separation effect (Test A in Table 3).

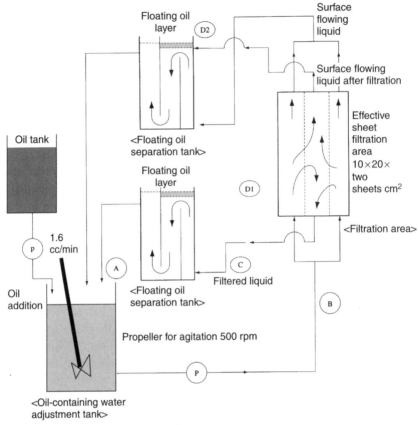

Oil tank

1.6 cc/min

Oil addition

Floating oil layer

Surface flowing liquid

Surface flowing liquid after filtration

Effective sheet filtration area 10×20× two sheets cm²

<Filtration area>

<Floating oil separation tank>

Floating oil layer

Filtered liquid

<Floating oil separation tank>

Propeller for agitation 500 rpm

<Oil-containing water adjustment tank>

Water-soluble surfactant solution, 20 ℓ

Supplemental
1: Flux of each stream
  Sheet surface flow   1 ⎤
  Sheet surface flow after filtration 1 ⎥— Total flux 3ℓ/min
  Filtrated flow   3 ⎦ Filtration pressure, 0.2 kgf/cm²

2: A–Oil content measurement point (result is shown in Fig. 5)
 B, C–Checkpoint for coalescence effect (condition of oil droplet)
 (Sample weighing point in Fig. 6)
 D1, D2–Location of floating oil separation

Fig. 4 A testing method for oil-water separation (the sheet described in Tables 2 and 3).

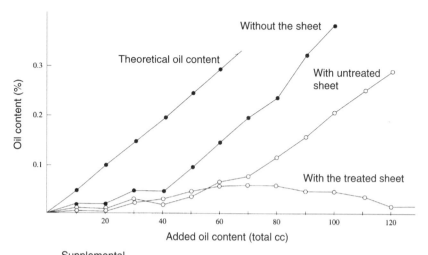

Supplemental
Oil content measurement: the Bobcock method (oil flotation by heating, volume measurement)

Fig. 5 Oil-water separation effect (the sheet in Tables 2 and 3 using the method described in Fig. 4).

separation (points D1 and D2 in the figure). Figure 7 shows the appearance of oil adsorption in the vicinity of the hydrogel in the sheet.

4 CONCLUSION

Application of gels in oil-water separation has been summarized here. Although only fundamental examples have been introduced, commercial products and instruments based on these concepts are already available. These are widely used for washing metal surfaces, managing lubricating oil in metal processing, and for other oil-containing water treatments [4]. Only the oil-water separation paper sheets were introduced here. However, the authors are currently developing fine oil-water separation media using hydrogel-containing porous ceramics. While the oil-water separation in the process control of metal processing was described, a safer treatment technique that can be used for oil content control of extracted soups in natural seasoning manufacturing is also desired. The authors are deeply indebted to the authors of the references cited.

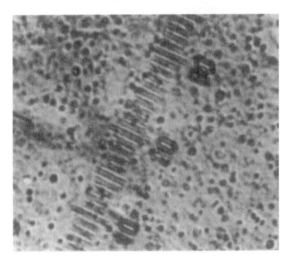

Before the passage (Fig. 4, point B)

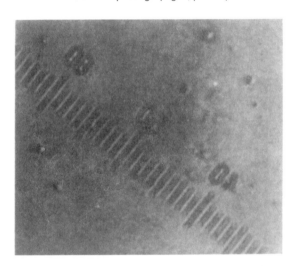

After the passage (Fig. 4, point C)

Supplemental
The smallest scale: 2.5 μm

Fig. 6   Appearance of the oil droplets before and after the passage through the oil-water separation sheet.

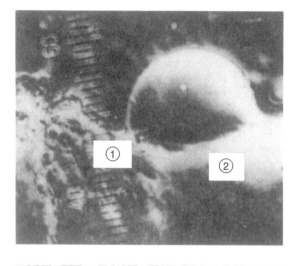

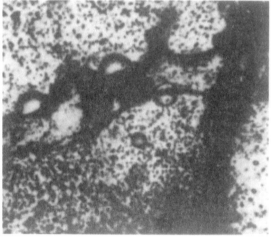

Supplemental
The smallest scale: 2.5 m

① Hydrogel ② Oleophilic part

Fig. 7   Appearance of the collected oil in the vicinity of the hydrogel.

REFERENCES

1 For example, Kuboshima, K. (1983). *Oil-water Separation Technology*, MOL (Ohm-sha), 21; Kuboshima, K. (1986). *J. Soc. Fiber Sci., Jpn.* **42**: 9; Special issue, (1973). Technology for waste oil treatment. *PPM* 12; (1976). *Introduction of Ocean Pollution Prevention*, Naruyama-do Shoten; Special Issue (1978). An approach for waste oil treatment. *PPM* 4; (1987) Tokkyo Kaiho 49915; (1987). Tokkyo Kaiho 279810; (1988). Tokkyo Kaiho 130797; (1991). Tokkyo Kaiho 65207; (1991). Tokkyo Kaiho 77603; (1992). Tokkyo Kaiho 20060; (1974). Tokkyo Kaiho 120261.

2 For example, Kuboshima, K. (1979). *Paper Pulp Technol. Times, Jpn.* **22**: 9; (1978). *J. Soc. Synthetic Paper* 37; (1980). *J. Soc. Synthetic Paper* 19.

3 (1986). Tokkyo Kaiho 12729.

4 Kuboshima, K. *Technical Report, Maruchi Industrial Research*, Hamamatsu, Japan.

Section 5

Application of Gels to Latent Heat Thermal Storage Media

HIROYUKI KAKIUCHI

1 INTRODUCTION

Among the first latent heat thermal storage media is a commercially available ice pillow made of gel [1]. Ice is used as the latent heat thermal storage medium, while softness is achieved by a hydrogel. Gelation inhibits thermal convection, which then maintains a low temperature for a prolonged period of time. The ice used as a latent heat thermal storage medium absorbs 80 cal/g from its surroundings. Those thermal storage media that utilize latent heat accompanying phase transition (such as from liquid to solid or from liquid to gas) are called latent heat thermal storage media. In addition to the latent heat type, thermal storage media include the heat capacity type and the chemical type. Table 1 lists types and classifications of thermal storage media [2].

Gels are used for latent heat thermal storage media mainly when the phase transition from solidification to melting state is utilized. When a material melts, it becomes a liquid. By gelling this liquid, practical

393

Table 1 Types and classification of thermal storage media.

1) Heat capacity type:	solid thermal storage media....	Nonmetallic type... bricks, soil etc.
		Metallic type... copper, magnesium and molten salt etc.
	Liquid thermal storage media....	Water, organic oils, brines, liquid metals etc.
	Gaseous thermal storage media....	High-pressure and temperature water vapor and high-temperature gas
2) Latent heat type:	inorganic solid-liquid phase transition: icewater	
		Hydration salts ($Na_2SO_4 \cdot 10\,H_2O$, $CH_3COONa \cdot 3H_2O$ etc.)
		Simple salts (LiH, LiF, NaOH etc.)
		Molten eutectic salts ($NaOH$-$NaNO_3$, $NaCl$-KCl-$MgCl_2$ etc.)
	Organic solid-liquid phase transition type:	Paraffin wax (n-decane etc.)
		Fatty acids (caprylic acid, stearic acid)
		High density polyethylene, urea etc.
		Transition heat of organic solid (pentaerythritol etc.)
	Gas-solid phase transition (steam accumulator etc.)	
	Solid-gas phase transition (heat of sublimation of dry ice)	
3) Concentration differential type:	absorbent type... Lithium bromide-water, water-ammonium etc.	
	adsorbent type... Zeolite-water vapor, silica gel-water vapor etc.	
4) Chemical type:	Hydration and decomposition of hydration salt ($CaO + H_2O = Ca(OH)_2$ etc.)	
	Formation of ammonium compounds ($FeCl_2 \cdot 6NH_2 = FeCl_2 \cdot 2NH_3 + 4NH_3$ etc.)	
	Metal hydrate decomposition ($MgH_2 = Mg + H_2$ etc.)	
	Formation of gaseous inclusion compounds	
5) Others:	Photochemical reaction (photoisomerization etc.), biomass (plants etc.)	

applications have been achieved. The characteristics of gels used for latent heat thermal storage media include:

1) water-retention;
2) dispersion stabilization or viscosity enhancement of aqueous media;
3) water supply to the environment; and
4) flexibility.

Ways in which these characteristics of gels are used in latent heat thermal storage media will be explained here.

2   INORGANIC HYDRATION SALT TYPE THERMAL STORAGE MEDIA

2.1   Phase Separation Phenomena

Inorganic hydration salt has large thermal storage capacity and is inexpensive. Therefore, many studies have been reported on applications for thermal storage media.

Many inorganic hydration salts show an unharmonic melting phenomenon (peritectic reaction). Upon solidification, they are formed by the reaction between the lower hydration salts and the liquid. The lower hydration salts of inorganic salts in aqueous solutions exhibit generally greater density than the liquid phase and precipitate at the bottom of the container. As shown in Fig. 1 (a) [3], they separate into the solid S and

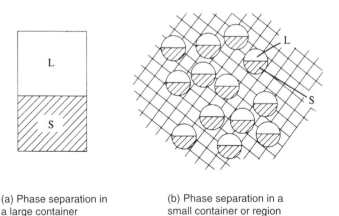

(a) Phase separation in   (b) Phase separation in a
a large container   small container or region

Fig. 1 Explanation of the thickener effect [3].

liquid L. This phenomenon is called phase separation. When phase separation takes place, the target hydration salt is formed around the lower hydration salt. Hence, the desired amount of the target salt cannot be obtained. If a viscosity enhancer called a thickener is added and the thermal storage media is divided into small regions, the contact area is increased and precipitation of lower hydration salt can be prevented. Table 2 lists representative combinations of an inorganic hydration salt and a thickener [4, 5].

2.2   Application of a Superabsorbent Polymer to Sodium Sulfate Decahydrate

2.2.1   Sodium Sulfate Decahydrate

Sodium sulfate decahydrate is a representative material of commercially used latent heat thermal storage media. Sodium sulfate decahydrate has a melting point at 32°C with the latent heat of melting of 60 cal/g. It is a grainy crystal below the melting temperature (see Fig. 2 (a)). When sodium sulfate decahydrate melts, it does not dissolve into the separated water at all, because solubility of sodium sulfate to water is negligible. Hence, as shown in Eq. (1), anhydrous sodium sulfate precipitates at the bottom of the liquid and separates into two phases with a saturated aqueous solution of sodium sulfate (see Fig. 2 (b)).

Melting:

$$Na_2SO_4 \cdot 10H_2O \text{ (S)} \rightarrow Na_2SO_4 \text{ (S)} + \text{saturated solution} \qquad (1)$$

When it is cooled as is, the surface of the precipitated anhydrous sodium sulfate returns to decahydro salt and the anhydrous sodium sulfate remains inside of the particle. As shown in Eq. (2), it phase separates into sodium

Table 2 Examples of inorganic hydration salt thickeners.

Thermal storage media	Thickener
$CaCl_2 \cdot 6H_2O$	Hydroxyethyl cellulose
$Ca(NO_3)_2 \cdot 4H_2O$	Poly(acrylic acid)
$NaCO_3 \cdot 10H_2O$	Oxidized polyethylene
$Na_2HPO_4 \cdot 12H_2O$	Starch glue
$Na_2S_2O_3 \cdot 5H_2O$	Wood pulp
$Na_2SO_4 \cdot 10H_2O$	Clay(bentonite, attapulgite)

(a) Grainy crystals   (b) Melt

(c) Recrystallized   (d) Gelled thermal storage medium

Fig. 2   Photographs of sodium sulfate decahydrate.

sulfate decahydrate, anhydrous sodium sulfate, and an unsaturated aqueous solution of sodium sulfate (see Fig. 2 (c)).

$$Na_2SO_4\,(S) + \text{saturated solution} \rightarrow Na_2SO_4\,(S) \cdot 10H_2O\,(S) + Na_2SO_4\,(S)$$
$$+ \text{unsaturated solution} \qquad (2)$$

2.2.2 Application of Superabsorbent Polymers

In order to form a small region as shown in Fig. 1 (b), thickeners, such as attapulgite clay that uses a capillary effect [6] and carboxymethyl cellulose [7], a water-soluble polymer are added. However, inorganic hydrated salts separate water when melting occurs and they incorporate water when solidification occurs. This cycle is repeated thousands of times. Thus, the sodium sulfate crystals grow and precipitate. In order to prevent the crystals of sodium sulfate from growing, addition of a crystal growth inhibitor to the water-soluble polymer is being evaluated [8]. Although a latent heat thermal storage medium using attapulgite clay as a thickener is commercially available [9] it has been suggested that thermal storage times may not meet the claims of the manufacturer [10].

A superabsorbent polymer was used as a thickener for sodium sulfate decahydrate [11]. There is also a report that a superabsorbent polymer having a water absorptivity of 600 times its volume was added at a 2.9 wt% amount. When sodium sulfate decahydrate repeated the melting-solidification cycle 100 times, there was no separation of water [12]. Even very small amounts of superabsorbent polymer can enhance viscosity and more than 90% of the water held is free water. Hence, use of this type of polymer as a thickener effectively takes advantage of water retention, water supply, dispersion stabilization, and the viscosity enhancement properties of superabsorbent polymers [13].

Currently, there is no well-accepted theory on preventing thermal storage media from decreasing in effectiveness when a superabsorbent polymer is used as the thickener. However, there is a report that the rate of crystal growth of sodium sulfate decahydrate and anhydrous sodium sulfate was smaller with a superabsorbent polymer than with attapulgite clay. If this is due to crystal growth only in this microenvironment, the concept shown in Fig. 1 (b) may be practically used. There are also some problems with the gelled sodium sulfate decahydrate [10]. Heat build-up is slow because there is no convection. There are also regions with fast and slow crystallization rates.

2.3 Application Examples

2.3.1 Thermal Storage Media for Floor Heating

A thermal storage media type floor heating system is an actual application example of sodium sulfate decahydrate [14]. A brief explanation of the system is given as follows. Beneath the floor, plastic capsules of a thermal storage medium are installed. Using electricity late at night, the thermal storage medium is melted by an electric heater. During the daytime, the floor will be maintained at around 25°C by the latent heat of the sodium sulfate decahydrate. According to Sumitomo Chemicals (Brand name: Sumithermal) [15], the heat capacity held in storage will not change even if storing and releasing are repeated 8000 times. It is also said that in the event the container is accidentally damaged, the thermal storage medium will not flow because it is a gel.

2.3.2 Thermal Storage Media for Air Conditioning

To help reduce the extreme demand on electricity during the daytime in summer, a thermal storage system with ice that uses late night electricity is now commercially available [16]. However, to make ice, the thermal medium must be cooled below the freezing point of water. This causes reductions in cooling machine efficiency. A mixed salt of sodium sulfate decahydrate that can hold heat by cold water at 3–5°C and has a melting point of ≈ 9°C has been reported recently [17]. This system uses sodium sulfate decahydrate as the main component.

To reduce the melting temperature of sodium sulfate decahydrate to 9°C, the melting point adjustment agents such as ammonium chloride, sodium chloride, and ammonium bromide are mixed. In this thermal storage medium, a superabsorbent polymer is also used as a thickener. If the mixture is not homogeneous enough, preparation of the thermal storage medium is affected—the heat stored is reported to be less than original designs specified. From this, it can be seen that formation of small regions as shown in Fig. 1 (b) is important. Figure 2 (d) illustrates a gelled thermal storage medium for air conditioning.

3 THERMAL STORAGE MEDIA FROM ORGANIC COMPOUNDS

There are many reports on the application of paraffin wax, fatty acids, and high density polyethylene as thermal storage systems. Unlike inorganic compounds, organic compounds do not have phase separation problems. Thus, it has been said that there is no need for gelation. However, recently, a thermal storage medium using a gelled paraffin that can maintain a solid structure even above the melting temperature of the paraffin was introduced [18]. In another example of a gelled system, a mixture of a paraffin and hydrocarbon rubber crosslinked by a hydrocarbon crosslinking agent has been reported [19]. Among the characteristics of this paraffin thermal storage medium, it will not fracture when processed into a sheet and it exhibits good flexibility. Thus, it utilizes one of the important properties of gels: it is between a solid and a liquid state. Although application gels are dominated by the hydrogels type, this is a rare and important commercial example of oleogels.

REFERENCES

1 Shimoda, A. (1996). *Kagaku* **50**: 75.
2 Inaba, H. (1996). *Reitoh* **71**: 443.
3 (1985). *Thermal storage and enhanced thermal storage technologies*, Thermal Storage and Enhanced Thermal Storage Technology Editorial Board Ed., IBC, p. 169.
4 (1985). *Ibid.*, p. 170.
5 Abhat, A. (1983). *Solar Energy* **30**: 313.
6 Brown, P.W. *et al.* (1986). *Solar Energy Mater.* **13**: 453.
7 (1983). Tokkyo Kokai 52996.
8 Marks, S.B. (1983). *Solar Energy* **30**: 45.
9 Hisamoto, S. (1993). *Reitoh* **68**: 763.
10 Kimura, K. (1996). *Reitoh* **71**: 458.
11 (1979). Tokkyo Kokai 16387.
12 Shin, B.C. *et al.* (1989) *Energy* **14**: 921.
13 (1983). Tokkyo Kokai 221395.
14 Yamaguchi, M. (1995). *Kagaku Kogaku Ronbunshu* **21**: 853.
15 *Technical Bulletin of 'Smithermal,''* Sumitomo Chemicals.
16 For example, (1993). *Kagaku Kogaku* **57**: 61.
17 Kakiuchi, H. *et al.* (1996). *Proc. 33th Symp. on Thermal Transfer, Jpn.*, pp. 471–472.
18 *Technical Bulletin on New Latent Heat Thermal Storage Media*, Mitsubishi Electric Wire Industries Co.
19 (1991). Tokkyo Kokai 66786.

Section 6
Application for Electrophoresis

YOSHINOBU BABA

1 INTRODUCTION

Electrophoresis is one of the most important biotechnologies used for DNA and protein analyses. It is an indispensable technique for human genome analysis, DNA sequencing, gene analysis, gene therapy, and molecular weight determination of proteins [1–3]. For electrophoresis, a gel or polymer solution is needed as the separation medium. By the molecular sieve effect of these matrix materials, biopolymers can be separated. Traditionally, gel electrophoresis using a flat plate gel was dominant [1, 2]. However, a capillary electrophoresis, which performs electrophoresis in a silica capillary with a diameter of $<100\,\mu m$, is now drawing attention [3–6]. For many years only polyacrylamide and agarose were used as electrophoresis gels. In the 1990s, the work on the human genome mapping project began and requirements for developing higher performance gels increased. The human genome mapping project made it necessary to develop a very high resolution gel in order to analyze a tremendous amount of genome information. Development of capillary array electrophoresis in which approximately 100 capillaries are placed in parallel and a gel that can be exchanged with the gel inside the capillary to

401

drve the microchip system is also becoming an important subject. To cope with these requirements, new gels have been developed recently and are now replacing the traditionally used ones [7–10].

The characteristics of these recently developed gels in addition to the traditional used ones and their application to capillary electrophoresis will be introduced here. Those readers who are interested in gel electrophoresis of these gels should refer to monographs [1, 2, 7] and a review [8].

2   POLYACRYLAMIDE AND AGAROSE GELS

Gel electrophoresis is the technique used to separate biopolymers, such as DNA and proteins, by applying voltage to both ends of the gel plate (see Fig. 1(a)). Separation can be adjusted by controlling the gel concentration in order to adjust the size of the biopolymers. For example, in DNA analysis, a gel containing 5–8% polyacrylamide is used for DNA with several tens–several hundred basis pair units; 3.5% polyacrylamide or 1–2% agarose is used for DNA with several hundreds–several thousands base pair units; and 0.3–1% agarose is used for several thousands–several tens of thousands base pair units.

On the other hand, capillary electrophoresis uses a fine capillary instead of a gel plate and separate biopolymers. The capillary used is made of a high purity silica with an inner diameter of 50–100 μm, an outer diameter of 150–400 μm, and a length of 10–50 cm as shown in Fig. 1 (b). The surface of the capillary is coated in order to resist breakage. In order to analyze biopolymers, a fine capillary is filed with a gel or a polymer solution [3–6]. For capillary electrophoresis, polyacrylamide is mainly used. Both ends of the capillary are placed in a container that contains a buffer solution in which high voltage electrodes are also immersed. When a sample is injected the one end of the capillary and anode electrode (the left-hand side in the figure) are both inserted in the sample tube. A voltage of 100–200 V/cm is then applied for 1–10 s. This operation injects the sample into the capillary electrophoretically. After injecting the sample into the capillary, the capillary and the electrode are returned to a buffer solution. When the sample side is the anode and high voltage is applied at 100–500 V/cm, the sample migrates to the cathode and separation is then achieved.

The detection is done online by a uv detector or laser-induced fluorescence detector.

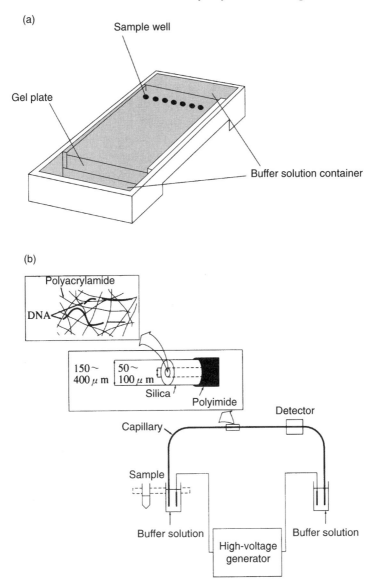

Fig. 1 Instruments for gel electrophoresis (a) and capillary electrophoresis (b).

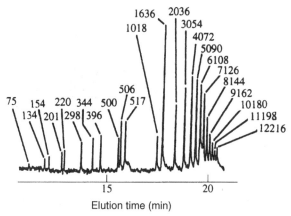

Elution time (min)

The number in the figure: the basic pair unit of DNA

Fig. 2 Separation of double strand DNA using capillary gel electrophoresis with polyacrylamide gel [11].

Figure 2 shows the double strand DNA obtained using capillary electrophoresis with polyacrylamide [11]. As can be seen from the figure, almost all base pair units from 75 to 12,216 base pair unit are separated. Furthermore, the separation time was only 20 min. A 500 base pair unit fragment is completely separated from a 506 base pair unit – with only 6 units between them.

Figure 3 illustrates an example of DNA sequencing using capillary gel electrophoresis with polyacrylamide [12]. From this figure, it can be seen that DNA gene information of up to 300 base pair units can be read in just 10 min. If the same sequencing is performed by the traditional method, it will take 2–3 h. It is shown, therefore, that capillary electrophoresis was more than ten times as fast as the other method. The precision of DNA sequencing also improved due to the high resolution of polyacrylamide.

3 NEWLY DEVELOPED GELS

Although traditionally used polyacrylamide and agarose were quite useful gels for electrophoresis, several new gels for higher resolution and functions have been developed [7–10]. The representative example among those gels is commercialized as Hydrolink™. In this series of gels, the one for DNA sequencing consists of a copolymer of acrylamide,

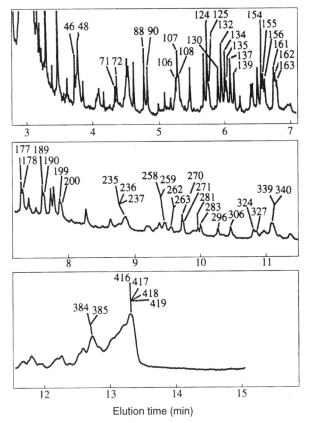

The number in the figure: the base pair unit of DNA

Fig. 3 Separation of DNA sequence reaction products using capillary gel electrophoresis with polyacrylamide gel [12].

N-methylolacrylamide and bisacrylamidomethyl ether; for double strand DNA sequencing it is made of a copolymer of N,N-dimethylacrylamide and ethylene glycol dimethacrylate. The Hydrolink^{TM} gel for DNA sequencing has been shown to increase the reading ability of the commercial automatic sequencer from 500 base pair unit to more than 1000 base pair unit. Traditionally, polyacrylamide gel was used as the standard gel in an automatic DNA sequencer. However, Hydrolink^{TM} is now used much of the time.

It is known that polyacrylamide is not very stable when undergoing hydrolysis. It is not desirable to use high-resolution analysis to hydrolyse the gel polymer. To overcome this shortcoming, new gels made of poly(N-acryloylaminoethoxy ethanol) and poly(N-acryloylamino propanol) have been developed [8, 9, 13]. These gels are more than 500 times as stable as polyacrylamide gel during hydrolysis and they permit extremely high resolution in DNA analysis. Figure 4 shows an example of a double strand DNA analysis using poly(N-acryloylaminoethoxy ethanol). As seen from the figure, the base pair units from 75 to 12,314 are nearly completely separated.

In capillary gel electrophoresis, the gel used can be replaced, which was not possible with the traditional disposable electrophoresis gel. This allows an automated system to be constructed and costs to be reduced, an important consideration given the tremendous amount of information being acquired on human genome DNA. A new thermoresponsive gel is being evaluated for use with capillaries [14]. A thermoresponsive gel reduces its viscosity suddenly above a certain temperature. Hence, it can be used as a DNA sequencing gel at room temperature and, then, easily removed from the capillary after measurement simply by increasing the temperature.

Figure 5 is an example of DNA sequencing using capillary gel electrophoresis and a thermoresponsive gel. It is obvious from this figure

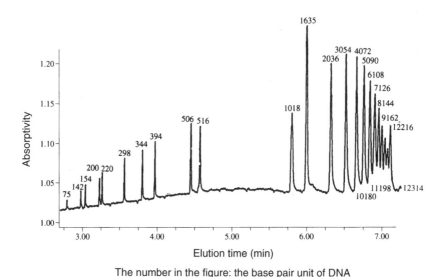

Fig. 4 Separation of a double strand DNA using capillary electrophoresis with poly(N-acryloylaminoethoxy ethanol) [13].

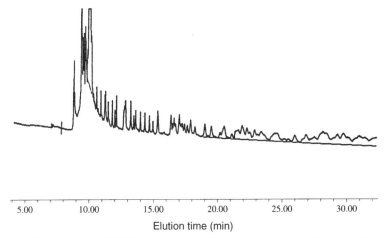

5.00   10.00   15.00   20.00   25.00   30.00

Elution time (min)

Fig. 5 Separation of DNA sequence reaction products using capillary gel electrophoresis with a thermoresponsive gel.

that DNA sequencing is feasible at room temperature and that the resolution was comparable to that of the traditional gel. After measurement, the gel could be easily removed from the capillary by increasing the temperature.

4 CONCLUSION

Polyacrylamide and agarose have been widely used for electrophoresis, in particular, in DNA sequencing, gene analysis, and gene therapy. Recently, high-performance gels that can replace traditional gels have been developed. By using these gels for capillary electrophoresis, analysis performance has improved dramatically. However, the study of electrophoresis gels has just begun. In the future, it is hoped that new development of gels accelerates application of capillary gels to fast sequencing of human genomes, high performance genome mapping, genetic screening, DNA analysis, and genetic analysis for illnesses. It is further hoped that such development significantly contributes to the human genome projects.

REFERENCES

1 Sambrook, J., Fritsch, E.F., and Maniatis, T. (1989). *Molecular Cloning*, 2nd ed., Chap. 6, Cold Spring Harbor: Cold Spring Harbor Laboratory Press.

2 Richwood, D. and Hames, B.D. (eds). (1990). *Gel Electrophoresis of Nucleic Acids,* 2nd ed., Oxford: IRL Press.

3 Baba, Y. (1997). *Capillary Gel Electrophoresis and Related Technique,* Amsterdam: Elsevier.

4 Baba, Y. and Tsuhako M. (1992). *Trends Anal. Chem.* **11**: 280.

5 Baba, Y. (1996). *J. Chromatogr.* **B687**: 271.

6 Dolnik, V. (1994). *J. Microcol. Sep.* **6**: 315.

7 Molinari, R.J., Connors, M., and Shorr, R.G.L. (1993). *Adv. Electrophoresis* **6**: 43.

8 Chiari, M. and Righetti, P.G. (1995). *Electrophoresis* **16**: 1815.

9 Simo-Alfonso, E., Gelfi, C., Sebastino, R., Citterio, A., and Tighetti, P.G. (1996). *Electrophoresis* **17**: 723.

10 Yoshida, R., Uchida, K., Kaneko, Y., Sakai, K., Kikuchi, A., Sakurai, Y., and Okano, T. (1995). *Nature* **374**: 240.

11 Baba, Y., Tomisaki, R., Sumita, C., Tsuhako, M., Miki, T. and Ogihara, T. (1994). *Biomedical Chromatogr.* **8**: 291.

12 Tomisaki, R., Baba, Y., Tsuhako, M., Takahashi, S., Murakami, K., Anszawa, T. and Kambara, H. (1994) *Anal. Sci.* **10**: 817.

13 Chiari, M., Nesi, M. and Righetti, P.G. (1994). *Electrophoresis* **15**: 616.

14 Baba, Y., Inoue, H., Iwakoshi, H., Hosoya, K. and Tanaka, N. (submitted).

CHAPTER 8

Electric and Electronic Industries

Chapter contents

Section 1
Communication Cables

1 INTRODUCTION

When communication cables, for example, fiberoptic cable and copper wire cable, are used underground, underwater or in air, water penetrates into the cable when the external sheath is damaged, thereby causing problems. In order to prevent water from penetrating into the cable, a layer of superabsorbent polymer is placed between the cable outer layer (sheath) and the cable core. If there is damage to the external sheath and water penetrates, the superabsorbent polymer absorbs the water and swells, thus protecting the cable core. The water that first penetrates the cable is used to form a hydrogel and is fixed to prevent water from running along the cable line. If water manages to enter through the interface between the cable core and the outer layer, serious damage will result. Figure 1 schematically illustrates the structure and mechanism for prevention of water penetration.

Communication cables have evolved into multiple strands, of finer diameter and higher performance to meet the demands of the growing information society. This has resulted in more sensitive electronics devices and the interruption in communications due to water penetration has

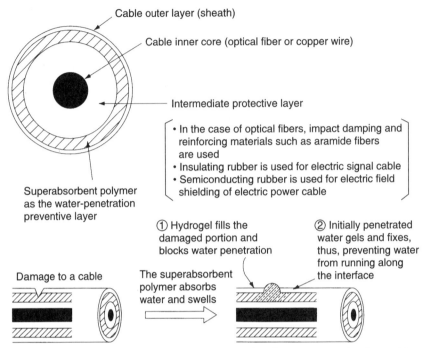

Fig. 1 Mechanism to prevent water penetration into communication cables.

become a serious problem. Thus, many communication cables have incorporated highly sensitive water-prevention systems. To achieve higher density and reliability, there is active competition between electronic wiremakers, electronics material makers, and resin makers.

Several examples of communication cables that utilize superabsorbent polymers are presented here along with a mechanism for water-penetration prevention.

2 PROPERTIES REQUIRED FOR SUPERABSORBENT POLYMERS FOR COMMUNICATION CABLES

Because communication cables are installed in air, underground, and underwater, oil slick or seawater can often penetrate the cable when the sheath of the outer layer is damaged. Thus a water-absorbing layer quickly absorbs water thereby preventing it from running within the sheath leading to further degradation of the cable and other connected devices. Superabsorbent polymers used for communication cables are quite different from those used for disposable diapers in that they are required to last for several to several tens of years. Furthermore, while they must be water-absorbing polymers, they must not be degradable by biological and other factors. Should they become a hydrogel as a result of contact with water, they must be stable and maintain sufficient gel strength for a prolonged period of time. If they are used to surround optical fibers, it is important that hydrogen gas, which damages light-transmitting capabilities, will not be generated.

Inexpensive and high water-absorptivity synthetic polymers (the water absorptivity of more than 100 cc/g), such as sodium polyacrylate, are used as superabsorbent polymers. To meet the forementioned requirements, polymers specifically designed for communication cables with high water absorptivity of salt water have recently been developed. As monomers, sodium acrylate and methoxy poly(ethylene glycol methacrylate) (average addition of ethylene oxide is (9 mol) are frequently used. Poly-(ethylene glycol diacrylate) (average addition of ethylene oxide is 8 mol) is used as a crosslinking agent [1].

3 WATER-ABSORBING WRAPAROUND TAPES

The method widely employed at present to introduce superabsorbent polymers to communication cable is to use a wraparound tape on either one side or both sides of which a superabsorbent polymer is coated. Figure 2 shows the structure of wraparound tape and Fig. 3 illustrates the

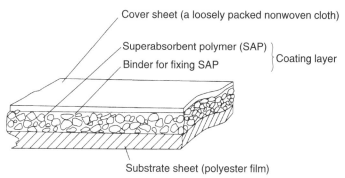

Fig. 2 Structure of a wraparound water-absorbing tape.

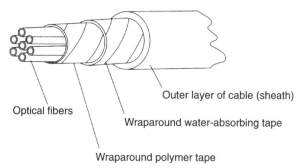

Outer layer of cable (sheath)

Optical fibers

Wraparound water-absorbing tape

Wraparound polymer tape

Fig. 3 An example of fiberoptic cable in which a wraparound water-absorbing tape is incorporated.

fiberoptic cable with the wraparound tape used as the water-absorbing layer.

The wraparound tape is made of a substrate sheet, such as a plastic film or nonwoven cloth, on which a superabsorbent polymer is coated. The superabsorbent polymer powder is mixed with an adhesive material, such as an organic polymer binder or a water-soluble resin, and coated onto the sheet. The sheet is then dried, fixed, and cut into strips. When the tape is in contact with water, the superabsorbent polymer particles absorb the water, swell, and become large hydrogel particles to prevent water penetration.

As the substrate sheet, a thin plastic film or tightly packed nonwoven cloth, such as polyester spanbond nonwoven cloth of approximately 0.15 mm, is used. A synthetic rubber, such as styrene-butadiene rubber or thermoplastic polyurethane elastomer, is used as a binder. In addition, a surfactant for hydrophilicity, an antioxidant for prevention of thermodeformation, and a silica-type inorganic filler for prevention of tackiness are used. For the superabsorbent polymer particles, various synthetic polymers, for example, polyacrylate and polyvinyl-type superabsorbent polymers, can be used. For this application, the particle sizes are an important parameter, because they polymer is required to be within the coating layer, and as the absorption rate is no retarded, quickly protrude from the layer when swelling.

Thus, for this purpose particle size from 0.05 to 0.5 mm will be convenient. Several percent of finer particles with diameters less than 0.05 mm could be added, as they are believed to protect the more delicate structures inside the cable from water penetration.

The basic structure of the wraparound tape is further improved by the addition of a cloth cover for handling purposes. If the superabsorbent coating layer is exposed, it tends to be sticky by absorbing water when the cable is manufactured. It is also possible that superabsorbent polymer particles fall from the tape during the connection of cables and adsorbs onto the cable core, resulting in the loss of transmission characteristics. Hence, recently, a thin loosely packed nonwoven cloth weighing 10–$30\,g/m^2$ and less than 0.1 mm in thickness is placed and thermally fused to form a three-layer structure in order to improve handling. The advantage of the water-absorbing layer of the wraparound tape is its excellent handling and processability. However, more than 0.3 mm is thicker than desired as the outer diameter of the fiberoptic cable tends to be thick [2–4].

4   WATER-ABSORBING MATERIALS MADE OF FIBERS, STRINGS, AND NARROW-WIDTH TAPES

Superabsorbent polymer fibers and synthetic or natural polymer fibers on which superabsorbent polymer particles are adhered using adhesives can also be used as water-blocking materials. These fibrous materials are bundled as they are with the wires and filled inside the sheath. There is also an ultrathin plastic film on which a superabsorbent polymer is coated to make a flat yarn-like narrow-width tape. This tape is used in the bundle with the wires inside the sheath. Although these materials are relatively low cost and easy to use, they are linear, and as such they tend to stretch and the coefficient of thermal expansion is also large. These properties are opposite those of the fiber core, which has high strength and a low coefficient of thermal expansion. Hence, it is difficult to design the tension during the construction and gathering. Further difficulties are how to ensure their long-term durability and reliability [5].

5   JELLY-LIKE WATER-ABSORBING MATERIALS

Jelly-like materials can be made of a resin from petroleum or isobutylene resin to which a mineral oil, metal soap and refined oil are added. Jelly-like water-absorbing materials are made by dispersing the powder of a superabsorbent polymer in this jelly-like material. When it contacts water, the superabsorbent polymer particles in the matrix absorb water and swell, these swollen particles leave the matrix and become a block of hydrogel. The mechanism to prevent water penetration is almost the same as other water-absorbing materials. There is no problem due to the difference in the coefficient of thermal expansion as in the previously mentioned fibrous materials. It can also block fine gaps within the cable, and act as impact-damping material because it has some fluidity. On the other hand, this jelly-like material is sticky and heavy at room temperature. During terminal treatment operation, such as cable connection, it might be necessary to use a thinner to remove the jelly-like material. Figure 4 shows the construction of fiberoptic cable in which this jelly-like material is incorporated. [6].

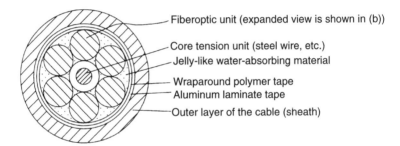

(a) A cross-sectional view of a fiberoptic cable in which a jelly-like water-absorbing material is incorporated

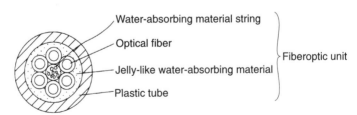

(b) A cross-sectional view of the fiberoptic unit used in (a)

Fig. 4 Example of the structure of a fiberoptic cable in which a jelly-like water-absorbing material is incorporated.

6   DIRECT-COATING TYPE WATER-ABSORBING MATERIALS

A superabsorbent polymer is coated directly on the spacer within the cable core and dried to form a water-absorbing layer without the use of wraparound tape. This method has an advantage of making a thinner-diameter cable. Figure 5 shows the incorporation of a water-blocking

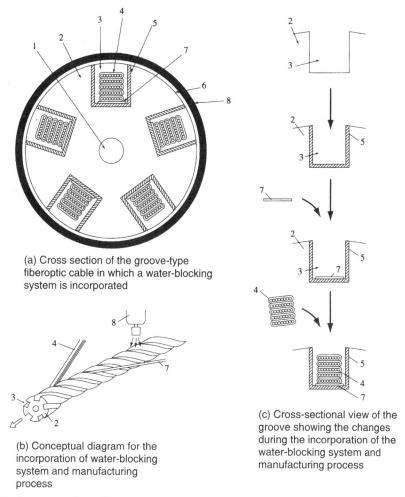

(a) Cross section of the groove-type fiberoptic cable in which a water-blocking system is incorporated

(b) Conceptual diagram for the incorporation of water-blocking system and manufacturing process

(c) Cross-sectional view of the groove showing the changes during the incorporation of the water-blocking system and manufacturing process

Fig. 5   Incorporation of water-blocking system into a multiwired, groove-type fiberoptic cable.

mechanism in a multiwired groove-type fiberoptic cable. Figure 5 (a) is the cross section of this cable, and Fig. 5 (b) is the conceptual diagram of the water-blocking mechanism and manufacturing process. Figure 5 (c) is the cross section of the change of the groove during the incorporation of the water-blocking system and other cable manufacturing processes.

In the middle of the communication cable, there is a spacer **2** made of a plastic. In the middle of this is the core-tension material **1** made of a steel wire. This steel wire prevents the optical fiber from fracturing due to the tension that is exerted on the cable as a whole. Around the core-tension material, multiple grooves **3** are formed. In each groove **3**, optical fibers that are laid in a tape manner **4** are stacked. This tape-like bundle of optical fibers consists of eight single-mode optical fibers, which are lined in a horizontal manner. Each optical fiber has a diameter of 0.25 mm and is covered with an UV-curable polymer. These fibers are lined in a horizontal manner and further covered by the UV-curable polymer giving a ribbon-like appearance. The fiberoptic bundle formed in this way has excellent processability, for example, cutting and connecting. Each groove has several layers of this ribbon-like bundle.

In this construction, a water-absorbing material **5** is adhered on the inner wall of the groove **3**, which provides better utilization of the space than the water-absorbing wraparound tape. This water-absorbing material is made of the powder of a superabsorbent polymer, such as polyacrylate, which is dispersed and adhered to the wall by water-soluble organic binder, such as poly(ethylene oxide)(PEO).

Incorporation of the water-absorbing material **5** onto the inner wall of the groove **3** is done by spraying the suspension of superabsorbent polymer powder in a methanol/poly(ethylene oxide) mixed solution onto the wall by a spray **8** (see Fig. 5 (b)). Poly(ethylene oxide), which is used as the binder, is water-soluble, hence, when water penetrates into the cable, it dissolves in the water, thus exposing the superabsorbent polymer particles. The exposed particle then absorb water and swell, filling the gap between the fiber and groove, and stopping the penetrating water. In the process shown in Fig. 5(b), after the water-absorbing material **5** is sprayed on, a water-soluble film **7** is adhered onto the bottom of the groove prior to drying the sprayed powder. This water-soluble film is smooth on the surface, and the bottom surface of the film, which is in contact with the binder solvent, dissolves to the solvent and achieves a good contact with the bottom of the groove. Despite the introduction of the water-absorbing layer in which superabsorbent polymer is dispersed, the bottom of the groove becomes flat. This is convenient for ribbon-like fiberoptic material **4** to be placed in a stable manner and to prevent the degradation of transmissivity. The contact of the spacer and optical fiber is flat. Therefore, the stress caused by the difference in the coefficient of thermal expansion accompanying temperature changes can be minimized. Finally, a wrap-around tape **6** made of a polyester tape with a thickness of 0.038 mm is used to form a bundle. The outer layer (sheath) is applied to make a communication cable shown in Fig. 5 (a). In this example. The water-absorbing material is fixed onto the spacer for optical fibers by a water-soluble binder. Hence, even during temperature changes, it will not compress the optical fiber. Instead of using a thick water-absorbing wraparound tape, a thin wraparound tape is used. Therefore, the cable can be thinner and high-density fiber bundles can be made.

7 APPLICATION TO POWER CABLES AND OTHERS, CONCLUSION

In this subsection, water-absorbing materials and a water-penetration prevention system for fiberoptic application were described. However, similar systems are also incorporated in copper wire signal cables. There is also a similar trend in the area of power cables, control cables, and structural support cables. For example, the main cable for supporting a long suspension bridge uses a strand made of several tens of high-strength, zinc-plated steel wires as a core. Its outside is coated by corrosion-prevention paint or covered with fiber-reinforced plastics. In this case, prevention of water penetration is as equally important as the case for communication cables where operation must be maintained. In these structural support cables, the aforementioned water-absorbing wraparound tape is often used inside the resin cover. The consumption of super-absorbent polymers in the area of water-absorbing materials for cable is steadily increasing as a high value-added asset, and strong future growth is expected [8–10].

REFERENCES

1 (1996). Tokkyo Kaiho 283697.
2 (1996). Tokkyo Kaiho 192486.
3 (1996). Tokkyo Kaiho 92525.

4 (1996). Tokkyo Kaiho 62473.
5 (1996). Tokkyo Kaiho 15586.
6 (1993). Jitsuyo Tokkyo Kaiho 11113.
7 (1995). Tokkyo Kaiho 306339.
8 (1993). Tokkyo Kaiho 234429.
9 (1994). Tokkyo Kaiho 281849.
10 (1996). Tokkyo Kaiho 277505.

Section 2
Batteries

1 PRIMARY BATTERIES

1.1 Introduction

Chemical batteries consist of three basic elements, namely, a cathode, an anode and an electrolyte. The energy obtained by the chemical reaction of the active material is obtained by the external circuitry as electric energy. While auxiliary batteries use reversible chemical reactions and thus are rechargeable, the chemical reactions of the primary batteries are irreversible and thus only the release of electricity is possible. For the active materials of the primary batteries, the following are used: solid metal oxides, such as manganese dioxide, mercury oxide, silver oxide, and nickel hydroxide; halogen compounds, such as silver chloride, copper chloride, and fluorinated graphite; and liquid oxidizing agents, such as air, oxygen, halogen, and thionyl chloride. The solid cathode of actual batteries is a composite of conducting carbon particles, such as graphite with a small amount of a binder, acetylene black, ketchen black or cokes, and, if necessary, an electrolyte. These materials are mixed and processed to make an electrode. As the anode, metals such as zinc (zinc can, zinc plate and an amalgam of zinc powder) and lithium are generally used. Depending on the type of anode, the electrolyte maybe either an aqueous electrolyte or a nonaqueous electrolyte. Examples of batteries that use an aqueous electrolyte are the mercury oxide-zinc battery (Ruben battery), the silver-silver chloride battery, the manganese dry battery (Leclanché battery), the alkali-manganese dry battery. Batteries that use a nonaqueous electrolyte are the manganese dioxide-lithium battery and the fluorinated graphite-lithium battery. As in the case of a dry battery in which a gel is used as an electrolyte, there are many examples that use gel technology for electrodes and electrolytes. Generally, if the polymer functions as a cathode, an anode or an electrolyte in chemical batteries, then the battery is called a polymer battery and is distinguished from other batteries. The batteries with a polymer gel as an electrolyte are also called polymer batteries. Recently, a film-like solid battery using an elastic gel that exhibits ionic conductivity comparable to electrolyte solution has been developed. As a battery that falls into the new category of polymer battery, its performance and characteristics are drawing attention. Gel technology helps to solidify the electrode and is going to play a significant role in the development of battery technology.

1.2   Dry Batteries and Gels

The manganese battery is the most familiar battery because of its wide use in daily life. The current dry battery uses manganese dioxide in the cathode, zinc (zinc can) in the anode, and a mixture of ammonium chloride and zinc chloride as the electrolyte. It was invented by Leclanché in 1868, and commercialized in the 1880s using a semisolid electrolyte obtained by adding a gelation agent, such as dextrin, and a moisture-absorbing agent, such as calcium chloride and glycerin. It is not an exaggeration to state that the dry battery was commercialized for the first time when gel electrolyte was developed. The basic Structure of the Leclanché battery is shown as follows:

$$\oplus C(\text{Carbon}) + MnO_2 | 6M \ NH_4Cl(2M \ ZnCl_2) \ (\text{Aqueous solution})|$$

$$Zn \ (\text{Mercury Treatment}) \ominus \quad (1)$$

Zinc chloride in the electrolyte has the shortcoming of decomposing starch gel and changing it into sol. Hence, current batteries use crosslinked starch or carboxymethyl cellulose (CMC) in place of a starch gel [1]. Although the open circuitry voltage is 1.5 V, when it is used at a high current load, the operating voltage becomes 1.1 to 1.2 V because of the slow reaction at the cathode. By contrast, the alkali manganese battery has both excellent current load characteristics and energy density:

$$\oplus C + MnO_2|6M \ KOH(ZnO \ Saturated) \ Aqueous \ solution|Zn \ominus \quad (2)$$

The voltage varies depending on the type of polarization reduction agent. However, the open circuitry voltage is the same as the manganese battery at 1.5 V.

There are various electrolyte gelation methods. For dry battery production, paper-lining is used. In this method, the paper is impregnated with a gelation agent that consists of a high-density and low-density layers, which act as separators.

Electrolyte solidification has been used to change the shape of batteries (for example, paper batteries), and for the more traditional purpose of preventing leakage in dry batteries. The flat battery incorporated into the instant camera consists of a conducting polymer film on which are stacked the cathode made of a printed manganese dioxide, the anode made of a printed zinc powder, and a gelled electrolyte. Four additional batteries are stacked to make a 6-V laminated structure dry battery [2]. This is an excellent example of the use of a thin battery.

1.3 Polymer Electrolytes and Lithium Batteries

The lithium battery is indispensable for watches, portable calculators, and cameras. It has contributed to the development of compact electronics as a result of its high voltage and high energy density. There has been significant progress in the development of the lithium ion-conducting solid electrolyte as well as the lithium battery. As lithium ion-conducting solid electrolytes, super ionic-conducting glass of lithium ion (LISICON), iodolithium (LiI), and solid solutions of polymers, such as poly(ethylene oxide) poly(vinylidene fluoride), and a lithium salt have been introduced. Among many organic and inorganic solid electrolytes, some have already been used for primary batteries.

The lithium-iodide battery 3 is made of an iodide adduct as the cathode, lithium metal as the anode, and a solid electrolyte that utilizes LiI formed when the cathode and anode contact. As the electricity is released, the lithium ion dissolved from the anode reacts with iodide in the cathode and a new LiI layer is formed at the cathode/anode interface. Because the specific conductivity of LiI is approximately 10^{-5} S/cm, the application is limited to pacemakers, hearing aids and other low-current applications. At present, there are no batteries that can replace the LiI battery because it has no leakage problems. Studies are currently underway on organic polymer solid electrolytes with higher ionic conductivity. Solid polymer electrolytes typically are solid solutions of matrix polymers and electrolyte salts. They were discovered by Wright *et al.* [4]. In 1979, Armand proposed the use of a polymer electrolyte as the solid electrolyte for the lithium battery [5]. However, the ionic conductivity was only comparable to LiI at that time.

In order to use polymer electrolytes for high-current batteries, the electrolyte must possess ionic conductivity comparable to other electrolytes at 10^{-3} to 10^{-2} S/cm. However, for memory backup batteries with a small-current application, the resistance can be 100 to 1000 times higher. Thus, if the ionic conductivity is more than 10^{-5} C/cm. It is sufficient as this value is estimated assuming that the thickness of the solid electrolyte is similar to the distance between electrodes of ordinary batteries at 25μm. Theoretically, if the electrode distance can be smaller, an electrolyte with a higher resistance can be used. This is possible only when the resistance of the electrode/electrolyte interface can be negligibly small in comparison to the electrolyte resistance. At present, solid electrolytes with much higher ionic conductivities than LiI at approximately 10^{-4} to 10^{-3} S/cm have been developed. However, lowering the resistance of the electrode/electrolyte interface is still an important problem to be solved.

A salt must dissolve, dissociate, and readily diffuse in a polymer matrix. The ethyleneimine and ethylene oxide repeating units (ionic dissociation group) in the polymer chains accelerate ionic dissociation. Whether the ionic dissociation group in the polymer structure can be classified as a main-chain type or a side-chain type depends on its location. Thus it is necessary to reduce the degree of crystallinity to increase the diffusivity of ions. As ethylene oxide readily crystallizes, instead of ethylene oxide a copolymer of ethylene oxide and propylene oxide can be used as an ionic dissociation group to suppress crystallization. In the side-chain type, it is effective to use the main chain with low glass-transition temperature T_g in order to reduce the T_g of polymer electrolytes. For the main-chain type, introduction of crosslinking structure is effective in suppressing crystallization [7]. Polymer electrolytes with high salt solubility made of poly)vinyl alcohol) [8] and selective cationic ion conductor by fixing anions on the polymer chain [9] have

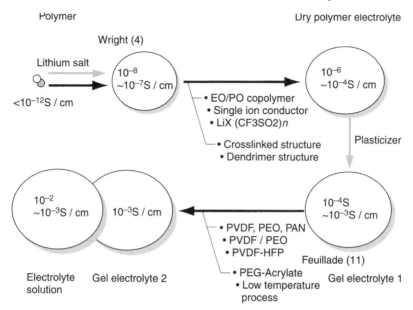

Polymer   Dry polymer electrolyte

Fig. 1 Trend of polymer electrolyte development.

been evaluated. Figure 1 shows the historical trend of the polymer electrolyte for batteries in terms of ion conductivity. The highest value obtained to date [6] at 10^{-4} S/cm is by the solid solution (salt-in-polymer type) of dendrimer called P(EO/MEEGE) and an electrolyte salt with the structure of $LiN(SO_2CF_3)_2$.

The ionic conductivity of the salt-in-polymer type polymer electrolyte can be expressed by the Williams-Landel-Ferry (WLF) equation derived from the free volume theory. In the relatively high-temperature regime above T_g, the ionic conductivity is comparable to solutions, hence, the battery can be used for high-current applications. A Canadian electric power company, Hydro Quebec Co., introduced a large battery made of PEO in which an electrolyte salt is dissolved for use at 60°C. When various high-boiling polar solvents are added to a dry polymer electrolyte, the polymer absorbs the solvent and gels at the same time, increasing the ionic conductivity at room temperature. The high-boiling solvent can be regarded as a plasticizer. A thermoplastic nonaqueous electrolyte gel in which an electrolyte liquid acts as a plasticizer for the polymer matrix was discovered by Feuillade and Perche [11] at almost the same time that Wright discovered the ionic conductivity of polymers. Studies on this gel had been made in connection with condenser and electrochromic devices (ECD). Thermoplastic nonaqueous electrolyte gels can be readily prepared by dissolving a polymer into a heated electrolyte and subsequently cooled. At room temperature, it becomes an elastic gel and exhibits ionic conductivity comparable to an electrolyte solution. Recent reports indicate that the advantages of the gel with polyacrylonitrile as the polymer matrix are due to the weak interaction between the matrix polymer and the Li ions [12].

1.4   Nonaqueous Electrolyte Gels

Nonprotic solvents are used for the electrolyte in lithium batteries. The molar ionic conductivity of the electrolyte increases linearly as the increase in electrolyte concentration is below $0.1 \, mol/dmm^3$. However, in the high concentration of electrolyte above $1 \, mol/dmm^3$, it is treated as a partially dissociated electrolyte [13]. Thus, the relationship between the electrolyte concentration and molar ionic conductivity varies depending on the type of electrolyte and solvent. As electrolytes for lithium batteries, detailed studies have been conducted on the mixture of a cyclic carbonate, such as propylene carbonate (PC) and ethylene carbonate (EC), and a glyme, such as 1,2-dimethyoxy ethane (DME) and diethoxy ethane (DEE). Glymes have strong basicity and solvate the lithium ion easily. When a glyme is mixed into PC or EC, the specific dielectric constant of the solvent reduces and the association constant increases. The glyme that solvates the lithium ion exhibits a behavior similar to the molten salt. This is because the solvent molecules become deficient to the electrolyte ions as the concentration of the electrolyte salt increases. A high concentration electrolyte exhibits thixotropy. Thus, if the glyme functional group changes into higher ether, propoxy and buthoxy, instead of methoxy and ethoxy, the electrolyte gels. Nonaqueous electrolyte gels can only be prepared by this type of ionic bonding of the solvent. In polymer electrolytes, gel structures can be classified in reversible network structure by physical association and coordination bonding, and permanent network structure by covalent bonding [14].

Among lyogels that contain nonaqueous solvent, elastic gels are especially noteworthy for battery applications. An elastic gel made of a polymer electrolyte can be prepared by applying heat or light onto the mixture of three components, namely, a network polymer from a polymer, a crosslinkable oligometer or a monomer, and electrolyte salt and solvent. It can exhibit the ionic conductivity comparable to solutions. The mechanical properties of this gel are a solid whereas its thermodynamic properties are those of a liquid. Polymer electrolyte gel will turn into a xerogel by drying to remove the solvent.

Feuillade and Perche [11] reported that an organic electrolyte that was gelled by poly(vinyl acetal) exhibited a specific conductivity of 10^{-4} S/cm. As mentioned before, a thermoplastic gel can be readily prepared by dissolving poly(ethylene oxide)(PEO), poly(vinylidene fluoride) (PVDF) or polyacrylonitrile (PAN) into an organic electrolyte and cooling [15]. These gels are either semisolid or solid and also thermoreversible. The temperature dependence of the ionic conductivity can be expressed by the Vogel-Tamman-Fulcher (VTF) equation:

$$\sigma(T) = AT^{-1/2} \exp\{-B/(T - T_g)\} \tag{3}$$

Appetecchi *et al.* [17] reported that instead of PEO, PVDF, and PAN, then used poly(methyl methacrylate)(PMMA), which has lower dielectric constant than these three polymers, as a gelation agent. An electrolyte gel with PMMA as a base was evaluated as the electrode for a fluorinated graphite-lithium battery (BR2320 coin-type battery) [18]. The gel was prepared by heating 1 mol/dm^{-3} of LiBF$_4$-PC electrolyte at 90°C to which 15% of PMMA was dissolved and cooled. The specific conductivity was half that of electrolyte solutions. The higher the molecular weight, the higher the viscosity and the easier the solidification. The conductivity decreases in proportion to the amount of PMMA. With a viscosity of 5×10^4 cP, the ionic conductivity was 2×10^{-3} S/cm, whereas with 1×10^5 cP, which showed no fluidity, the ionic conductivity was 1×10^{-3} S/cm.

When a monomer or oligomer is polymerized in an electrolyte, a covalent-type gel with high elasticity can be obtained. Various polymerization methods, such as vinyl polymerization, urethane reaction, polyene-thiol reaction and ring-opening reaction of a lactone, can be used to form gels [19, 20]. If an acrylic monomer is polymerized in an electrolyte, a high ionic conductive gel with high creep resistance and high modulus can be obtained due to the phase transition of the gel. Figure 2 shows a gel film obtained by photopolymerizing poly(ethylene glycol acrylate) in an electrolyte, and as can be seen, the film is twice its original size when stretched. Ionic conductive film prepared from a polymer latex exhibited higher mechanical strength [21]. Table 1 lists ionic conductivity of various polymer electrolyte gels.

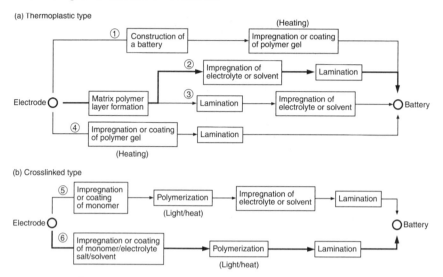

Fig. 2   Process of manufacturing polymer batteries.

Table 1   Ionic conductivity of polymer electrolytes.

	Electrolyte		Ionic conductivity (S/cm)	Temperature °C
Polymer matrix	Electrolyte salt	Solvent		
P(Vdf-HFP)	LiPF$_6$	EC/PC	3×10^{-3}	22
P(Vdf-HFP)	LiPF$_6$	EC/DMC	3×10^{-3}	22
PAN	LiClO$_4$	EC/PC	2.9×10^{-3}	20
PAN	LiAsF$_6$	BL	6.1×10^{-3}	20
PAN	LiN (CF$_3$SO$_2$)$_2$	BL/EC	4.0×10^{-3}	20
PAN	LiCF$_3$SO$_3$	EC/PC	1.4×10^{-3}	20
PEG-DA	LiCF$_3$SO$_3$	PC	1×10^{-3}	Room temperature
PEG-A/TMPA	LiBF$_4$	PC/DME	3×10^{-3}	25
PEG-A/TMPA	LiN (CF$_3$SO$_2$)$_2$	EC/DME	4.6×10^{-3}	25
PEG-A/TMPA	LiPF$_6$	EC type	6.4×10^{-3}	25
SBR	Li-salt	BL/DME	1.4×10^{-4}	Room temperature

P(Vdf-HFP): copolymer of vinylfluoride and hexafluoropropylene, SBR: styrene-butadiene rubber, PAN:polyacrylonitrile, PEG-DA: poly(ethylene glycol diacrylate), PEG-A/TMPA: poly(ethylene glycol monoacrylate)/trimethylolpropane acrylate

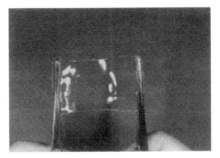

Fig. 3 Electrode gel film.

1.5 Polymer Film Primary Batteries

Similar to dry polymer electrolyte, various batteries have been developed using electrolyte gels. Among non-aqueous types the film battery in which lithium metal is used as the anode is drawing attention. While there are various proposals on plasticized gels and crosslinked gels, a realistic process is shown in Fig. 3. Details on the gel-preparation methods will be described later when auxiliary batteries are discussed. Figure 4 shows the polarization characteristics of lithium in an acrylate gel. Unlike the dry-type polymer system, there is no significant polarization at the gel/lithium

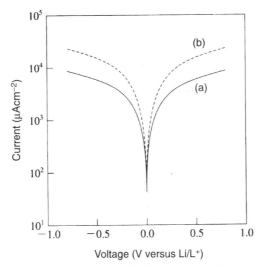

Fig. 4 Polarization characteristics of lithium electrode in an electrolyte gel.

interface. However, if the gel is used as the battery element, the interfacial resistance with a gel is considered to be problematic.

Recently, so-called paper batteries have been developed. An example is the film battery used for a hobby in which scissors are used. This battery is a three-layered structure that includes the cathode, a solid electrolyte layer, and aluminium foil, formed on a transparent film substrate. The battery is limited to microcurrent applications [22] because it produces a low voltage of 0.6 V and a maximum current of $0.004\,mA/cm^2$. It is presumed that a lithium ion conductive solid electrode based on acryloyl modified poly(ethylene glycol) is used as the electrolyte. Although aluminium is used for the anode, the active material is considered to be lithium [23].

Yuasa Corporation established a joint venture, ACEP Co., with HydroQuebec Co. in 1990 and began commercialization of a polymer film primary battery [24]. It is presumed that manganese dioxide is for the cathode, lithium foil is for the anode, and a plasticized polymer electrolyte using poly(ethylene glycol acrylate) is the matrix. The problematic polarization at the cathode and electrode interface is reduced to one-fourth that of ordinary batteries by utilizing the Plasticized S.P.E. System as the electrolyte. By utilizing a polymer electrolyte, the large exothermic reaction reaching 200 to 300°C between the lithium and electrolyte solution and the manganese dioxide and electrolyte solution has been significantly inhibited. The paper battery has a nominal voltage of 3 V, capacity 150 mAh (cutoff voltage 2.0 V), weight 2.6 g, and is $54\,mm \times 86\,mm \times 0.5\,mm$ in size. Although the composition of the electrolyte in the actual battery is unknown, there is no change in appearance until 140°C, moreover, it is so sage that it can be cut by

Table 2 Polymer primary batteries.

	Construction of battery			Voltage (V)	Volumetric energy or size	Weight (g)	Ref.
No.	Cathode	Anode	Electrolyte				
1	MnO$_2$	Li	PEO Crosslinked Plasticized	3	450 Wh/ℓ ($86 \times 54 \times 0.5$ mm)	0.5	24)
2	MnO$_2$	Al(Li)	PEO Crosslinked Dry polymer	1.5	29.5 mAh ($85 \times 54 \times 0.5$ mm)	3.4	22)
3	Fluorinated graphite	Li	PMMA Gel (LiBF$_4$/PC)	2.57	> 100 mAh (BR2320)		18)

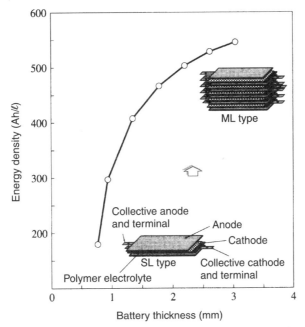

Fig. 5   The relationship between the thickness and energy density of a thin primary battery.

scissors. Table 2 summarizes the performance of the paper battery. Figure 5 plots the battery thickness on the abscissa and energy density on the ordinate using a model calculation for a stacked battery. Thus the thinner the batteries, the worse the efficiency and the lower the energy density; as the battery increased in thickness, the energy density increased, approaching an asymptotic value at approximately 3 mm.

The application of electrolyte gels to primary batteries has entered the practical application stage, and it is now possible to manufacture almost paper-thin batteries. In order for these batteries to be commercially acceptable, there must be improvement in current properties, establishment of manufacturing processes, and reduction of cost. The application of electrolyte gels to auxiliary batteries will be introduced later. For auxiliary batteries, however, higher reliability is required than for primary battery applications. The salt-in-polymer type solid electrolyte is also undergoing changes. Angell et al.[25] proposed a new xonxept, polymer-in-salt type electrolyte, which is not entering a new development phase.

2   AUXILIARY BATTERIES

2.1   Introduction

It was important to solidify electrolytes to prevent leakage and to reduce the thickness of primary batteries. However, for auxiliary batteries, it is important to improve safety, increase energy and, in particular, for large power supply, reduce cost. The acceleration of the development of auxiliary batteries in the past several years was due to the JOULE Project sponsored by the Commission of European Communications (CEC) and the Electric Car Development Project sponsored by the United States Advanced Battery Consortium. Similarly in Japan, research is underway on the research and development on electric power storage by dispersion type batteries under the New Sunshine Project supported by the Ministry of Industrial Technology/NEDO [26]. There are many more examples of the solidification of auxiliary batteries than there are of primary batteries, and a wide variety of electrolytes from inorganic to organic materials is available. A film auxiliary battery can be made by forming TiS_2 as the cathode and Cu as the anode on both sides of a paper electrolyte [27]. The paper electrolyte is made by dispersing a copper ion conductive inorganic compound into styrene/butadiene rubber and coating it on both sides of the paper. Although this is a small-capacity battery of 0.6 V, it is flexible and processable. For example, if a lithium ion conductor is used for the inorganic solid electrolyte, it can be made into a lithium battery. Of the dry polymer electrolytes, polyphosphazene solid electrolyte exhibits the highest ionic conductivity. When this electrolyte is used with amorphous vanadium pentoxide as the cathode and lithium metal as the anode, a paper auxiliary battery [28] provided a voltage of 3 V at the ultrathin thickness of 0.2 mm. Lithium is used as the anode because it has excellent repeatability. The ionic conductivity of polymer electrolytes can be increased by simply increasing the use temperature without relying on the special polymer structure. Thus a high-temperature operating, large power supply has been developed using TiS_2 as the cathode and lithium as the anode [29]. The most common solid solution for an electrode consists of a polymer matrix, poly(ethylene oxide), and an electrolyte salt. Although the ionic conductivity at room temperature is 10^{-7} S/cm, it increases to 10^{-4} S/cm above 100°C. From a differential scanning calorimetric (DSC) study, the exotherm of the reaction of the lithium metal with the polymer electrolyte was found to be extremely low compared to that with an ordinary electrolyte solution. Auxiliary batteries are thinner or safer, and are being developed further using gels that exhibit ionic conductivity comparable to solid electrolyte batteries even at room temperature. Structurally, auxiliary batteries are basically the same as primary batteries. However, their durability under the repeated operation depends on the materials used. In addition, hydrogels are almost never used for solid-state auxiliary batteries because no generation of gases is allowed. The requirements for polymer electrolyte for use in primary batteries are reduction of impedance, good shelf-life, and good mechanical strength as an elastic body. For auxiliary batteries, an additional requirement is chemical stability towards oxidation/reduction.

2.2 Acrylate Gels

As discussed previously in the section on primary batteries, an acrylate is often used as a crosslinkable monomer to form a polymer matrix for a nonaqueous electrolyte. The salt-in-polymer type polymer electrolyte made by dissolving trifluoroethylene glycol methacrylate into methoxy poly(ethylene glycol methacrylate) forms a comb-like network structure at the covalently bonded portion. The ethylene oxide chain consists of 22 monomer units, and the ionic conductivity at room temperature is reported to be 10^{-6} S/cm. If sodium thiocyanate is used [30] instead of lithium trifluoromethane sulfonate, the ionic conductivity reduced to 10^{-9} S/cm. However, by adding 50 wt% of PC as a plasticizer, the ionic conductivity reaches 10^{-4} S/cm.

A high-modulus, ionic-conducting transparent gel can be prepared by mixing an acrylate monomer into an electrolyte and, if necessary, a polymerization initiator, followed by heating or irradiation [31]. The irradiation source can be UV light, visible light, an electron beam or γ-ray. Using acrylate monomers, various polymer network structures, such as comb and ladder-like structures, can be obtained by changing the number of acrylic functional groups. By polymerizing a bifunctional acrylate compound, such as poly(ethylene glycol diacrylate), an ionic-conducting gel with a polyethylene-like structure as a matrix can be prepared [32]. Balance Co. developed a polymer auxiliary battery by combining this electrolyte with V_3O_8 as the cathode and lithium metal as the anode.

A polyethylene-like structure is considered to consist of more than 10 ethylene oxide units in the main chain and ladder-like polymer networks. The battery is manufactured by coating both the cathode and the anode with the monomer, later irradiating by electron beam, and then laminating both.

The authors have succeeded in developing a high ion-conducting gel that possesses a comb-like network structure by UV-curing a mixture of monofunctional and trifunctional acrylate [33]. An ion-conducting gel is a viscoelastic material that consists of an electrolyte salt, solvent, and monomer. Figure 1 shows the gel-preparation scheme in which the basic components are a monomer, electrolyte salt, and solvent. When the composition of these compounds changes, the properties of the gel change greatly. A gel can be obtained when methylbenzoylformate is added by 0.5 wt% as a photoinitiator in a mixture of these three components. This is done by UV radiation from a high-pressure UV lamp of $100 \, \text{mW/cm}^2$ intensity onto this mixture for 30 s. The electrolyte used was $LiBF_4$, the monomer was a $1:9$ mixture of ethoxy poly(ethylene glycol acrylate) and trimethylolpropane triacrylate, and the solvent was a $7:3$ mixture of propylene carbonate and dimethoxy ethane. Figure 2 is a ternary diagram of the ionic conductivity with the concentration of three components. The area within the dotted line is the formation of homogeneous gels. Each line corresponds to the ionic conductivity in the range from 10^{-7} to 10^{-3} S/cm. The ionic conductivity increases in the order of (e), (d), (c), (b), and (a). Precipitation of the electrolyte or formation of the

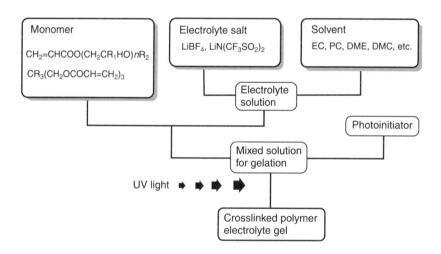

Fig. 1 Gel-preparation scheme.

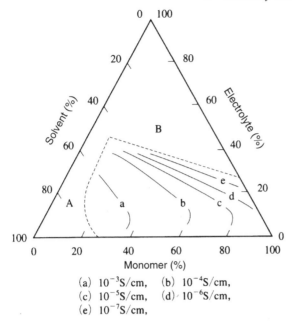

(a) 10^{-3}S/cm,   (b) 10^{-4}S/cm,
(c) 10^{-5}S/cm,   (d) 10^{-6}S/cm,
(e) 10^{-7}S/cm,

Fig. 2 A ternary diagram of ionic conductivity.

flowing gel or sol is indicated by the area outside the dotted line. Although the ionic conductivity is sensitive and changes depending on the amount of solvent and monomer, it is rather insensitive to the concentration changes of the electrolyte. Figure 3 is the ternary diagram of the viscoelastic properties of this gel composition. Similar to Fig. 2, the area within the dotted line is the area of solid gel formation. Each line corresponds to the modulus in the range from 10^2 to 10^6 dynes/cm^2. The modulus decreases in the order of (e), (d), (c), (b), and (a). The lower the concentration of the monomer and electrolyte or the higher the concentration of the solvent, the lower the modulus. From Figs. 2 and 3, it can be concluded that the softer the gel, the higher the ionic conductivity.

However, a unique aspect of this result is that in the high ion-conductivity area, a high creep resistance and modulus can be obtained despite the low polymer concentration.

Figure 4 plots the electrolyte concentration on the abscissa and the ionic conductivity on the ordinate. The data were obtained for the gel at the point a in Fig. 2, where the composition was 13 wt% monomer, 67 wt% solvent, and 20 wt% electrolyte salt. The line (a) is the ionic

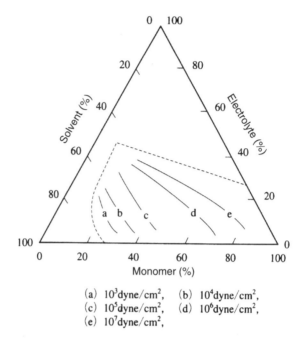

(a) 10^3dyne/cm^2, (b) 10^4dyne/cm^2,
(c) 10^5dyne/cm^2, (d) 10^6dyne/cm^2,
(e) 10^7dyne/cm^2,

Fig. 3 A ternary diagram of the moduli of gels.

conductivity prior to the gelation, sand the line (b) is that after gelation. The ionic conductivity shows a maximum at 1 mol/kg of the electrolyte concentration. Above this concentration, it tends to be lower.

However, the ratio of lines (a) and (b) is smaller when the electrolyte concentration is higher. Hence, the high electrolyte concentration is not necessarily disadvantageous. The modulus tends to be higher with higher electrolyte concentration. By polyene-thiol reaction, gels with similar properties can also be prepared [34]. Instead of the photoinitiator, thermal polymerization initiators, such as di-t-butylperoxide, can be used for gelation, although the rate of reaction is slower than photopolymerization. While the costs of UV and thermal curing are low in comparison to the electron-beam curing, the side effects from the residual polymerization initiator and monomer can be problematic. Thus, it is important to evaluate the selection of the polymerization initiator, and the process and maintenance to complete the polymerization. The residual monomer in the system can be detected by the infrared (IR) absorption band at $1638\,\text{cm}^{-1}$ or by analysis using photodifferential scanning calorimetry

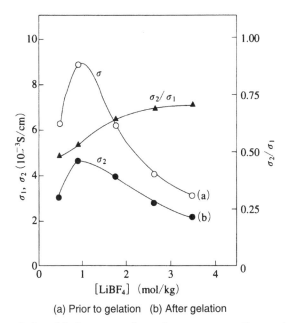

(a) Prior to gelation   (b) After gelation

Fig. 4   The relationship between electrolyte concentration and ionic conductivity.

(photo-DSC). The residual monomer in a battery forms a heterogeneous high-resistance electrode/electrolyte interface due to the oxidation-reduction effect that occurs during charge and discharge of the battery.

2.3 Performance of Auxiliary Batteries

A polymer/lithium battery was manufactured using the composition shown at point a in Fig. 3. A paper-like lithium auxiliary battery was manufactured with a stainless steel foil-supported polyaniline sheet as the cathode and a copper-foil-supported lithium foil as the anode [35]. Polyaniline was synthesized by electrochemical polymerization and the polyaniline film began to swell after the charge/discharge cycle even when it was pressed initially, exhibiting a gel-like appearance; the density approaches 0.4. The electrolyte gel was prepared by irradiating UV light onto the cathode and separator made of porous polyolefin film impregnated by a electrolyte mixture. These elements were stacked and placed in an aluminium laminate container to form a sheet battery. Many conducting polymers exhibit a gel-like appearance due to absorbing solvents, and polyaniline is no exception. However, the electrochemical interface between the gel and polyaniline is good. The thickness of the battery element is 280μm and the total thickness including the packaging material is approximately 500μm. It is possible to discharge with a voltage of 3.7 V, capacity of 15 mAh, and continuous discharge of 20 mA. The repeatability reduced 60% of the initial value after 200 cycles. The conducting polymers are comparable to inorganic materials in terms of energy density per unit weight. However, the energy density per unit volume is quite low at 40 mAh/g (electrode). To cope with this problem, an example of a high-density polymer battery was constructed and presented in the following.

A lithium-ion auxiliary battery can be constructed by an organic/inorganic composite electrode made of polyaniline and vanadium pentoxide as the cathode, graphite as the anode, and ag el as the electrolyte [36]. Figure 5 shows the structure of this battery. In order to overcome the disadvantage of the bulkiness of the polymer electrode, an organic/inorganic hybrid material (V_2O_5PANI $= 9/1$) was used. This cathode is a totally new type of dual-ion transport electrode in which both anion and cation are transported in the film at the same time. By using trifluoromethane sulfonoimide instead of $LiBF_4$ as the electrolyte salt for the

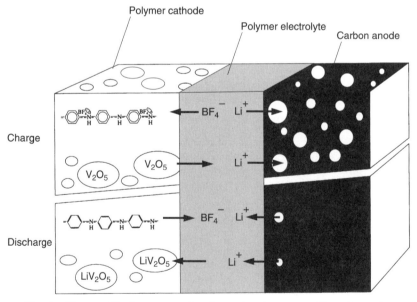

Fig. 5   Charge/discharge mechanism of polymer auxiliary batteries.

polymer electrolyte, the ionic conductivity of the gel increases. Figure 6 shows the temperature dependence of the ionic conductivity of the gel. A behavior similar to the electrolyte solution has been observed. Figure 7 shows the current property of the gel. Even at a high rate of 1 CmA the capacity was more than 80% of the result when the discharge capacity at 0.2 CmA is taken as 100%. Figure 8 shows the discharge capacity at various temperatures when the result at 25°C is taken as 100%. Even at −10°C, the efficiency is over 60%.

2.4 Characteristics of Polymer Auxiliary Batteries

As in other gels, the ion-conduction mechanism of an acrylic gel is similar to that of electrolyte solution, and the acrylic gel is formed by incorporating a solvent. The polymers used as the three main components of batteries must have high ionic diffusivity as they accompany large volume changes. As a result, processing methods are limited. This also

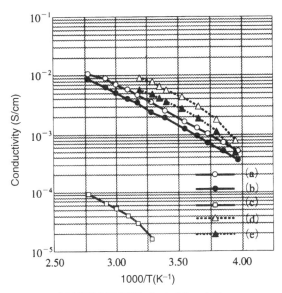

(a) LiBF$_4$/PC-DME electrolyte solution
(b) PEGA electrolyte gel of (a)
(c) PEO-PPO copolymer dry electrolyte
(d) LiN(CF$_3$SO$_2$)$_2$/EC-DME electrolyte solution
(e) PEGA electrolyte gel of (d)

Fig. 6 The temperature dependence of ionic conductivity.

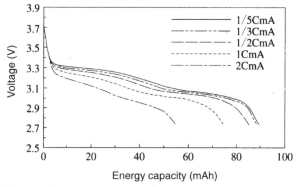

Fig. 7 Discharge curves of a polymer auxiliary battery.

applies to PVDF, which is not used because it forms a fluorinate compound with lithium at the anode where lithium is used. However, it is indispensible as a binder material in lithium batteries, and once again is attracting attention as a polymer matrix for polymer electrolyte.

It is known that plasticized rubber, which is made of a copolymer of PVDF and HFP and impregnated with electrolyte solution, shows high ionic conductivity [37]. Thus by increasing the ratio of HFP to PVDF, the ionic conductivity and processability can be improved. A research team as Bellcore Co. Reported that if the HFP content is increased in the HFP/PVDF copolymer used as the polymer matrix, the melting point of the copolymer decreases. At 12% HFP, the copolymer can be processed at 120 to 130°C. A gel is prepared using an electrolyte composition of $1\,mol/\ell LiPF_6/EC/DMC$, and this gel electrolyte was used for the 4-V

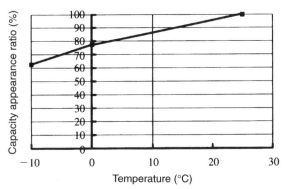

Fig. 8 Temperature characteristics of a polymer auxiliary battery.

lithium ion polymer auxiliary battery [38]. For the cathode, $Li_xMn_2O_4$ was used, and for the anode, mesomicrobeads manufactured by Osaka Gas were used. Both the cathode and anode electrodes were impregnated with polymer electrolyte, dried, and then laminated under heat; later, the electrolyte was filled. The research team at Bellcore Co. Pointed out that the batteries with polymer gels enhanced safety, lowered fabrication cost, had higher shape flexibility, and easier scalability. Although the safety data is incomplete, the lithium ion polymer battery using $LiCo_2$ as the cathode is believed to be safer than the auxiliary battery that used electrolyte solution [39]. If PAN was used instead of PVDF and $LiPF_6/EC/PC$ is the electrolyte, the gel exhibited a self-extinguishing property in a specific concentration range of $LiPF_6$. The safety of the auxiliary battery that used these materials was reported to be high (see Table 1) [40].

The safety of lithium batteries against internal short circuits and exothermic reactions is achieved mainly by lowering the melting point of the separator. However, in polymer batteries, this separator is not used except for safety and process control. In the system in which a separator is used, the separator itself can be a gel [41]. The safety and current performance often are in opposition to each other, and whether high current performance is maintained is sometimes questionable. In any case, the anode is made of carbons. The batteries that use gels have quite different application conditions and evaluation methods than those of the high-temperature operating power storage batteries described earlier in this section. The anti-flammable properties of the electrolyte solution itself is important for the safety of batteries made of gels.

2.5 Towards Increased Energy Density of Polymer Batteries and Future Trends

A polymer electrolyte is effective in dendrite formation of lithium. By using a polymer electrolyte, lithium, which is a high-energy material, can be used freely. From this viewpoint, the solidification technique using polymers has drawn attention. Recently, in a totally different approach, a polymer electrolyte gel is used as an element for a high-energy-density battery. Studies are being conducted on a sodium/sulfur battery (Na-S battery) that utilizes a polymeric sulfur. In this case, as the system in which this battery is used is large, high temperature is required. Doeff *et al.* [42] proposed a new, room-temperature operating battery using polymerization of an organic sulfur compound instead of inorganic sulfur.

Table 1 Lists of polymer auxiliary batteries.

No.	Construction of battery			Voltage (V)	Energy per volume or size	Energy per density or weight	Reference
	Cathode	Anode	Electrolyte				
1	$LiMn_2O_4$	C	PVDF-HFP Gel ($LiPF_6$/EC/DMC)	4	260 Wh/ℓ	110 Wh/kg	38
2	$LiCoO_2$	C	PEO (crosslinked) Plasticized	3.6	130 mAh (86 × 54 × 0.5 mm)	(4 g)	39
3	$LiCoO_2$	C	PAN Gel ($LiPF_6$/EC/PC)	3.6	—	—	40
4	V_2O_5/PANI	C	PEGA/TMPA Gel	2.5–3.7	168 Wh/ℓ (83 × 52 × 3 mm)	84 Wh/kg	36
5	V_6O_{13}	Li	PEGDA Gel	2.0–3.5	126 Ah/ℓ	68.3 Ah/kg	35
6	PANI	Li	PEGA/TMPA Gel ($LiBF_4$/PC/DME)	2.5–3.7	20 Wh/ℓ (Thickness 0.5 mm)	9.6 Wh/kg	35
7	RSSR/PANI	Li	PAN Gel ($LiBF_4$/EC/PC)	2.0–3.7	—	180 mAh/g Cathode	34
8	V_2O_5 (Amorphous)	Li	PEEGE Dry Polymer	2.0–3.9	3 mAh (45 × 45 × 0.2 mm)	(0.95 g)	28

For the anode, Li was used instead of NA, and for the electrolyte, a lithium ion-conducting polymer solid electrolyte was used instead of the ceramic solid electrolyte normally used for the Na-S battery. The organic sulfur compound in the polymer electrolyte precipitates on the cathode surface during recharging by polymerizing, and dissolves into the solid electrolyte during discharge. The shortcoming of this new battery is the lack of the repeatability of the organic sulfur compound. Sotobe and Koyama have succeeded in drastically improving the repeatability by using polyaniline on the cathode and a gel for the electrolyte [43]. The energy density per cathode weight reached 180 mAh/g.

The application of batteries, in particular, auxiliary batteries, is determined at present time by power consumption and operation time of electronic devices rather than size or shape. However, there has been steady progress in reducing the power consumption of electronic devices and the potential for paper batteries is improving. These batteries have excellent potential for safe operation as a high-power energy source for electric power driven automobiles.

Although gelation technology was not discussed in this subsection, it is attracting attention as an essential technology for synthesizing high-energy materials [44] in addition to the solidification of electrolytes. Electrolyte gel technology is already an advanced area for application in batteries and thus in the not so distant future batteries using these gels will become commercially available. Surprisingly, basic data by electrolytes for batteries is rather scarce, but fundamental studies are desired because of their high potential.

REFERENCES

1 Ohta, A. and Izawa, H. (1983). Demand for battery separators. *Battery Material Symposium* **1**: 253.
2 Yoshizawa, S. (ed.). (1986). *Batteries*, Kodansha Scientific.
3 Yamamoto, K. and Matsunaga, (1990). *Polymer Batteries*, Kyoritsu Publ.
4 Wright, P.V. (1975). *Br. Polymer* **7**: 319.
5 Armand, M.B. (1983). *Solid State Ionics* **9/10**: 1115.
6 Nishimoto, J., Furuya, N., and Watanabe, M. (1995). *Proc. 62nd Meeting of the Soc. Electrochemistry*, 3112.
7 Ito, K., Dodo, M., and Ohno, H. (1994). *Solid State Ionics* **68**: 117.
8 Yamamoto, T., Inami, M., and Kanbara, T. (1994). *Chem. Mater.* **6**: 44.
9 Zhou, G.B., Khan, I.M., and Smid, J. (1986). *Polym. J.* **18**: 661.
10 Gauthier, M., Fauteux, D., Vassort, G., Belanger, A., Rigaud, M., Armand, M.B., and Deroo, D. (1985). *J. Electrochem. Soc.* **132**: 1333.

11 Feuillade, G. and Perche, Ph. (1975). *J. Appl. Electrochem.* **5**: 63.
12 Voice, A.M., Southall, J.P., Rogers, V., Matthews, K.H., Davies, G.R., McLintyre, J.E., and Ward, I.M. (1994). *Polymer* **35**: 3363.
13 Izutsu, K. (1995). *Electrochemistry of Non-aqueous Electrolytes*, Baifu-kan.
14 Flory, P.J. (1953). *Principles of Polymer Chemistry*, Ithaca: Cornell University Press.
15 (1986). Tokkyo Koho 23947.
16 G.G. Cameron, Ingram, M.D., and Sorrle, G.A. (1987). *J. Chem. Soc., Faraday Trans. J.* **83**: 3345.
17 Appetecchi, G.B., Croce, F., and Scrosati, B. (1995). *Electrochim. Acta* **40**: 991.
18 Iijima, T., Toyoguchi, Y., and Eda, N. (1985). *Denki Kagaku* **53**: 619.
19 US Pat. 4,709,002.
20 US Pat. 4,709,002.
21 (1993). *Trigger* **3**: 20.
22 Fujiwara, Y. (1989). *Polyfile, Jpn.* **12**: 36.
23 (1988). Tokkyo Kaiho 130473.
24 Noda, T., Kato, S., Aibara, Y., Ashida, K., and Murata, K. (1994). Yuasa-Jiho **77**: October.
25 Angell, C.A., Liu, C., and Sanchez, E. (1993). *Nature* **362**: 137.
26 Kuwabara, K. (1997). *Kogyo Zairyo* **1**: 27.
27 Sotomura, T.G., Kondo, A., and Iwaki, T. (1989). *Prog. Batteries & Solar Cells* **8**: 163.
28 Nakazoe, I. (1989). *Polyfile, Jpn.* **12**: 36.
29 Gauthier, M., Fauteux, D., Vassort, G., Belanger, A, Rigaud, M., Armand, M.B., and Deroo, D. (1985). *J. Electrochem. Soc.* **132**: 1333.
30 Xia, D.W., Soltz, D., and Smid, J. (1984). *Solid State Ionics* **14**: 221.
31 (1986) US Pat. 4,908,283.
32 (1987). US Pat. 4,830,939.
33 Kabata, T., Fujii, T., Kimura, O., Ohsawa, T., Samura, T., Matsuda, Y., and Watanabe, M. (1993). *Poly. Adv. Tech.* **4**: 205.
34 (1990) US Pat. 5,223,353.
35 Ohsawa, T., Kimura, T., Kabata, T., Samura, T., and Yoshino, K. (1992). *J. Electronic Information and Commun. Soc., Jpn.* **8**: 391.
36 Ohsawa, T., Kimura, O., Kabata, T., Katagiri, N., Fujii, T., and Hayashi, Y. (1994). *The Electrochem. Soc. Proc.* **28**: 481.
37 Armand, M.B. (1985). *NATO ASI Sor. E.* **101**: 63.
38 Schmutz, C., Tarascon, J.M., Sozdz, A.S., Watten, P.C., and Shokoohi, F.K. (1994). *The Electrochem. Soc. Proc.* **28**: 330.
39 Noda, K., Katoh, S., Aibara, Y., Ashida, K., and Murata, K. (1994). Yuasa-jiho **77**: October.
40 Akashi, H., Tanaka, K., and Sekai, K. (1997). *Kogyo Zairyo* **1**: 1.
41 Toriyama, J., Okada, M., Yasuda, H., and Fujita, O. (1996). *Denchi Toron-kai* **3A32**: 239.
42 Doeff, M.M., Lerner, M.M., Visco, S.J., and DeJonghe, L.C. (1992). *J. Electrochem. Soc.* **139**: 2077.
43 Sotobe, M. and Koyama, N. (1996). *Denchi Toron-kai* **1A27**: 99.
44 Livage, J. (1991). *Chem. Mater.* **3**: 578.

Section 3
Fuel Cells

1 WHAT IS A FUEL CELL?

A fuel cell device is employed to obtain electric power from the chemical reaction between an oxidizing and reducing agent. In auxiliary batteries, the oxidizing and reducing agents (so-called active materials) are enclosed within the battery. When the active materials cease to be active, no electricity is generated and it becomes necessary to recharge (reactivate) the battery. Thus, this is a type of energy-storage device. By comparison, the oxidizing and reducing agents in the fuel cell are continuously supplied externally, hence, reactivation is unnecessary and continuous generation of electricity is feasible. As a result, the fuel cell is a power generator rather than a rechargeable battery.

Figure 1 explains the fundamental principle of an oxygen/hydrogen fuel cell. A single cell is composed of an electrolyte membrane with electrodes on both of its sides. Hydrogen is separated into a proton and an electron on the catalyst that is placed on the hydrogen electrode. The proton travels through the separation membrane and reacts with the oxygen and the electron that has passed through the external circuitry on the oxygen electrode to form water. This external electron provides

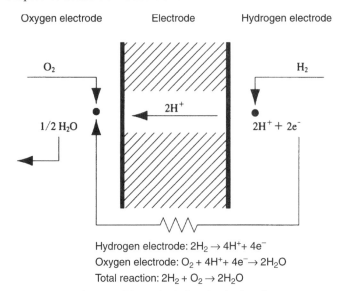

Hydrogen electrode: $2H_2 \rightarrow 4H^+ + 4e^-$
Oxygen electrode: $O_2 + 4H^+ + 4e^- \rightarrow 2H_2O$
Total reaction: $2H_2 + O_2 \rightarrow 2H_2O$

Fig. 1 The principle of a fuel cell.

electricity, and the voltage obtained from the oxygen/hydrogen reaction is 1.23 V. Hence, if a 100-V fuel cell is desired, 82 single cells are connected in a series.

The basic characteristics of fuel cells are as follows:

i)   unlike with the electricity obtained by burning, there is no restriction by Carnot's cycle (even in the latest power plants, the efficiency is said to be approximately 40%) and thus they are high efficient (the theoretical efficiency of oxygen/hydrogen reaction at 25°C is 83%); and

ii) their operation is clean because there are no reaction by products other than water.

Depending on the materials used for the separation membrane, the ionic conductivity is different as is the temperature at which the desirable ionic conductivity is obtained. Fuel cells can be classified as alkaline, phosphate, molten carbonate, solid oxides, and polymer electrolyte types. These all exhibit different characteristics, and research and development have been focusing on these differences. Readers interested in these aspects of fuel cells are referred to the works by Fueki and Takahashi [1], and Takahashi [2]. Discharge characteristics of various fuel cells are shown in Fig. 2. [3]

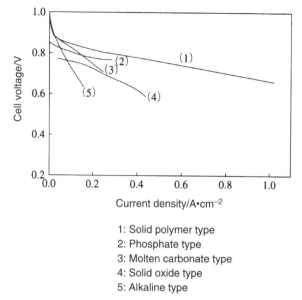

Fig. 2 Voltage versus current density curves for various fuel cells.

1: Solid polymer type
2: Phosphate type
3: Molten carbonate type
4: Solid oxide type
5: Alkaline type

2 POLYMER ELECTROLYTE FUEL CELLS

The fuel cells that use electrolyte gels are polymer electrolyte fuel cells (PEFCs). These fuel cells were used in the Gemini spacecraft during the 1960s and also the first to be commercially available.

At that time, a mixed electrolyte membrane made of fluorocarbon and poly(styrene sulfonic acid) was used. Later, as development continued for aerospace and military purposes, in 1972, the perfluorocarbon membrane Nafion® was introduced and it drastically improved chemical resistance.

In 1986, excellent fuel cell performance was confirmed as shown in Fig. 3 using a membrane developed by Dow Chemical Co. [4]. Then, in 1992, with the discovery that the amount of electrode-supported platinum catalyst could be reduced substantially, commercial application became possible, and accordingly, the pace of development accelerated.

In addition to the two previously mentioned PEFCs have the following characteristics:

i)   low-temperature operation (from room temperature to 100°C);
ii) high-power output density;
iii) no leakage of electrolyte;

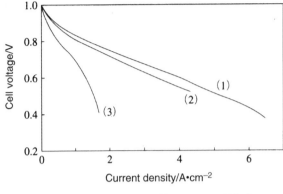

1: Dow membrane, H_2/O_2, 3.9 atm, 102 C
2: Dow membrane, H_2/O_2, 3.9 atm, 85 C
3: Nafion® 117, H_2/O_2, 3.9 atm, 85 C

Fig. 3   Characteristics of the fuel cell developed by Ballard Power Systems, Canada, using the Dow membrane.

iv)   resistance to differential pressure; and
v)   the availability of a wide selection of durable materials.

Especially because of their rapid response due to low-temperature operation, high-power density, and the fact that they are clean, application of fuel cells to automobiles has strong potential [5, 6]. When used in automobiles, air instead of oxygen and hydrogen-rich gas from methanol instead of hydrogen may be used.

3   DEVELOPMENT OF POLYMER ELECTROLYTE FUEL CELLS

In the United States, General Motors Co.(GM) has developed a small electric car that uses a modified methanol gas fuel as part of a project sponsored by the US Department of Energy (DOE). Chrysler and Corp. And Allied Signal Co. Also developed a small electric car in 1993 with hydrogen as the fuel under the same DOE project. In Canada, Ballard Power Systems succeeded in test driving a 120-kW PEFC powered bus using hydrogen fuel in a project sponsored by the British Colombia government. In Germany, Siemens has developed 34-kW PEFC for submarines. In Japan, as part of the New Sunshine Project, 1-kW level fuel cells have been developed. This project entered its second phase in 1996 with the goal of developing power systems of several tens kilowatts. Mazda Corp. used PEFC developed by Ballard Power Systems test drive a small car using an oxygen/hydrogen system. Fuji Electric has also succeeded in developing 5-kW fuel cells. In many research laboratories, research from the fundamental to applied, is being actively conducted.

4 SOLID ELECTROLYTES

The solid electrolytes in fuel cells must function as a separation membrane so that oxygen and hydrogen will not mix, and as a proton conductor. The polymer electrolyte membrane that has an acid side group exhibits these two functions. However, in order to exhibit conductivity, it must contain water because the acid group must dissociate. Hence, solid electrolyte in broad sense is the gel that contains water.

The perfluorocarbon sulfonic acid membrane is generally used. At present, Nafion® from Du Pont, Dow membrane from Dow Chemical, Asiplex® membrane from Asahi Chemicals, and Flemion® from Asahi Glass are commercially available. Figure 4 shows their basic structures; their exchange capacity is approximately 1.00 meq/g.

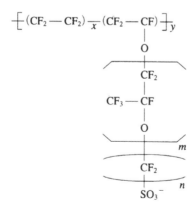

Nafion® membrane: $m \geq 1$, $n = 2$, $x = 5 \sim 13.5$
Dow membrane: $m = 0$, $n = 2$
Asiplex® membrane: $m = 0, 3$, $n = 2 \sim 5$
Flemion® membrane: $m = 0, 1$, $n = 1 \sim 5$

Fig. 4 The structure of various perfluorocarbon sulfonic acid membranes.

Unlike commercial ion exchange resin, the perfluorocarbon sulfonic acid membrane is not chemically crosslinked, but physically crosslinked by the crystalline region of the tetrafluoroethylene backbone. Its glass transition temperature is approximately 130°C [7]. The sulfonic acid groups form clusters, which consist of several tens of sulfonic acid groups with several hundred water molecules of approximately 3 to 4 nm in diameter [8, 9]. When the perfluorocarbon sulfonic acid membrane is used as the separation membrane for a fuel cell, it is important that it be impervious to oxygen and hydrogen permeation,proton transport, and the diffusion of water. It has been suggested that proton and water pass through the cluster and that gaseous components pass through the cluster interface and the amorphous portion of the polymer [10, 11].

The higher the exchange capacity and the thinner the membrane thickness, the greater the proton conductivity; this is a desirable property. Actually, there has been a report of a membrane with exchange capacity of 1.25 meq/g and a membrane thickness of 50 μm. However, while this appears to be encouraging, there was an increase in the permeation of the gaseous components, which led to the reduction of cell voltage caused by a chemical short circuit. In addition, there was a reduction of mechanical strength. Hence, it is important to carefully select the membrane properties based on need.

As shown in Fig. 5, if a dry gas is used, the membrane dries and the proton conductivity also decreases [12]. In particular, in order to prevent drying of the hydrogen electrode side, a moistened gas is used. Detailed

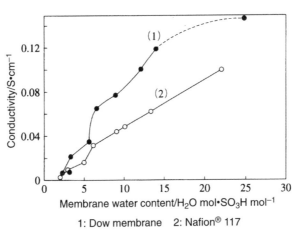

1: Dow membrane   2: Nafion® 117

Fig. 5   The dependence of the proton conductivity of perfluorocarbon sulfonic acid membranes on water content.

studies have been made of the relationship between the humidity of the atmosphere and the water content of the membrane and also between the water content of the membrane and the diffusion coefficient of water [11–13]. On the other hand, the generated water accumulates on the oxygen electrode side, blocking the path of the oxygen gas, which can cause the cell voltage to suddenly decrease or become unstable. As the conditions during the operation are complex, it is important to understand precisely the basic properties of the electrolyte membrane and cope with the situation.

5 PROBLEMS OF POLYMER ELECTROLYTE FUEL CELLS

Although polymer electrolyte has many interesting properties and the fuel cells using such polymer electrolyte offer great potential, some problems remain to be solved. They include:

i) management of water in the cell and between the cells;
ii) catalyst poisoning of the platinum catalyst by a small amount of CO impurity in the fuel;
iii) lifetime of the membrane and catalyst;
iv) the performance of a modifier if methanol is used as a fuel (current modifier operates around 200°C and the time necessary for this temperature rise may become a limiting factor); and
v) reduction of the cost of the single cell as a whole (in particular, reduction of the amount of platinum catalyst).

Two methods have been studied that address the problem of catalyst poisoning by CO: Mixing of a small amount of oxygen in the fuel [14, 15], and developing a catalyst that is resistant to catalyst poisoning [16, 17]. It has been demonstrated that it is possible to reduce the use of platinum catalyst to less than one-tenth of the traditional amount used by studying the reacting interface [18].

While the perfluorocarbon sulfonic acid membrane has provided satisfactory properties, further improvement is also desired for the membranes that:

i) can be used above 100°C (at high temperature, the catalyst activity increases and CO poisoning reduces. The temperature above 150°C can be used to heat water);
ii) exhibits high proton conductivity; and
iii) are low in cost.

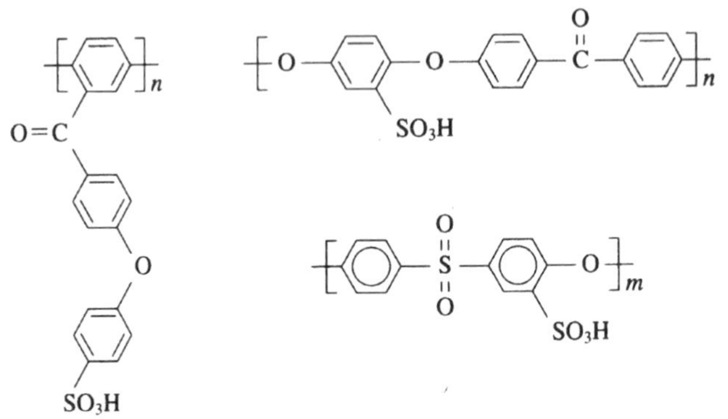

Fig. 6   Examples of newly synthesized sulfonic acid membranes.

Sulfonic-acid-modified heat-resistant polymers can be used to develop membranes that can be used above 100°C. For example, poly-benzimidazole modified with sulfonic acid [19] or doped with sulfuric acid or phosphoric acid [20, 21] shows good proton conductivity above 150°C.

Figure 6 shows poly(ether ether ketone) [19] or poly(acryl ether sulfone) [22] modified by sulfonic acid that have also been synthesized.

As low-cost solid electrolyte membranes, radiation-grafted membranes are currently being evaluated. These include: Teflon (PTFE)-fluoroethylene-propylene (FEP) membranes [23, 24], FEP-polystyrene membranes [25, 26], and PTFE membranes on which monomers with functional groups are grafted [27]. Using radiation grafting, a higher proton-conductance membrane has also been synthesized [28]. The chemical stability of these membranes is an important issue, and their long-term reliability can be demonstrated, it will be cost competitive.

6   CONCLUSION

Of all the fuel cells, polymer electrolyte fuel cells have very interesting characteristics, and they owe a great deal to solid electrolyte gels. In the future, establishment of a fuel supply infrastructure and progress in solid electrolyte technology will make possible relatively small-scale power generation for automobiles, computers, portable telephones, homes, buildings, schools, and hospitals, all to be derived from fuel cells.

REFERENCES

1   Fueki, K. and Takahashi, M. (1987). *Design Technology for Fuel Cells*, Science Forum.

2   Takahashi, T. (1992). *Chemistry One Point: 8, Fuel Cells*, Kyoritsu Publ.

3   Parthasarathy, A., Srinivasan, S., and Appleby, A.J. (1992). *J. Electroanal. Chem.* **339**:101.

4   Prater, K.B. (1990). *J. Power Sources* **29**:239.

5   Lemons, R.A. (1990). *J. Power Sources* **29**:251.

6   Barbir, F. (1995). *NATO ASI Ser. E.* **295**:241.

7   Okuyama, K. and Nishikawa, F. (1994). *Nippon Kagakukai-shi* 1087.

8   Gierke, T.D., Munn, G.E., and Wilson, F.C. (1981). *J. Polym. Sci., Polym. Phys. Ed.* **19**:1687.

9   Hsu, W.Y. and Gierke, T.D. (1983). *J. Membrane Sci.* **13**:307.

10   Ogumi, Z., Takehara, Z., and Yoshizawa, S. (1984). *J. Electrochem. Soc., Electrochem. Sci. Technol.* **13**:769.

11   Okuyama, K. and Nishikawa, F. (1994). *Nippon Kagakukai-shi* 1091.

12   Zawodzinski, Jr., T.A., Springer, T.E., Davey, T., Jestel, R., Lopez, C., Valerio, J., and Gottesfeld, S. (1993). *J. Electrochem. Soc.* **140**:1981.

13   Hinatsu, J.T., Mizuhata, M., and Takenaka, H. (1994). *J. Electrochem. Soc.* **141**:1493.

14   Gottesfeld, S. and Pafford, J. (1988). *J. Electrochem. Soc.* **135**:2651.

15   Lemons, R.A. (1994). *J. Power Sources* **29**:251.

16   Kawatsu, S. and Iwase, M. (1994). *Proc. 35th Battery Symp., Jpn.* pp. 299–300.

17   Schmidt, V.M., Brockerhoff, P., Hohlein, B., Menzer, R., and Stimming, U. (1994). *J. Power Sources* **49**:299.

18   Wilson, M.S. and Gottsfeld, S. (1992). *J. Electrochem. Soc.* **139**:L28.

19   Kobayashi, T., Mutsugawa, M., Sanui, K., Ogata, N., and Watanabe, M. (1994). *Polym. Preprints, Jpn.* **43**:736.

20   Savinell, R.F., Yeager, E., Tryk, D., Landau, U., Wainright, J.S., Gervasio, D., Cahan, B., Litt, M., Rogers, C., and Scherson, D. (1993). The electrolyte challenge for a direct methanol-air polymer electrolyte fuel cell operating at temperatures up to 200°C. *NASA Conf. Publ. Data*, **3228**, 167.

21   Wainright, J.S., Wang, J.T., and Savinell, R.F. (1994). Acid doped polybenzimidazoles: a new polymer electrolyte. *ECS Extended abstracts*, Spring Mtg., San Francisco, CA, **94-1**:618.

22   Nolte, R., Ledjett, K., Bauer, M., and Mulhaupt, R. (1993). Partially sulfonated poly (arylene ether sulfone)-A versatile proton conducting membrane material for modern energy conversion technologies. *J. Membrane Sci.* **83**:211.

23   Rouilly, M.V., Kotz, E.R., Haas, O., Schere, G.G, and Chapiro, A. (1993). Proton exchange membranes prepared by simultaneous radiation grafting of styrene on to Teflon-FEP films. Synthesis and characterization. *J. Membrane Sci.* **81**:89.

24   Gupta, B., Buchi, F.N., and Schere, G.G. (1993). Materials research aspects of organic solid conductors. *Solid State Ionics* **61**:213.

25   Buchi, F.N., Gupta B., Halim, J., Haas, O., and Scherer, G.G. (1994). A new class of partially fluorinated fuel cell membranes. *ECS Extended Abstracts*, Spring Mtg., San Francisco, CA, **94-1**:615.

26   Buchi, F.N., Gupta, B., Haas, O., and Scherer, G.G. (1995). (1995). Study of radiation-grafted FEP-g-polystyrene membranes as polymer electrolytes in fuel cells. *Electrochim. Acta* **40**:345.

27 Guzman-Garcia, A.G., Pintauro, P.N., Verbrugge, M.W., and Schneider, E.W. (1992). Analysis of radiation-grafted membranes for fuel cell electrolytes. *J. Appl. Electrochem.* **22**:204.

28 Okuyama, K. and Nishikawa, F. (1996). *Nippon Kagakukai-shi* 830.

Section 4
Sensors

1 ENZYME SENSORS

1.1 Fixation of Enzymes onto Gels and Construction of Sensors

Because enzymes are excellent biofunctional catalysts that catalyze specific chemical reactions in the body at moderate conditions and show high substrate specificity, it is possible to develop various artificial devices by engineering applications. However, enzymes must be used in moderate conditions because otherwise they will denature and their functions can not be utilized fully. In general, enzyme reactions take place in aqueous solutions. However, if enzymes must be dissolved in reaction systems, there will be many limitations in their use in engineering applications. Therefore, depending on the application purposes, fixed enzymes, namely, those that are fixed on appropriate substrate surfaces, have been utilized [1, 2]. Upon fixation, it is important that the fixed enzyme not denature, maintain a high enzyme activity, and will not leak or desorb from the substrate. Enzyme fixation methods can be divided largely into bonding onto a substrate, crosslinking, and inclusion methods. In the inclusion method, polymer gels are used mainly as the fixation matrices,

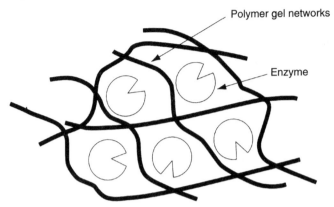

Fig. 1 Fixation of an enzyme into a hydrogel by the inclusion method.

and the fixation method is shown in Fig. 1. The enzyme coexists at the time of polymerization so that it may be entrapped in the fine networks of the polymer gel.

In the inclusion method, polyacrylamide is often used. An enzyme-fixed polymer gel can readily be obtained by adding a crosslinking agent, N,N′-methylene-bis-acrylamide, and an acrylamide monomer solution in which an accelerator is included to the enzyme solution, which is the polymerized. In this way, it is possible to form an enzyme-fixed membrane that prohibits leakage of the enzyme out of the gel by forming gel networks with an appropriate crosslinking density. This is achieved by adjusting the monomer and crosslinking agent concentration. However, in order to improve the fixation strength, enzymes are often crosslinked with the polymer using a compound such as glutaraldehyde which reacts with the amino residues of proteins. When a polymer gel is used as the fixation matrix,low molecular weight solute or reaction products can enter or leave the gel freely while the enzyme is fixed. Hence, such systems are used in such material production devices as bioreactors and enzyme sensors. As bioreactors are described in this monograph in part II, Chapter 2, Section 5, here enzyme sensors, in particular, glucose sensors, will be introduced.

1.2   Oxygen and Hydrogen Peroxide-detecting Glucose Sensors

One of the most actively studied enzyme sensors is an enzyme electrolyte for glucose analysis using glucose oxidase (GOD).the idea of constructing

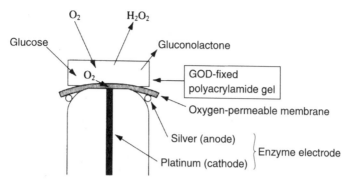

Fig. 2 The structure of the glucose sensor constructed for the first time by Updike and Hicks.

an artificial device by combining a biofunctional molecular element, such as an enzyme, and a transducer was proposed by Clark and Lyons [3]. However, Updike and Hicks [4] in 1967 constructed the first enzyme electrolyte, a glucose sensor that utilized a fixed enzyme. As shown in Fig. 2, this enzyme electrolyte used a glucose-oxidase-fixed membrane obtained by the inclusion method in polyacrylamide gel. This membrane was physically fixed into the electrolyte on which an oxygen-permeable membrane is adhered. The principle of glucose sensing by this sensor is as follows. As shown in Fig. 3, GOD is a catalyst used to oxidize glucose and to produce gluconolactone.

As the reaction proceeds, the coenzyme near the activity center of GOD, that is, oxidized flavin adenine dinucleotide (FAD), changes into reduced flavin adenine dinucleotide ($FADH_2$), which is oxidized by the oxygen in the system and turns into FAD again (see Figs. 3 and 4).

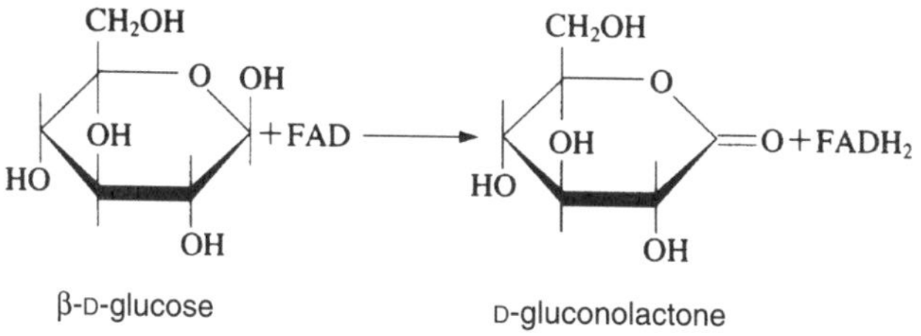

$$FADH_2 + O_2 \longrightarrow FAD + H_2O_2$$

Fig. 3 The enzyme reaction of glucose oxidase (shown in the figure) as the reaction of coenzyme $FAD/FADH_2$ in glucose oxidase.

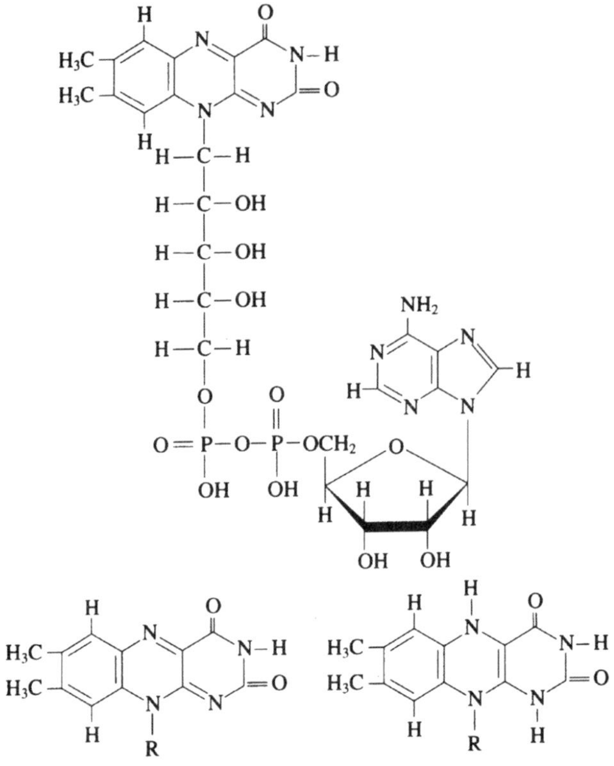

Fig. 4 The structure of oxidized coenzyme flavin adenine dinucleotide (FAD) and reduced coenzyme (FADH$_2$)

Because the oxygen in the system, that is, in the polyacrylamide gel, accompanies this oxidation of FADH, the permeation of oxygen to the electrode decreases. Hence, it is now possible to determine the glucose concentration by measuring the reduction of oxygen concentration in the system using the oxygen electrode. It is also possible to determine the glucose concentration by oxidizing the hydrogen peroxide produced by the enzyme reaction and measuring the increase in the oxidation current [5–7].

When an oxygen or hydrogen peroxide enzyme-detecting electrode is used, the oxidation reaction uses oxygen as the electron acceptor. Hence, oxygen is indispensable for the enzyme reaction. However, the disadvantage of this method is the strong dependency of the current response on the concentration of the dissolved oxygen. In clinical blood sugar analysis, glucose concentration in the blood is relatively high. For a healthy man, it is approximately 5 mM, while for diabetics, it reaches as high as 20 mM.

This causes oxygen deficiency during the measurement of an accurate glucose concentration can not be determined. To cope with this difficulty, the sample was diluted, and to reduce oxygen and the oxidation of hydrogen peroxide, a relatively high voltage is necessary. However, this leads to complications caused by the active molecules of the other electrode. For example, determination of hydrogen peroxide is carried out normally at 0.7 V (versus SCE) using a metal electrode. At this voltage, other materials in the blood, such as ascorbic acid and uric acid, also influence the sensor output.

1.3 Low Molecular Weight Molecule Mediator-type Glucose Sensors

In order to solve the aforementioned disadvantage of the oxygen and hydrogen peroxide glucose-detecting sensor, a second-generation glucose sensor, that is, a mediator-type glucose sensor, has been developed [8]. During the enzyme reaction of GOD, the enzyme functions as the electron acceptor to oxide $FADH_2$ into FAD again. If this oxidation reaction of $FADH_2$ can be carried out electrochemically, it is then possible to determine the glucose concentration by observing the oxidation current. $FADH_2$ Unfortunately, there is no direct electron transfer from to the electrode because $FAD/FADH_2$ of GOD is included deep inside the enzyme protein. Therefore, a redox (oxidation-reduction) molecule is introduced in the system and through it, $FADH_2$ is oxidized indirectly. This redox molecule is the replacement artificial electron acceptor, and because it mediates the electron transfer between the electrode and $FADH_2$, it is called a mediator (see Fig. 5). An oxidation-type redox molecule receives an electron from $FADH_2$ and oxidizes $FADH_2$ into FAD. The redox molecule itself becomes a reduction type that is oxidized by the electrode to return to the original oxidation type. Thus, as a result of the current from the oxidation reaction of the mediator by the electrode, amperemetric glucose sensing becomes possible, and unlike oxygen, it will not be consumed by the progression of the enzyme reaction. The mediator functions repeatedly under no oxygen and high glucose concentration to oxidize $FADH_2$ into FAD. Furthermore, because the standard

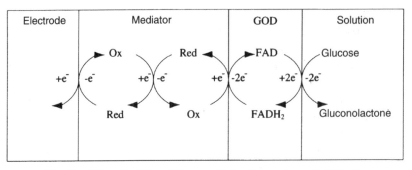

Fig. 5 The principle of the mediator-type glucose sensor.

oxidation/reduction potential ($E^{0\prime}$) of $FADH_2/FAD$ is $-0.4\,V$ (versus SCE), if the redox molecule has sufficiently higher $E^{0\prime}$, then it can function fully as a mediator. Hence, as it is not necessary to use high voltage to influence the electrode active compounds in the blood, the improvement in sensor performance is expected.

Traditionally, ferrocene derivatives [8–10] and quinones [11, 12] have been used as mediators. For these small molecular weight mediators, the electron transfer from GOD to the electrode is achieved by the diffusion of the mediator molecules to the electrode. If the mediator molecules are fixed by inclusion into the polymer gel in a manner similar to enzymes, it is possible for these mediator molecules to leak into the measurement solution after repeated use. This phenomenon can occur even if these molecules are somewhat water insoluble, which leads to the gradual reduction of electrode activity. In particular, for the development of implanted sensors, the fixation of mediator molecules onto the electrode is a serious problem.

1.4 Polymer Mediator-type Glucose Sensors

In order to develop high value-added glucose sensors with long-term stability and biocompatibility, systems with polymer mediators with redox active groups have recently been studied. For example, long-term stability was improved as follows. The mediator molecules can be chemically fixed to the polymer substrate, which traditionally has been used as the enzyme-fixation matrix. This polymer functions as a polymer matrix and a mediator, allowing the preparation of an enzyme electrode without leakage of the enzyme and mediator. If a hydrophilic polymer mediator that swells by water to form a hydrogel is used, the solute is distributed in the enzyme-fixed membrane. This solute diffuses to the vicinity of the enzyme causing an enzyme reaction in a manner similar to the forementioned polyacrylamide gel. The electron-transfer process from glucoseoxidase, which is a biopolymer, to the electrode through the polymer mediator is also attracting attention. In the case of a polymer mediator, the mediator molecule itself can not diffuse and transfer electrons to the electrode because the redox site is fixed onto the polymer chain. Hence, in this case, the electron transfer is considered to take place as a result of electron hopping in which the electron exchange reaction between the redox sites occurs consecutively. Electron hopping occurs when the polymer main chain is flexible and the redox site is locally mobile. Then the neighboring redox sites approach each other by the conformational changes of the polymer main chain and thus the electron exchange reaction takes place [13, 14]. Using this fast-charge-transfer polymer mediator, a more effective mediation reaction can be expected.

So far, it has been reported that redox polymers, such as polysiloxane or polyarrylamide on which ferrocene is introduced as a side chain, are effective mediators [15–25]. Other effective mediators include poly(vinyl pyridine) or poly(vinyl imidazole) on which osmium complex is introduced in the side chain. However, the structure-mediator function correlation has been poorly understood. Here, the glucose sensor was used as an example of enzyme sensors, and as shown in Table 1, sensors using enzymes other than glucose oxidase have also been actively studied. It is now possible to detect electrochemically materials in the body using an electrode with an enzyme and a polymer gel.

Table 1 Other Enzyme Sensors that Utilize Polymer Gels.

Materials detected	Enzyme	Literature
L-α-glycelophosphoric acid•L-lactic acid	L-α-glycerolphosphate oxidase•lactate oxidase	26)
Nicotinamide adenine dinucleotide	Horseradish peroxidase	27)
Choline•acetylcholine	Choline oxidase•acetylcholine esterase	28)
Creatine•creatinine	Creatineamide hydrase•creatinineamide hydrase	29)
Cholesterol	Cholesterol oxidase•horseradish peroxidase	30)

2 HUMIDITY SENSORS

2.1 Introduction

There are many humidity sensors that use polymers, and depending on the detection principle, a polymer with different structure and properties is chosen. As shown in Table 1, there are two major types of humidity sensors: one that uses the polymer with electrical property changes and is subject to weight changes upon humidity changes, and another that is subject to optical property changes due to moisture absorption, which has recently been studied. The sensors that utilize the changes in electrical properties can be further divided into those that measure electrical resistance and those that measure electrical capacitance. Sensors with weight changes utilize either a quartz oscillator or a surface acoustic wave (SAW) generator. The sensor that measures electrical capacitance uses hydrophobic polymer with the water uptake of less than 2 to 3% even in the high humidity region, but the sensor that measures electrical resistance uses a hydrophilic polymer. In this case, the water uptake often becomes quite high reaching more than 100%.

Here, the humidity sensors that utilize hydrophilic polymers will be described as they relate to gels. In addition, sensors that use a quartz oscillator will also be discussed.

The solid polymers used for resistance-type humidity sensors with acid and alkaline groups, such as sulfonic acid or quaternary ammonium, absorb water depending on the humidity of the atmosphere. As the electrical resistance decreases in proportion to the absorbed water, the humidity can be determined by measuring the electrical resistance. If the current applied is dc, then polarization or electrolysis may occur. Thus, generally, ac impedance is measured. Figure 1 shows the chemical structures of representative polymers that can be used for humidity sensors; these are all hydrophobic polymers. In addition, they are either

Table 1 Humidity sensors that utilized polymers.

Humidity sensor	Method using electrical properties	Electrical-resistance measurement sensor
		Electrical-capacitance measurement sensor
	Method using weight changes	Quartz-oscillator sensor
		Surface acoustic wave (SAW) generator sensor

Polycation

Poly(N,N-dimethyl-s,5-dimethylene piperidium chloride)(DPiC)

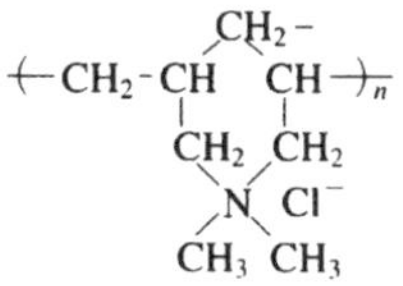

$$\left[CH_2-CH\overset{\displaystyle CH_2-}{\underset{\displaystyle \begin{array}{c} CH_2 \ \ CH_2 \\ \diagdown N \diagup \ Cl^- \\ CH_3 \ \ CH_3 \end{array}}{|}}CH\right]_n$$

Poly(2-hydroxy-3-methacryloyloxypropyl trimethylammonium chloride)(HMPTAC)

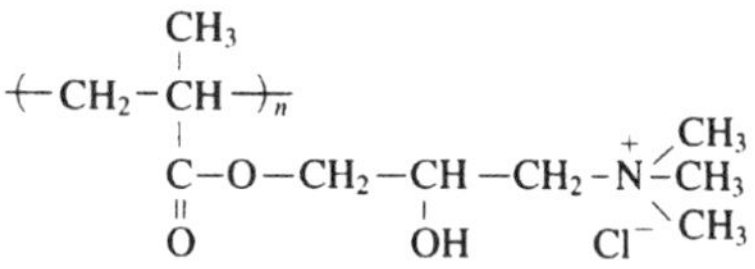

$$\left[CH_2-\underset{\underset{\displaystyle O}{\overset{\displaystyle |}{C}}\text{-}O-CH_2-\underset{\underset{\displaystyle OH}{|}}{CH}-CH_2-\overset{+}{N}\overset{CH_3}{\underset{Cl^-\diagdown CH_3}{\diagup}}CH_3}{\overset{\displaystyle CH_3}{\underset{\displaystyle |}{CH}}}\right]_n$$

Quaternarized poly(4-vinyl pyridine)

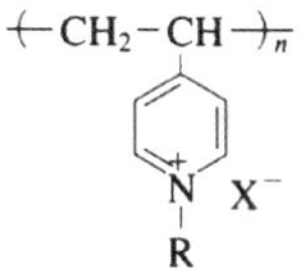

$$\left[CH_2-CH\right]_n$$

R = alkyl
X = halogen

Poly(vinyl benzyl trimethylammonium chloride)

$$\left[CH_2-CH\right]_n$$

Cl^-

$\underset{CH_3\ CH_3}{\overset{+}{N}}\text{-}CH_3$

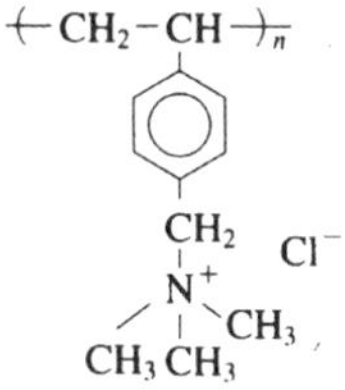

Polyanion

Poly(styrene sulfonic acid)

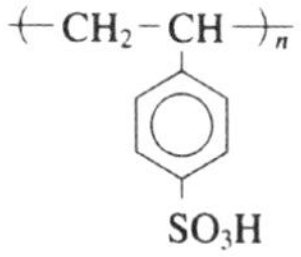

$$\left[CH_2-CH\right]_n$$

SO_3H

Poly(2-acrylamide-2-methylpropane sulfonic acid)(AMPS)

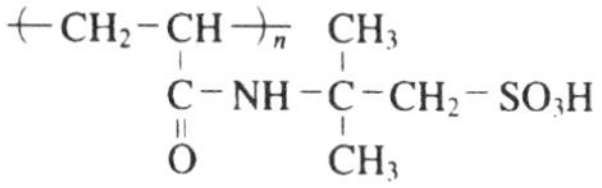

$$\left[CH_2-CH\right]_n \quad \underset{\underset{\displaystyle O}{\overset{\displaystyle |}{C}}-NH-\underset{\underset{\displaystyle CH_3}{|}}{\overset{\overset{\displaystyle CH_3}{|}}{C}}-CH_2-SO_3H}{}$$

Fig. 1   Representative polymers that are used for electrical-resistance humidity sensors.

polycations that possess quaternary ammonium or polyanions that possess sulfonic acid. The resistance of these polymers greatly changes in response to humidity changes. As shown in Fig. 2, a comb-shaped gold electrode is either evaporated or fused onto a ceramic substrate. Then a thin polymer film is coated onto the electrode either by spin coating the polymer solution or dip coating the ceramic into the solution. The impedance of such element decreases exponentially as shown in the open circle in Fig. 3.

However, as the hydrophilic polymer alone would dissolve in water, it has a drawback in lack of stability under a high humidity atmosphere. In order to cope with this problem, various polymers are used. They include ionen [31] made of methylene and quaternary ammonium groups in the main chain, a mixture of perfluorosulfonic acid ionomer and phosphate pentoxide [32], sulfonated polystyrene [33], Nafion membrane that has perfluorosulfone group [34], and a fluorinated polymer that possesses a perfluorosulfone group and a perfluorocarboxyl group [35]. An example of evaporating the electrode onto the stable porous ceramic or polymer to which hydrophilic polymer is impregnated has been reported [36, 37]. A unique approach is to use a humidity-sensitive membrane made by plasma treatment. For example, on a thermally table silicone polymer, a mixture of tetramethylsilane and ammonia is polymerized [38]. Hexamethyldisloxane and ammonia are also polymerized on a glass plate to make a plasma-polymerized membrane [39].

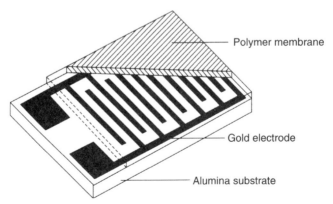

Fig. 2 An electrical-resistance humidity sensor.

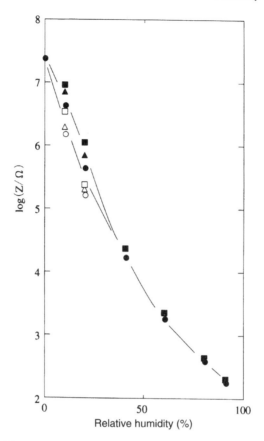

(○, ●) HMPTAC homopolymer
(△, ▲) Crosslinked HMPTAC polymer
(□, ■) IPN, closed symbols: humidity-increasing process,
        open symbols: Humidity-decreasing process

Fig. 3 The temperature dependence of the impedance of an electrical-resistance humidity sensor.

2.2 Sensors That Utilized Copolymers

A cationic or anionic hydrophilic vinyl monomer is polymerized with a hydrophobic vinyl monomer. This polymer is thinly coated onto a substrate using an appropriate solvent to make a widely used sensor as shown in Fig. 2. For example, a copolymer made of a vinyl monomer and reactive cationic monomer is coated on a substrate and has been made into a commercially available sensor [40]. There are reports on the moisture-sensitivity characteristics and response time of the sensors that used copolymers made of hydrophilic vinyl monomers with quaternary ammonium, sulfone or carboxylic groups and hydrophobic monomer, such as methyl methacrylate or styrene [41–43]. However, sensors that use polymers are inferior to those described later in terms of moisture-condensation property and long-term stability under high humidity.

2.3 Sensors with Crosslinked Hydrophilic Polymers

There are two methods of forming a water insoluble, crosslinked hydrophilic membrane on the substrate shown in Fig. 2. One is to copolymerize a hydrophilic vinyl monomer with a crosslinking monomer with more than two functionalities on the substrate.

The other method is to coat a polymer with a hydrophilic group or the group that can become hydrophilic on the substrate, followed by crosslinking by light or radiation. Compared to the former method of polymerizing monomers, the reproducibility of the crosslinking method is better and many practical examples exist.

As an example of the former, a humidity-sensitive membrane was made by photopolymerizing styrene sulfonate and N,N′-methylene-bis-acrylamide on a substrate [44].

As an example of the latter method, the copolymer of 4-vinyl pyridine or 2-dimethylaminoethyl methacrylate and a crosslinking monomer, 4′-methacryloyloxy calcon, is quaternarized by 1-bromopropane. This copolymer was coated on a substrate, followed by photocrosslinking [45]. Another example includes the crosslinking of poly(dimethyldiaryl-lammonium chloride) by γ-radiation by a cobalt 60 source [46]. As shown in Fig. 4, it has been reported that a mixture of poly(vinyl pyridine) and a linear hydrocarbon with both ends halogenated was coated onto a substrate and thermally reacted [47]. Similarly, crosslinking of poly(chloromethyl styrene) with N,N,N′, N′-tetramethyl-1,6-hexanediamine was also reported [48]. A unique moisture-sensitive material [49] can be made as shown in Fig. 5 by the sol-gel method. After hydrolysis of organosilane, if such hydrophilic groups as sulfone or quaternary ammonium are introduced onto the three-dimensionally crosslinked organopolysilsesquioxane, a water-resistant humidity-responsive membranae can be made. By using a substrate with OH groups on the surface, the water resistance will further

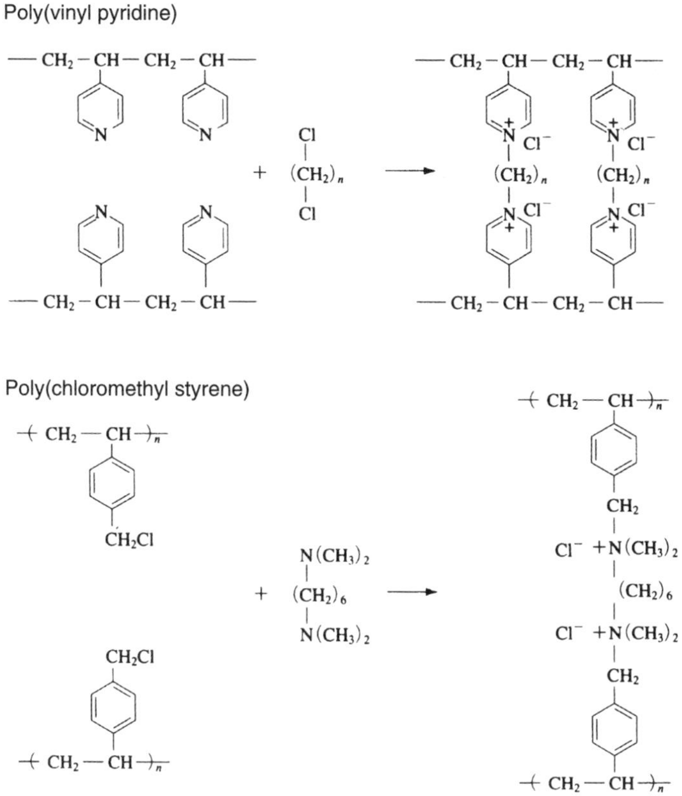

Poly(vinyl pyridine)

Poly(chloromethyl styrene)

Fig. 4 Simultaneous quaternarization and crosslinking of poly(vinyl pyridine) and poly(chloromethyl styrene).

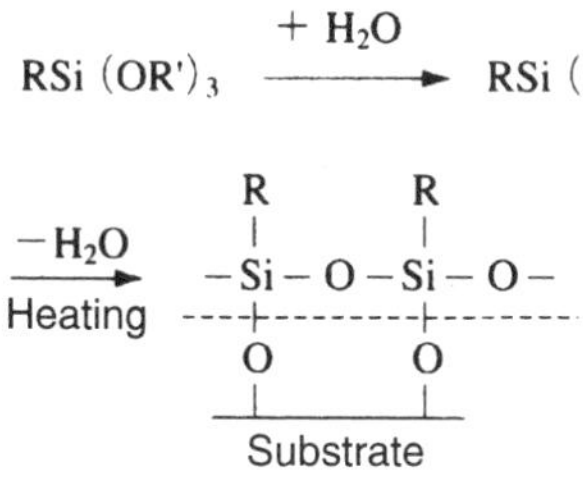

Fig. 5 Synthesis of crosslinked organopolysiloxanes.

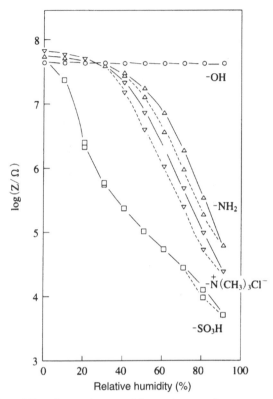

Fig. 6 The humidity dependence of impedance of organopolysiloxane to which hydrophilic groups are introduced.

improve. Figure 6 shows the humidity dependence of impedance of several sensors made by this method.

2.4 Sensors Utilizing Interpenetrating Polymer Networks

Two polymers with different properties can be mixed by the interpenetrating polymer network (IPN) method and used as a modification of the forementioned crosslinking approach. As shown in Fig. 7, a hydrophilic polymer and hydrophobic polymer are separately crosslinked to form physical interpenetration. The hydrophilic polymer absorbs water but will not dissolve in water or deform greatly due to the inhibition effect of the hydrophobic polymer. As an example, diisocyanate crosslinked HMPTAC polymer as the hydrophilic polymer and melamine resin as the crosslinked

Crosslinked hydrophobic polymer

Crosslinked hydrophobic polymer

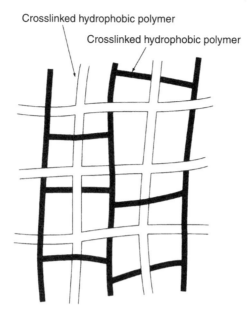

Fig. 7 Interpreting polymer networks (IPN).

hydrophobic resin were used to prepare membranes that are insoluble to any solvent. If, instead of melamine resin, ethylene glycol dimethacrylate (EGDMA) is used, a better water-resistant sensors could be manufactured [50, 51]. Figure 3 shows the humidity dependence of the impedance of the humidity sensor made of HMPTAC-EGDMA IPN membrane. For comparison, the figure also plots the impedance of the raw materials, uncrosslinked HMPTAC polymer membrane, and the membrane cross-linked only by isocyanate. This IPN sensor did not change its character-istics even after immersion in water for 24 h.

2.5 Sensors Utilizing Graft Polymers

This method uses a hydrophobic polymer with sufficient water and solvent resistance as the main chain to which hydrophilic monomers are chemi-cally grafted by a catalyst, light or radiation. The polymer will be insoluble in water yet its electrical resistance is moisture dependent. This graft copolymer will be used for the sensor. As for the hydrophobic polymer, poly(tetrafluoroethylene)(PTFE) film or a porous poly(ethylene)(PE) film is used. The hydrophilic polymers can be chosen from poly(styrene sulfonate) [52] or quaternarized poly(vinyl pyridine) [53, 54]. However, grafting these polymers directly onto PTFE film will be difficult. As shown in Fig. 8, the former can be made by grafting polystyrene first, followed by sulfonation of the benzene group. The latter can be made by grafting poly(vinyl pyridine), followed by quaternarization of the pyridine ring. A gold electrode was evaporated onto these graft copolymers as shown in Fig. 9 to manufacture a surface-exposed or sandwiched element. When the initial PTFE film is thick (0.5 mm), polystyrene or poly(vinyl pyridine) is grafted only to the surface. Thus, only a surface-exposed sensor can be manufactured.

When the thickness of the PTFE film is thin (0.05 mm), the graft polymer layer enters inside the film and spreads throughout the entire interior of the PTFE film. After the film is sulfonated or quaternarized, gold electrodes can be evaporated on both sides of the film to form a sandwiched element. Figure 10 shows the humidity-sensing curves of the surface-exposed sensor using a thick PTFE film on which poly(styrene sulfonic acid) is grafted. The salt exhibits higher impedance compared to the acid. The measurement was done at various frequencies. Assuming a parallel equivalence circuitry of the resistance R_p and capacitance C_p, complex impedance analysis is performed. Figure 11 shows the R_p against the amount of water absorbed. It can be found that the R_p depends only on the amount of absorbed water while it is independent of the type of cations. Hence, it can be concluded that the electrical conductivity is likely from only the protons from the absorbed water molecules.

PTFE-graft-quaternarized poly(vinyl pyridine) is a useful material for studying electrical conductivity mechanisms. After grafting, poly(vinyl pyridine) can be quaternarized with an alkyl group of different size, which is added to the pyridinium group or halogen atoms that are the counter ion of the pyridinium group. By changing the size of the alkyl group, the greater the number of carbons in the alkyl group, the higher the impedance. This result agrees with the results that the longer the alkyl group, the fewer the number of water molecules per pyridinium group. When the type of ion is changed, the humidity-sensing characteristics change as shown in Fig. 12; R_p obtained from the complex impedance of this element is plotted against the number of water molecules in Fig. 13. When the number of water molecules is small, R_p obviously changed depending on the type of anionic group. However, if the same anionic

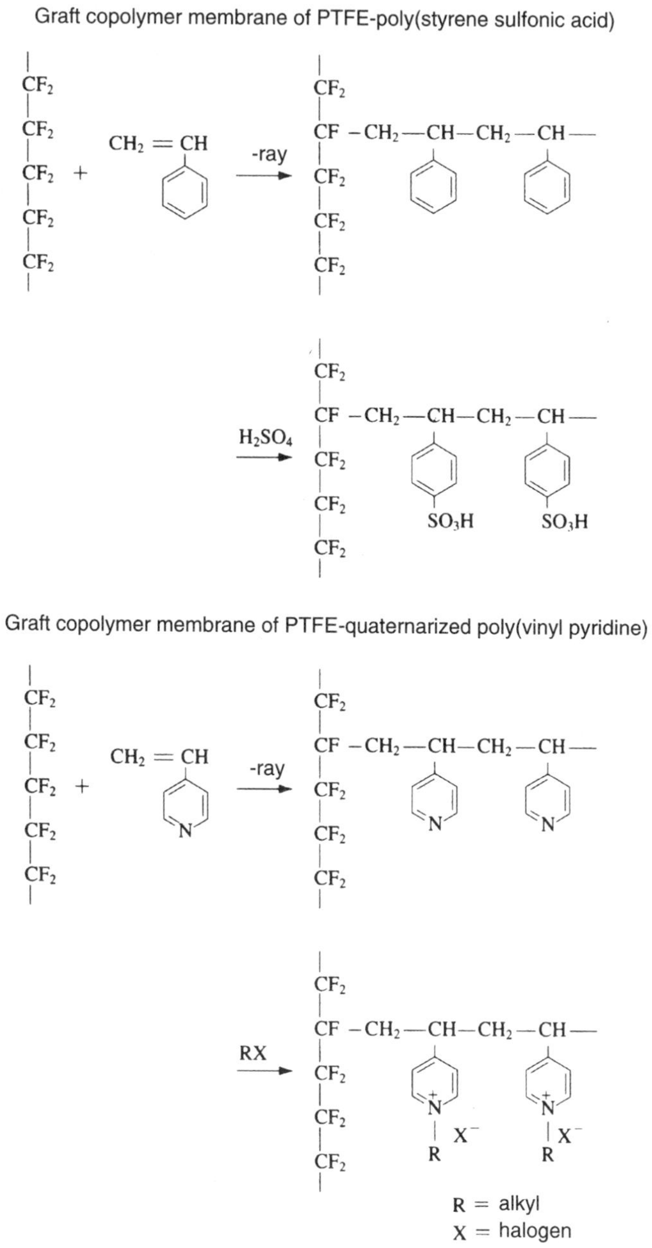
Graft copolymer membrane of PTFE-poly(styrene sulfonic acid)
-ray
H2SO4
SO3H
Graft copolymer membrane of PTFE-quaternarized poly(vinyl pyridine)
-ray
RX
R = alkyl
X = halogen

Fig. 8 Graft polymerization onto PTFE film.

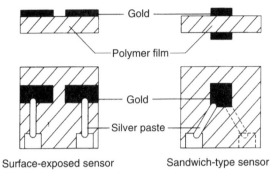

Surface-exposed sensor   Sandwich-type sensor

Fig. 9 Humidity sensors using a graft copolymer film.

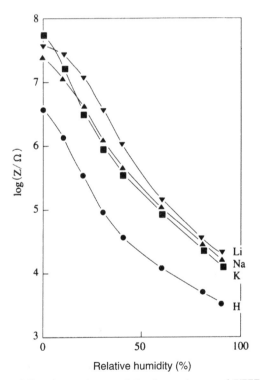

Fig. 10 The humidity dependence of the impedance of PTFE membrane on which surface poly(styrene sulfonate) was grafted.

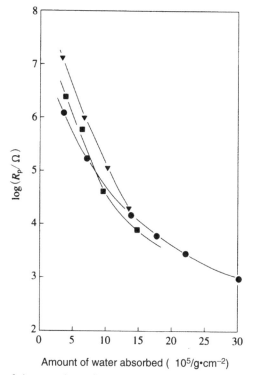

Fig. 11 R_p of the previous figure versus the amount of absorbed water.

group is used, R_p depends only on the number of water molecules and is independent of the size of the alkyl group. The differences probably diminish at high water content because the proton becomes the main carrier while in the low water content regime the halogenated anion is the carrier.

If an ordinary polyethylene membrane is used instead of PTFE to graft poly(vinyl pyridine) and to quaternarize it, the response time is too slow to be useful. Unlike for the PTFE, the reaction proceeded homogeneously, and hydrophobic polyethylene and hydrophilic quaternarized vinyl pyridine are well mixed. Consequently, the diffusion of water is slow.

In order to overcome this shortcoming, porous polyethylene can be used as the graft substrate. In this case, if the grafting monomer penetrates in all areas of pores and grafts only with the wall surface, it is ideal.

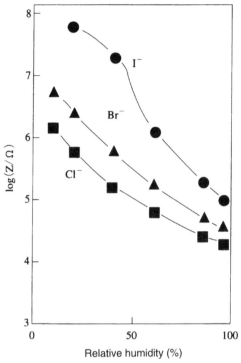

Fig. 12 Humidity-sensitivity characteristics of PTFE-graft-quaternarized poly(vinyl pyridine).

As a grafting monomer, 2-acrylamide-2-methylpropane sulfonate (AMPS) [55] or 2-hydroxy-3-methacryl-oxypropyl trimethylammonium chloride (HMPTAC) [56, 57] can be used. On both sides of the grafted porous membranes, gold electrode is evaporated to manufacture a sandwich-structured sensor. Although it is different from grafting, a membrane with similar characteristics as the rafted membrane can be made by sulfonating the wall surface of the porous polyethylene membrane [58]. The grafted polymer membrane shows excellent water resistance and is resistant to deformation. However, it take time to perform graft polymerization.

2.6 Quartz Oscillator Method

A quartz oscillator can be used to measure the weight increase of a membrane and to determine humidity. A humidity sensor was made by

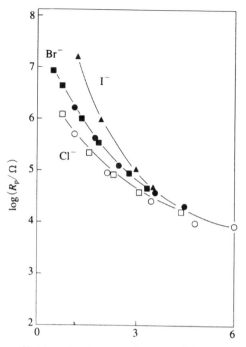

Fig. 13 R_p of the previous figure versus the amount of absorbed water.

coating a thin film of hydrophilic polymer on a quartz oscillator. From the resonance frequency changes, the amount of water absorbed can be calculated. As a hydrophilic polymer, Randin and Zülig [59] used a copolymer of 2-acrylamide-2-methylpropane sulfonic acid and 2-hydroxyethyl methacrylate, cellulose acetate and epoxy resin. Redeva *et al.* [60] used membranes obtained by glow discharge polymerization of acetonitrile, maronic acid dinitrile, o-phthalic acid dinitrile, and hexaethyldisiloxane.

REFERENCES

1 Chibata, I. (1986). *Fixed Biocatalysts*, Kodan-sha Scientific.
2 Uozumi, T. and Ohta, T. (1993). *Enzyme Engineering*, Maruzen Publ.
3 Clark, L.C. and Lyons, C. (1962). *Ann. N.Y. Acad. Sci.* **102**: 29.
4 Updike, S.J. and Hicks, G.P. (1967). *Nature* **214**: 986.
5 Bourdillon, C., Bourgeois, J.P., and Thomas, D. (1980). *J. Am. Chem. Soc.* **102**: 4231.

6 Sittampalam, G. and Wilson, G.S. (1982). *J. Chem. Edu.* **59**: 71.

7 Okahata, Y., Tsuruta, T., Ijiro, K., and Ariga, K. (1989). *Thin Solid Films* **180**: 65.

8 Cass, A.E.G., Davis, G., Francis, G.D., Hill, H.A.O., Aston, W.J., Higgins, I.J., Plotkin, E.V., Scott, L.D.L., and Turner, A.P.F. (1984). *Anal. Chem.* **56**: 667.

9 Lange, M.A. and Chambers, J.Q. (1985). *Anal. Chim. Acta* **175**: 89.

10 Wang, J., Wu, L., Lu, Z., Li, R., and Sanchez, J. (1990). *Anal. Chim. Acta* **228**: 251.

11 Ikeda, T., Hamada, H., Miki, K., and Senda, M. (1985). *Agric. Biol. Chem.* **49**: 541.

12 Mottga, N. and Guadalupe, A.R. (1994). *Anal. Chem.* **66**: 566.

13 Watanabe, M., Nagasaka, H., Sanui, K., and Ogata, N. (1992). *Electrochim. Acta* **37**: B:1521.

14 Watanabe, M., Nagasaka, H., and Ogata, N. (1995). *J. Phys. Chem.* **99**: 12294.

15 Degani, Y. and Heller, A. (1989). *J. Am. Chem. Soc.* **111**: 2357.

16 Hale, P.D., Inagaki, T., Karan, H.I., Okamoto, Y., and Skotheim, T.A. (1989). *J. Am. Chem. Soc.* **111**: 3482.

17 Gregg, B.A. and Heller, A. (1990). *Anal. Chem.* **62**: 258.

18 Gregg, B.A. and Heller, A. (1991). *J. Phys. Chem.* **95**: 5970.

19 Hale, P.D., Boguslavsky, L.I., Inagaki, T., Karan, H.I., Lee, H.S., Skotheim, T.A., and Okamoto, Y. (1991). *Anal. Chem.* **63**: 677.

20 Calvo, E.J., Danilowicz, C., and Diaz, L. (1993). *J. Chem. Soc., Faraday Trans.* **89**: 377.

21 Ohara, T.J., Rajagopalan, R., and Heller, A. (1993). *Anal. Chem.* **65**: 3512.

22 Tatsuma, T., Saito, K., and Oyama, N. (1994). *Anal. Chem.* **66**: 1002.

23 Kaku, T., Karan, H.I., and Okamoto, Y. (1994). *Anal. Chem.* **66**: 1231.

24 De Lumlet-Woodyear, T., Rocca, P., Lindsay, J., Dror, Y., Freeman, A., and Heller, A. (1994). *Anal. Chem.* **67**: 1332.

25 Nagasaka, H., Saito, T., Hatakeyama, H., and Watanabe, M. (1995). *Denki Kagaku* **63**: 1088.

26 Katakis, I. and Heller, A. (1992). *Anal. Chem.* **64**: 1008.

27 Vreeke, M., Maidan, R., and Heller, A. (1992). *Anal. Chem.* **64**: 3084.

28 Garguilo, M.G., Huynh, N., Proctor, A., and Michael, A.C. (1993). *Anal. Chem.* **65**: 523.

29 Schneider, J., Grudig, B., Renneberg, R., Cammann, K., Madeas, M.B., Buck, R.P., and Vorlop, K.D. (1996). *Anal. Chim. Acta* **325**: 161.

30 Boguslavsky, L., Kalash, H., Xu, Z., Beckles, D., Geng, L., Skotheim, T., Laurinavicius, V., and Lee, H.S. (1995). *Anal. Chim. Acta* **311**: 15.

31 Noguchi, H., Uchida, Y., Nomura, A., and Mori, S. (1989). A highly reliable humidity sensor using ionene polymers. *J. Mater. Sci. Lett.* **8**: 1278.

32 Huang, H. and Dasgupta, P.K. (1991). Perfluorosulfonate ionomer-phosphorus pentoxide composite thin films as amperometric sensors for water. *Anal. Chem.* **63**: 1570.

33 Xin, Y. and Wang, S. (1994). An investigation of sulfonated polysulfone humidity-sensitive materials. *Sensors & Actuators A.* **40**: 147.

34 Yamanaka, A., Kodera, T., Fujikawa, K., and Kita, H. (1988). *Denki Kagaku* **56**: 200.

35 Huang, P.H. (1988). Halogenated polymeric humidity sensors. *Sensors & Actuators* **13**: 329.

36 Sakai, Y., Sadoaka, Y., Omura, H., and Watanabe, N. (1984). *Kobunshi Ronbunshu* **41**: 205.

37 Sakai, Y., Sadoaka, Y., Okumura, S., and Ikeuchi, K. (1984). *Kobunshi Ronbunshu* **41**: 209.

38 Inagaki, N. and Suzuki, K. (1984). *Kobunshi Ronbunshu* **41**: 215.

39 Sadhir, R.K. and Sanjana, Z.N. (1991). Plasma deposited thin films suitable as moisture sensors. *J. Mater. Sci.* **26**: 4261.

40 Takaoka, Y., Maebashi, Y., Kobayashi, S. Bando, K., Tokkyo Kokai 16467 (1983).

41 Tsuchitani, S., Sugawara, T., Kinjo, N., and Ohara, S. (1985). Humidity sensor using ionic copolymer. *Sensors & Actuators* **210**.

42 Tsuchitani, S., Sugawara, T., Kinjo, N., Ohara, S., and Tsunoda, T. (1988). A humidity sensor using ionic copolymer and its application to a humidity-temperature sensor module. *Sensors & Actuators* **15**: 375.

43 Takahashi, K., Motegi, K., Rakuma, T., Itagaki, T., and Suzsi, T. (1988). *Sanyo Technical Review* **20**: 114.

44 Hijikigawa, M., Miyoshi, S., Sugihara, T., and Jinda, A. (1983). A thin-film resistance humidity sensor. *Sensors & Actuators* **4**: 307.

45 Ohtsuki, S. and Dozen, Y. (1988). *Kobunshi Ronbunshu* **45**: 549.

46 Rauen, K.L., Smigh, D.A., Heineman, W.R., Johnson, J., Seguin, R., and Stoughton, P. (1993). Humidity sensor based on conductivity measurements of a poly (dimethyl-diallylammonium chloride) polymer film. *Sensors & Actuators* **17**: 61.

47 Sakai, Y., Sadaoka, Y., and Matsuguchi, M. (1989). Humidity sensor durable at high humidity using simultaneously crosslinked and quaternarized poly (chloromethyl styrene). *J. Electron Soc.* **136**: 171.

48 Sakai, Y., Sadaoka, Y., and Matsuguchi, M., and Sakai, H. (1995). Humidity sensor durable at high humidity using simultaneously crosslinked and quaternarized poly (chloromethyl styrene). *Sensors & Actuators* **24**: 689.

49 Sakai, Y., Sadaoka, Y., and Matsuguchi, M., Moriga, N., and Shimada, M. (1989). Humidity sensors based on organopoly-siloxane having hydrophilic groups. *Sensors & Actuators* **16**: 359.

50 Sakai, Y., Sadaoka, Y., and Matsuguchi, M., Hiramatsu, I., and Hirayama, K. (1990). Humidity sensor composed of interpenetrating polymer networks of hydrophilic and hydrophobic polymers. *Proc. 3rd Internat. Mtg. Chemical Sensors*, pp. 273–276.

51 Sakai, Y., Sadaoka, Y., and Matsuguchi, M., and Hiramatsu, K. (1993). Humidity sensor composed of interpenetrating polymer networks of hydrophilic and hydrophobic methacrylate polymers. *J. Electrochem. Soc.* **140**: 432.

52 Sakai, Y., Sadaoka, Y., and Ikeuchi, K. (1986). Humidity sensors composed of grafted copolymers. *Sensors & Actuators* **9**: 125.

53 Sakai, Y., Sadaoka, Y., and Fukumoto, H. (1988). Humidity sensors and water-resistive polymeric materials. *Sensors & Actuators* **13**: 243.

54 Sakai, Y., Sadaoka, Y., and Matsuguchi, M., Kanakura, Y., and Tamura, M. (1991). A humidity sensor using polytetra-fluoroethylene-graft-quaternarized-polyvinylpyridine. *J. Electrochem. Soc.* **138**: 2475.

55 Sakai, Y., Rao, V.L., Sadaoka, Y., and Matsuguchi, M. (1987). Humidity sensor composed of microporous film of polyethylene-graft-poly-(2-acrylamido-2-methyl-propane sulfonate). *Polym. Bull.* **18**: 501.

56 Sakai, Y., Sadaoka, Y., matsuguchi, M., and Rao, V.L. (1989). Humidity sensor using microporous film of polymethylene-graft-poly-(2-hydroxy-3-methacryloxypropyl trimethyl-ammonium chloride. *J. Mater. Sci.* **24**: 101.

57 Sakai, Y., Sadaoka, Y., Matsuguchi, M., Rao, V.L., and Kamigaki, M. (1989). A humidity sensor using graft copolymer with polyelectrolyte. *Polymer* **30**: 1068.

58 Sakai, Y. and Sadaoka, Y. (1985). Humidity sensors using sulfonated microporous polyethylene films. *Denki Kagaki* **53**: 150.

59 Randin, J.P. and Zülig, F. (1987). Relative humidity measurements using a coated piezoelectric quartz crystal sensor. *Sensors & Actuators* **11**: 319.

60 Radeva, E., Kobev, and Spassov, L. (1992). Study and application of glow discharge polymer layers as humidity sensors. *Sensors & Actuators B* **8**: 21.

CHAPTER 9

Sport and Leisure Activity Industries

Chapter contents

Section 1
Sporting Goods Applications

SEISUKE TOMITA

1 INTRODUCTION

Sporting goods sales in Japan reached their zenith in 1991 and have since decreased every year by several percent per year. The market in 1991 was 1.526 trillion yen and in 1995 it was 1.415 trillion yen. Table 1 lists the 1995 sales details of sporting goods sales, golf supplies were 298.5 billion yen.

Outdoor and marine sports goods are expected to experience growth. With the wide-ranging lifestyles, sales tends are seen. Further as can be seen in trendy sports (e.g., snowboarding and inline skating), trends are ever fluid and changing.

Sporting goods reflect well both the interests and the trends of consumers. Therefore, new activities are continuously being developed and as a result, incorporating new technologies, in particular new materials has also actively been pursued. Much of the revolution in sporting goods came as a result of new materials. This improved performance and expanded the market, which then eventually developed newer foci. This cycle was repeated again and again. A good example is the use of carbon fibers in tennis rackets.

480

Table 1   Sales of Sporting Goods in 1995.

Item	Sales (million yen)
Golf	298,570
Ski	203,345
Fishing	177,400
Athletic wear	147,190
Outdoor	101,800
Sports shoes	84,600
Tennis	79,815
Swimwear	53,944
Baseball	50,930
Marine sports	48,447
Snowboard	48,000
Cycling	25,400
Badminton	12,085
Martial arts	11,016
Ping-pong	8,668
Other	64,519
Total	1,415,729

White paper on the golf industry, 1996
Yano Economic Institute

Tennis rackets were traditionally made of wood. However, the market embraced carbon composites because of superior performance, the market for the grew, and finally the wooden racket was passé. Similar tends can be seen in shafts. These changes in raw materials increased the consumption of carbon fibers in the consumer market place where here to fore use was limited mostly to military applications. However, this increased consumption helped create industrial applications and develop further carbon fiber composite technology. Application of carbon fiber composites is not limited to tennis rackets and golf club shafts. Sporting goods that benefit from being very strong and lightweight are gradually being made of carbon fiber composites. Expensive goods, such as golf clubs, tend to be made of high-performance materials. Examples include titanium alloy casting for golf club heads and lightweight carbon composites for the club shaft. Figure 1 [2] illustrates the shift from steel to carbon shafts. Raw materials change and new technologies are introduced far more rapidly compared to other industries.

These examples show that new materials have the greatest impact on those sporting goods that require special functionality. While the role of

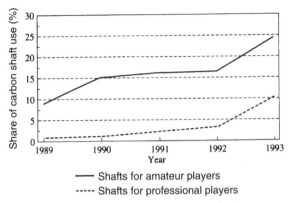

Fig. 1 Trends in golf club shaft.

gels in sporting goods has attracted attention in recent years, active applications must await the future. Whether gels influence sporting goods the way the carbon composite did will depend on future materials developments and research into characteristics.

Because gels possess both the modulus and the microstructure that emulate human tissues, they are expected to be valuable to sporting goods developers and various applications will be developed in the future. Currently, many items are under development but there are only a few examples of commercialized sporting goods that use gels. Those gels currently in use are not hydrogels like those in the body, but synthetic polymer gels. Because of their high water absorption capability, hydrophilic polymer-based hydrogels are being used in daily commodities and industrial materials as water absorbing materials. Sporting goods products do not at this time reflect this trend.

This subsection will touch on those sporting goods that use gels, including tennis shoes developed by the authors, and the technology under development from the patent literature will be introduced.

2   APPLICATION TO SPORT SHOES

The market for athletic shoes is very large. The sales in Japan in 1995 reached 154.4 billion yen (36 million pairs) [1]. In recent years, the performance of sport shoes has been improving and specialized shoes have been developed for specific sports. In particular, running shoes and tennis shoes used by athletes have been designed with special attention to friction and abrasion. These shoes are used very hard and ground friction is an important element to consider when improved performance is the goal. Sports that involve sudden starting and stopping impose severe burdens on the body and are thus the case of much injury.

During running, impact with the floor is measured at five times that of the runner's own weight in the vertical direction. Because this impact leads to injury and fatigue during activity, absorbing or relaxing this impact during running has been an important goal. Many methods have been reported to minimize this impact. Currently, methods used can be largely divided into the following—those that utilize special mechanisms and those that utilize properties of the material. In general, the former requires complex parts and, thus, the latter approach will be advantageous from the economical standpoint. However, current suppliers of sport shoes are manufacturers in developing countries, such as China and Indonesia, and obtaining and applying special materials may sometimes be problematic. Many complex mechanisms are involved in these shoes. They include attachment of an air bag at the sole to reduce impact, where a fluid is contained in the bag, or viscous fluid is moved through orifices in the sole in order to absorb impact. In the most common approach a soft form, such as a rubber sponge, is used. The materials used are often thermoplastics or thermoplastic elastomers, such as EVA and polyurethane. One of the requirements for shoes is that a lightweight material be used. Because soft forms have very low density, they are widely used in shoes. However, these impact absorption mechanisms offer only narrow property selection. They make use only of the softness of the materials and no special impact absorption mechanism is involved. Unless a special mechanism is added to absorb impact, mechanical design features are needed. Figure 2 shows an invention for impact absorption via a mechanical design [3]. It was necessary to develop impact absorption mechanisms that respond over wide property ranges based on material selection and viscoelastic properties. Using the impact absorption material developed by the authors as an example, application of a synthetic polymer gel to shoes will be explained.

(a)

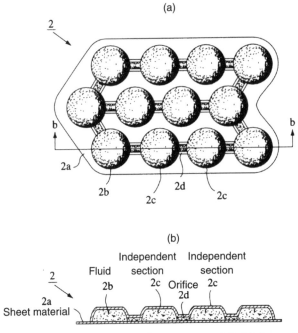

(b)

Fig. 2   Impact damping mechanism by passage of fluid through orifices.

3   DEVELOPMENT OF TENNIS SHOES

In order to use a gel in the heel part of the sole, it is necessary to solve problems that are unique to shoes. These include the shape stability of the gel, property changes caused by penetrants, changes in appearance when the soles get dirty, good processability by injection or compression molding, in addition to mechanical property requirements. Of course, there are problems related to cost and supply of materials. Because shows are not particularly expensive commodities, cost reductions and mass production issues are always important consideration in shoe production. From the performance viewpoint, properties that are difficult to quantify, such as comfort, are also important properties.

Movements of a human foot are complex and shoes must be designed to adjust to them. Therefore, materials must be placed in the desired locations. Loss of shape and property changes with time by plastic deformation, such as creep, must be avoided. Figure 3 shows the cross-sectional view of the sole of a shoe [4]. A gel was placed only at the heel

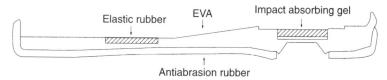

Fig. 3 Location of parts in tennis shoe.

to reduce heel impact. The front of the shoe has a more elastic rubber because this is where the initial dashing force is created. A gel material that has good shape retention makes it possible to design a sole that responds to even the delicate movements of the foot. Because shape stability is excellent, it can be molded into a spherical or hemispherical shape to make use of the mechanical properties that are influenced by shape. Because shoes are made by combining parts composed of different materials, property changes caused by the transfer of materials and wear and tear are important parameters to consider.

A gel that is commercially available has the trade name ELAMUS. It is made of a hydrogenated styrene-butadiene block copolymer that is synthesized by solution polymerization. It is swollen by a liquid plasticizer. To maintain shape, a crosslinked structure resembling that of human tissue is formed (see Fig. 4) [5]. The characteristics of this gel are that even if the same polymer main chain is used, various properties can be tailored by changing the low molecular weight material that swells the polymer. By selecting appropriate compatibility and miscibility to allow the polymer molecules to be mobile, gel properties can be tailored.

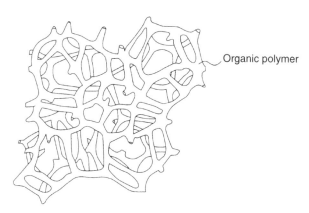

Fig. 4 Microstructure of ELAMUS.

Traditionally, forms made of vulcanized rubber with a large amount of plasticizer have been used. However, these gels exhibit much more finestructues than the superstructure of the pores of these rubbers and offer better stability and homogeneity of properties.

Because sporting goods are used by humans, they must be stable at the temperatures at which they will be used. If glass transition temperature (which is due to the molecular motion of polymer chains) is to be used as the energy absorption mechanism, the sudden change in modulus at the glass transition often causes unstable properties. If other viscoelastic properties are to be used for energy absorption, it is necessary either to copolymerize monomers or to mix polymers of different mobilities [6] (see Fig. 5) [7].

Figure 5 shows the temperature dependence of the viscoelastic properties of ELAMUS. It can be seen that excellent modulus and damping characteristics are obtained within the temperature range required for shoes. By applying impact damping characteristics to the heel portion of tennis shoes, impact during running has been reduced markedly.

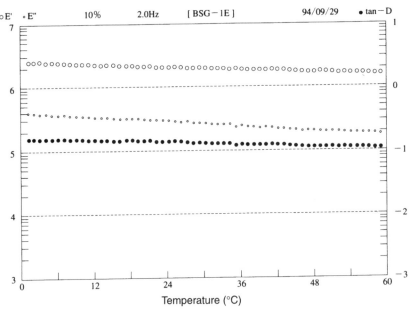

Fig. 5   The temperature dependence of viscoelastic properties of ELAMUS.

By utilizing the shape stability of the gel, the shape effect is also obtained (see Fig. 6) [8].

By utilizing the shape effect, deformation caused by body weight is made to work for comfort. This phenomenon creates a sole that feels soft. Softness is particularly felt at the beginning of deformation and a comparison with a flat bulk plate provides the necessary information.

There are many shoe development examples available. These include the use of a low molecular weight silicone oil to swell a silicone polymer [Tokko Kaiho 291110 (1986), trade name: Alpha Gel], a gel with polyurethane as the polymer component (product name: Solbosen), and a rubber form made of a high styrene, styrene-butadiene copolymer to

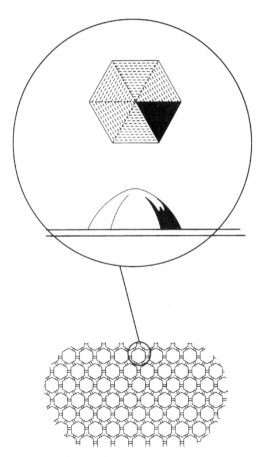

Fig. 6 The shape of an impact-absorbing material.

which a plasticizer with a high glass transition is impregnated [product name: ZDEL, Tokkyo Koho 5983 (1996)]. However, these gels often lack shape retention stability. Furthermore, the low molecular weight compound that swells the polymer diffuses out of the gel. As a result, it contaminates the surrounding parts of the shoes or reduces the degree of swelling. Hence, it seems to be a common approach to use it in a flexible bag, usually made of polyurethane.

Applying these impact-reducing or damping materials to sporting shoes is becoming common practice. Applications are now being used in the manufacture of walking shoes.

4 APPLICATION OF GELS TO AREAS OTHER THAN SHOES

Applications of gels to areas where impact damping is needed are being studied. The products already available commercially include external supporters (braces, etc.) for the arm and leg joints. In aggressive sports, such as basketball and volleyball, bagged gels are used to protect players from impact injuries that could result from falls.

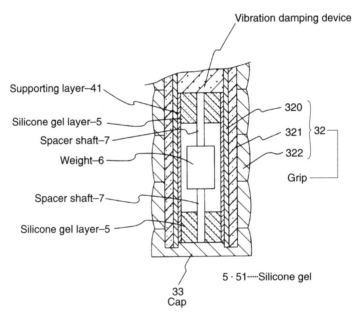

Fig. 7 Impact absorption of a racket.

An interesting impact damping proposal is geared toward preventing tennis elbow by incorporating a silicone gel with the shaft of the tennis racket (see Fig. 7) [9]. Traditionally, the shaft was wrapped with a viscoelastic or thin foam backing material and then taped. Benefits were minimal with this older system but contact with the ball felt different with the newer designs, which provides important feedback when playing tennis. However, if the benefits are significant, the number of impact damping applications will increase because there are many sports that will benefit.

A unique hydrogel application actually protects the body from sun during the summer. A gel in a hat or actually in a swimwear may be used to lower body temperature by using the latent heat of water vaporization, thereby allowing one to enjoy sports under a scorching sun. A hat for golfers has already become a very popular product.

5 PATENTED APPLICATION EXAMPLES

A search of patent literature yields examples of gel development or application to sporting goods, even if they are not yet commercialized. Of those examples found, the one that interests the author is selected and introduced here.

A hydrogel is prepared by a mechanochemical reaction using PVA as the polymer backbone. This is then used as the core of a golf ball in place of a ball made of strings. The string-wrapped, liquid-core ball uses a bag in which a dispersion of high-density particles in water is placed (see Fig. 8 for details). This bag is then wrapped in rubbery strings to improve feel upon impact. Although this ball is now seldom used, it had been very popular. Instead of a liquid core, the use of a gel that had high elasticity and handled easily was proposed. Unfortunately, the performance of the two-piece solid ball has been improved significantly but market lack of interest has caused it not be commercialized [10].

6 CONCLUSION

Application of gels to current sporting goods is dominated by mechanical properties alone, such as impact relaxation or damping. In the future, it is expected that the development and application of hydrogels, which are already well developed in medicine and daily commodities and are compatible with the body, will advance even further.

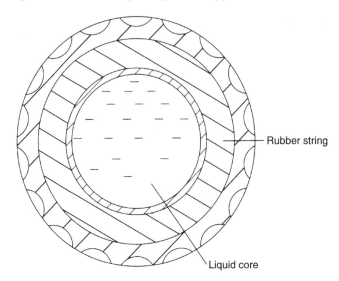

String-wrapped, liquid-core golf ball

Fig. 8   Construction of liquid core.

It is also tempting to dream about applications, such as the use of stimuli-responsive gels in sporting goods so that optimum performance can be achieved whatever the environment because the gels will be able to take any environment into account. The property of the material may be adjusted depending on the way one walks or runs, or the level of experience. In addition to the importance of comfort, the possibility of developing high-performance sporting shoes and goods also may be possible.

REFERENCES

1   Yano Economy Research Institute (ed.). (1995). *White Papers on the Sporting Goods Industries.*
2   Proctor, S.K. (1994). *Proc. World Sci. Congress of Golf.*
3   (1995). Tokkyo Kaiho 16104.
4   Internal document of Bridgestone Sports, Inc.
5   (1993). Tokkyo Kaiho 239256.
6   Ozaki, K. (19??). *J. Soc. Rubber, Jpn.*
7   Internal document of Bridgestone Sports, Inc.
8   Internal document of Bridgestone Sports, Inc.
9   (1991). Tokkyo Kaiho 49776.
10   (1987). Tokkyo Kaiho 112575.

Section 2
Artificial Snow

1 INTRODUCTION

Skiing is a major winter sport. Those Japanese who enjoy skiing number more than 10 million. The popularity of snowboarding among young people is also high. During skiing season, ski slopes are full of people on the weekends. Unfortunately, the period suitable for skiing is approximately four months. For the rest of the year only those who can visit high mountains or travel to the Southern Hemisphere can ski. There are many artificial skiing facilities for those who really like to ski during the off season. These include skiing on a grass slope or lawn, skiing on a plastic sheet, and skiing on golf balls embedded on a slope. Nonetheless, they hardly give one the feel of skiing on snow. Therefore, many facilities create snow all year-round by using cooling systems. One method of doing this was developed by a company in Australia in which a superabsorbent polymer was used. This method will be described here.

2 BRIEF EXPLANATION OF THE SUPERABSORBENT POLYMER METHOD

2.1 Snow Manufacturing Method

This method makes snow by freezing water-absorbed polymer gel on cooling pipes. The polymer used in this method is a poly(acrylic acid) type. To this polymer 100–120 times the water is absorbed and then sprayed and frozen on the cooling pipes installed on the slopes.

The gelled polymer will flows only very little even on these slopes. It will form snow on the slope much better than if water had been sprayed directly. However, the surface is hard and inappropriate for skiing if it is merely frozen. Hence, the ice surface must be scraped by a grooming machine and fine ice, namely, powdered snow then results (see Fig. 1).

2.2 Air Conditioning System for Cooling

2.2.1 Cooling System

In this method, the snow is kept cold by a coolant placed under the snow that then keeps the snow on the cooling pipes. The coolant is a liquid such as ethylene glycol solution or calcium chloride solution) or a cooling gas. These coolants offers better temperature distribution and are easy to deal with in the event of leaks. These coolants are chosen because leakages can be spotted with the naked eye and oxygen-related problems are nil. The basic cooling system is shown in Fig. 2.

Although snow that uses a polymer has a small freezing point window, when the polymer absorbs 100 times its weight it has a freezing

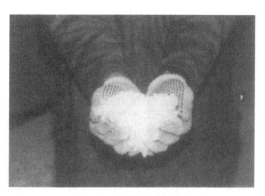

Fig. 1 Artificial snow.

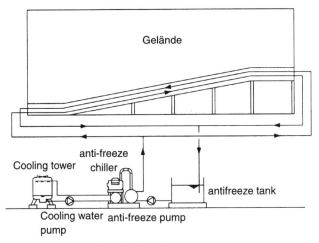

Fig. 2 Cooling system.

point of approximately 0°C. Therefore, by controlling cooling temperature, the snow surface temperature can always be maintained below the freezing point (the snow is actually maintained at ∼2°C). In other words, as long as the snow surface is maintained below the melting temperature, even if room temperature (in this case, the air temperature approximately 1–1.5 m above the snow surface) is slightly high, the snow can be maintained. It is even possible to increase room temperature in order to reduce thermal shock to the body.

The amount of cooling Q (W/m^2) to maintain snow per unit area is:

$$Q = \alpha(W/m)^2$$

where T_R is the room temperature (°C), T_s is the snow surface temperature (°C), and α is the surface thermal conductivity (W/m^{2}K).

As the room temperature increases, the amount of cooling increases and, accordingly, costs increase. Thus, room temperature during the summer months is set to at ≈ 10–15°C.

The average temperature of the cooling pipe is expressed as:

$$T_P = T_S - Q_d/\lambda$$

where d is the average thickness of the snow (m) and λ is the average thermal conductivity of the snow layer W/mK(d).

To set the cooling temperature high and operate the cooling device at a high efficiency, d must be small and λ must be large. While thickness of the powdered surface is desirable because it improves the feel of snow, thermal insulation increases by increased void content. Thus, λ decreases and costs are too high. The powdered surface snow is cost effective at when 3–5 cm.

2.2.2   Air Conditioning System

Air conditioning can be reduced due to cooling from the bottom of the snow. However, it is necessary to control the transport of water on the surface, such as from sublimation and frosting. Hence, moisture control is also an important element.

The surface snow temperature of this system is targeted at $-2°C$. The dew temperature is defined as the temperature below which water in air condenses. If the temperature of the air that is in contact with the snow falls below the dew temperature of the air, moisture transport from the air to the snow surfacer is in the form of condensation or frosting. If the temperature of the air is above the dew temperature, the moisture transport from snow to air in the form of sublimation takes place. For snow with polymer, the friction coefficient becomes large if sublimation proceeds as a consequence of the increased polymer content on the snow particle surface. Thus, it has an adverse effect on the sliding phenomenon. The dynamic friction coefficient will be discussed later. Conversely, if too much frost develops, the snow surface will harden. The surface also will melt due to the thermal insulation of the frost. When it freezes again, it turns into ice. The energy necessary to freeze the moisture in air into frost increases the load on the cooling equipment.

If the dew temperature is much higher than the snow surface temperature, there is the possibility of condensation on the ceiling and beams of the building occurring due to radiant snow. Accordingly, moisture control is essential in this system. In particular, during the summer months the dew temperature climbs to $>20°C$ and moisture control devices become indispensable. Figure 3 is an air conditioning flow diagram.

Moisture control systems can be divided into chemical and cooling methods. The chemical method reduces ambient humidity by absorbing moisture from air using an activated charcoal rotary or lithium chloride solution. In this method, it is necessary that the charcoal or absorption solution remove the absorbed moisture cyclically if continuous operation is needed. Energy is needed to recycle these materials. Figure 3 shows the

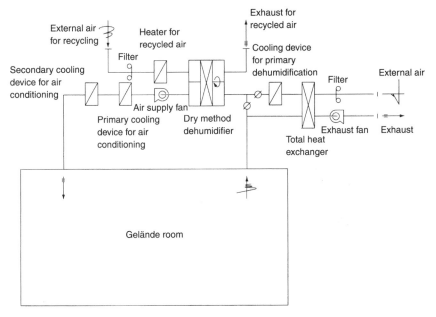

Fig. 3 Air conditioning system.

dry method (charcoal method) humidity reduction system. In this device, the dew temperature of the air at the exit can be set $< 10°C$.

The air is cooled by cooling the coil in order to condense moisture and remove it. If the coolant is water, the dew temperature at the exit is as high as $5–10°C$. By using other coolants or a coolant gas, it is possible to set the dew temperature at $0°C$. Below, $0°C$, the condensed water freezes and blocks air recirculation pathways.

In an actual facility, the primary dehumidification is done by cooling, followed by secondary dehumidification using the chemical method. With this method, the dewpoint of the air in the facility is controlled so that it is only slightly higher than the snow surface temperature. This is because a small amount of frosting is permissible but sublimation, which increases the friction coefficient, must be avoided. The frost is used to supplement the snow that was used by the skiers.

In winter, it is possible for the dew temperature external air to fall below the snow surface temperature. If external air is introduced as is, water sublimes from the snow surface and the friction coefficient increases. Hence, in the air conditioning system, moisture generated by breathing, for example, and reduction in humidity caused by introduction of external air will be balanced by either adding moisture or reducing the intake of external air. The snow surface temperature may also be lowered below the dewpoint of the external air.

These facility studies were performed in Japan after the technology was transferred from Australia. Some of the technologies have already been patented in Japan and abroad.

3   PROPERTIES OF POLYMER MIXED SNOW

3.1   Thermal Property Modification of Snow by Polymer

It is important to know the thermal conductivity of snow when designing an artificial skiing facility. In this subsection, the measured thermal conductivity of the snow layer will be introduced. As shown in Fig. 4, a thermal flow sensors and thermocouples are embedded in the snow layer to measure the heat flow and temperature distribution within the snow layer. Similarly, thermocouples are also installed on the snow surface to measure the temperature distribution at the snow surface as well as the representative room temperature 1 m above the snow surface. Figure 5 shows the temperature distribution and other values within the snow layer and the room. From these measurements, the thermal conductivity of the snow as measured to be 0.26 W/mK for powdered snow and the ice layer beneath the powdered snow was 1.23 W/mK.

3.2   The Friction Coefficient of Polymer Mixed Snow

Snow mixed with polymer is almost the same in appearance as natural snow. However, depending on the room air conditions, the snow surface conditions change. Ability to slide changes depending on snow conditions. Thus, the dynamic friction coefficient of the surface of the polymer mixed snow, artificial snow made with tap water, and natural snow was measured. We obtained the variation of the dynamic friction coefficient of artificial snow as well as natural snow, which is the target value for artificial snow.

3.2.1   Comparison of the Laboratory Measured Properties of Polymer-Mixed Snow and Snow Made from Tap Water

A polymer-mixed snow and snow from tap water were prepared on a horizontal experimental table in a laboratory. Using a pulling device, an

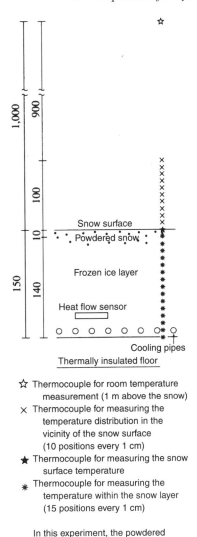

Fig. 4 Positions for thermal property measurement of snow.

experimental ski was pulled and the tensile force on the ski was determined by strain gauge to determine the dynamic friction coefficient (see Fig. 6). Figure 7 shows the time dependent dynamic friction coefficient of the powdered snow prepared by shaving the surface of the ice made from the polymer-mixed water and tap water. From this figure, it

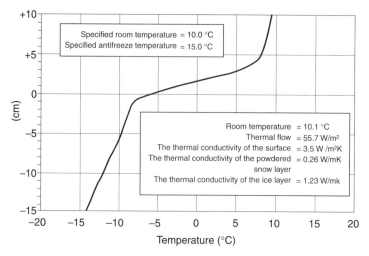

Fig. 5 Temperature distribution within and outside of the snow in the vertical direction.

is found that the dynamic friction coefficient of the polymer-mixed snow is generally greater than for the tap water snow. This result indicates that even if the polymer concentration is around 1%, ski resistance can be significant.

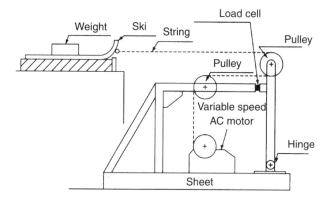

Fig. 6 An instrument for friction coefficient measurement

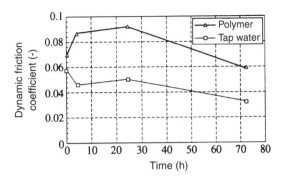

Fig. 7 Change of dynamic friction coefficient.

3.2.2 Comparison of the Measured Properties of Polymer-mixed Snow and Natural Snow in an Experimental Gerende and at a Certain Ski Resort

In an experimental gerende and on natural snow, the speed between two points was measured by a photoelectric speedometer. From the slope angle and the distance between the two points, the dynamic friction coefficient was determined. Two pairs of sliding skis were prepared. On one pair, only a stone ground surface was used. On the other pair, graphite-based wax was applied after stone grinding had been done.

The time dependent coefficient of the polymer-mixed artificial snow in the experimental gerende is shown in Fig. 8. In the figure, G

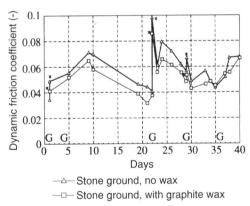

Fig. 8 Time dependent changes of the dynamic friction coefficient in an experimental gerende.

indicates that the grooming operation took place so as to make the surface powdery.

The date with the asterisk indicates that the measurement was taken immediately after grooming. As can be seen from the figure, the dynamic friction coefficient of the experimental gerende is within the range of 0.03–0.08, with day-to-day variations. The value measured immediately after grooming increased as compared to the value measured prior to the grooming. This is due to the increase in the front resistance of the ski due to the softened snow. It is also due to the increased contact area on the ski surface with the snow. It is well known that even in the natural snow, the resistance of fresh snow is greater than of compacted snow. The gradual increase in the dynamic friction coefficient without grooming from day 1–day 9 in Fig. 8 is possibly due to the poor humidity control in the experimental gerende and increased polymer concentration on the snow surface. Contrarily, when the humidity within the room is increased, frosting took place and the dynamic friction coefficient decreased as in day 9–day 22 in Fig. 8. Accordingly, in order to improve sliding, it is effective to increase the water content on the sliding surface and to decrease the relative concentration of the polymer. To maintain this condition, it has been confirmed that the humidity control is important.

Table 1 lists the experimental value of the dynamic friction coefficient of natural snow. The dynamic friction coefficient of natural snow also is within the range 0.05–0.1. Hence, if the condition of the artificial snow is properly controlled, the polymer-mixed snow can be comparable to natural snow with respect to the dynamic friction coefficient.

3.3   Grooming Machine

As described already, simply freezing a gelled polymer will not be appropriate for skiing, as the snow layer is icy plate. Thus, the surface

Table 1 Dynamic friction coefficient of natural snow.

Slope angle	Ski No. 1	Ski No. 2	Supplemental
5.4	0.060	0.048	compacted snow (snowed a day earlier)
7.9	0.066	–	compacted snow (snowed a day earlier)
10.4	0.098	0.068	New snow (measured during the snowing)
16.7	0.110	0.057	New snow (measured during the snowing)

Ski No. 1, stone ground, no wax
Ski No. 2, stone ground with wax

Fig. 9 Grooming machine

must be scratched and powdery snow needs to be created. To achieve this, mechanical force needs to be applied. It was also described that the humidity needs to be controlled in such a way that only slight frosting but not sublimation occurs, which creates better sliding properties. This frosting and aggregation of the snow particles creates a hardened snow surface. Thus, the snow must be returned to its powdery state daily. The instrument used to do this is called a grooming machine (see Fig. 9). The machine in the photographs equipped with low-pressure tyres for running on snow. It is also equipped with an oil-driven shaving device.

4 CONCLUSION

The main problem in using superabsorbent polymer for artificial snow is the increased friction coefficient of the snow. Methods to avoid this problem have been described here. The most effective way is to avoid the presence of the polymer on the snow surface. On the other hand, from the viewpoint of placing water on a slope, the following advantages of the polymer-mixed water exist:

i)   polymer gel does not easily flow on a slope; and
ii) when crushed ice or ice blocks are placed on a slope, there will be spaces between the crushed ice or between the ice particles and the cooling pipes, whereas with a polymer gel, these spaces can be eliminated.

Therefore, a method is proposed in which the lower layer ice is made with polymer-mixed water solution on which frosting created by a fine water mist is formed on top of the polymer ice plate. Then, this frost is made into powdered snow. This approach will offer benefits for future artificial ski facilities. When artificial ski facilities are constructed, complex optimization of the building along with cooling and air conditioning systems is necessary. The author has observed formation of water condensation in unexpected locations as well as structural deformation of buildings caused by the pressure of the formed ice even when every aspect was carefully handled. In the future, it is this author's desire to construct more comfortable and less expensive ski facilities using these experiences.

INDEX

503

ISBN 0-12-394963-7